Global Warming and the Risen LORD

*CHRISTIAN DISCIPLESHIP
AND CLIMATE CHANGE*

JIM BALL

Published in Washington, D.C., by the Evangelical Environmental Network (EEN).

EEN is a ministry that seeks to educate, inspire, and mobilize Christians in their effort to care for God's creation, to be faithful stewards of God's provision, and to advocate for actions and policies that honor God and protect the environment.

Web: http://www.creationcare.org
Contact: een@creationcare.org

Published in association with Russell Media, Boise, Idaho, providing editorial, design, digital development, print, fulfillment, distribution and marketing services to cause oriented organizations.

Web: http://www.russell-media.com
Contact: customerservice@russell-media.com

Unless otherwise noted, Scripture quotations are from the Holy Bible, New International Version. Copyright © 1973, 1978, 1984 by the International Bible Society, Inc. Used by permission of the Zondervan Publishing House. All rights reserved.

This book may be purchased in bulk for educational, business, ministry, or promotional use. For information please email een@creationcare.org.

ISBN (print):   978-0-9829300-1-4
ISBN (e-book):   978-0-9829300-2-1

Library of Congress Control Number: 2010940683

Printed in the United States of America using the most climate-friendly methods available, including 40% post- consumer waste recycled paper and environmentally friendly soy based inks.

ADVANCE PRAISE FOR
GLOBAL WARMING
and the RISEN LORD

"I was delighted to meet Jim Ball and discuss with him and others the Bible's message about God's care for the whole creation, and our part in that great work. Jim has taken a courageous stand in explaining and exploring the way in which the current crisis of a changing climate has to be taken seriously by all Christians who live under God's word. I am delighted to see this book as one further step along a transformative road."

> **Bishop Tom Wright,**
> **Professor of New Testament and Early Christianity,**
> **University of St Andrews**

- - -

"Addressing energy security is essential for maintaining the well-being of our families and nation. In his book my friend Jim Ball discusses how we can make that happen by focusing on clean sources of energy. Highly recommended."

> **Roberta Combs**
> **President, Christian Coalition of America**

- - -

"My friend Jim Ball has been at the forefront of climate change for nearly two decades, helping to show Christians the way, especially in the area of climate policy. And now he has added to that a book also filled with biblical and spiritual resources to help encourage and sustain us as we fight the good fight. Global warming is a huge challenge, but Jim reminds us that it is the Risen LORD in whom we put our hope. Essential biblical analysis for a troubled planet. An extraordinary accomplishment!"

> **The Rev. Richard Cizik,**
> **President and Founder,**
> **The New Evangelical Partnership for the Common Good**

"For me, fighting to overcome global warming is about people I've met like Srey from Cambodia or Mutinta of Malawi desperately trying to feed their families. Jim's book tells the stories of many such people and connects them to our story with the Risen Lord. He helps us understand that through the power of the resurrection we can overcome global warming. Inspiring and hopeful!"

The Rev. Dr. Jo Anne Lyon
General Superintendent, The Welseyan Church

- - -

"Rev. Ball helps us see the tremendous impact climate change will have on the world's poor, and as such it is something all Christians must respond to. But the book's most important contribution is helping us understand the incredible spiritual resources our faith in Christ provides as we work to overcome this challenge."

Bishop George McKinney
Founder and Pastor
St. Stephen's Cathedral, Church of God in Christ

- - -

"Jim Ball proves once again in this book why he is such an important and prophetic voice that challenges Christians to live up to their calling to steward the earth and to care for the poor. As part of an organization that focuses on ministering to the poorest of the poor around the world, I am thankful for this provocative work which clearly reminds us of our responsibilities in this area as well as pointing us toward the One who is both Creator and Redeemer. I recommend this book to all who want to be called to a higher realm of creation care."

Dave Evans,
President, Food for the Hungry

"As Jim's book ably summarizes, the science is clear and compelling that climate change is a profound challenge that must be met. But the challenge is not only scientific; it is also moral and spiritual. The voices of religious leaders need to be heard. Exercising faith can bring guidance, hope and inspiration. All three can be found in Global Warming and the Risen LORD."

Sir John Houghton,
Former Chair, Intergovernmental
Panel on Climate Change's Scientific Assessment

- - -

"Jim's book displays his passion for protecting the poor and vulnerable and provides a clarion call for Christians to do so by walking faithfully with the Risen LORD as He leads the way in overcoming global warming. The third part of the book offers a rich description of what actions can be taken to meet this challenge."

Dr. Ron Sider,
President, Evangelicals for Social Action,
author of Rich Christians in an Age of Hunger

- - -

"Jim Ball gives a unique Resurrection perspective on climate change. His Christ-centred critique brings challenge and hope in equal measure. I admire his work because like William Wilberforce his values and his activism are biblically based. When the Son of Man comes will he find faith on the earth? Yes, if people take Jim's message seriously."

The Rt Rev James Jones,
Bishop of Liverpool

"For too many of us, the issue of global warming is an unknown topic. Yet it haunts our future, a time when our grandchildren are likely to ask why so many of us chose to remain uninformed and silent. Jim Ball has done an excellent job of introducing us to global warming issues and why this is such a biblical, a thoroughly theological matter."

The Rev. Gordon MacDonald,
author and speaker

- - -

"As a longtime leader for creation care, Jim Ball is one of our most important religious voices on climate change. Jim's book provides both challenge and encouragement for this fight, and reminds us that climate change is the single greatest threat facing wildlife today."

Larry J. Schweiger,
President and CEO, National Wildlife Federation

- - -

"In leading a global Christian health ministry I have come to understand how deeply degraded environments affect the health of people living in the world's poorest communities. Dirty water, dirty air, eroded soils, deforestation and more add much to the burdens of poverty they face. It is clear that climate change is another real threat to people who cannot bear any more. Jim Ball opened my eyes to the fact that whatever the impacts of climate change, they will hit the poorest the hardest. We should not stand for that. No one has done more to awaken American Christians to the reality of global climate change than Jim. His new book is the next step in bringing his prophetic voice to this issue. We all should read it."

Michael J. Nyenhuis
President/CEO, MAP International

Global Warming

and

the

Risen

LORD

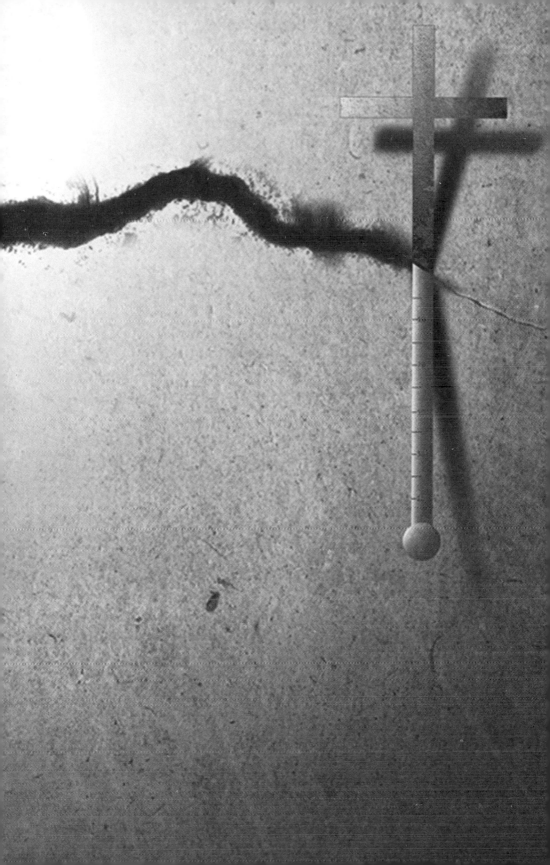

TABLE OF CONTENTS

PREFACE: A PERSONAL NOTE TO THE READER

In the fall of 1990, at the beginning of my Ph.D. studies, another Ph.D. student told me she was there to study a Christian approach to nature. At that time I couldn't grasp why anyone would do so, and I literally said to her, "Why would you waste your time doing that?" Thankfully, she didn't get defensive or brusque. Instead, she said, "Well, you're a Christian. Why don't you read the Bible in light of this question?"

And so I did. When I came to Colossians 1:15-20, I was convicted. If Christ creates all things and His blood reconciles all things, how can I treat the rest of creation with indifference or even contempt? For years I had missed the important theme in Scripture that we are called to be stewards, to care for the rest of God's creation.

Once I began to study environmental issues, I came across discussions about climate change or global warming. As a follower of Christ, I had long-held concerns about hunger, the poor, and peacemaking; I soon grasped the profound challenge global warming presented in these areas. My study of and involvement in this challenge has led me to conclude that climate change is not just an environmental problem. Rather it is its own category, one that touches on many other concerns.

If we found seven people who had never heard of climate change and gave them a full briefing on the effects of global warming, I doubt they would describe it primarily as an environmental problem. A military person would say it's a major national security issue. A development worker would say it is perhaps the most important relief and development issue of this century. One working in the energy industry would suggest this challenge could radically transform the energy sources of our economy. A health professional would argue it could be the most important public health threat faced in our lifetime. On the comments would go, as these seven people would see how climate change relates to their lives and areas of interest and expertise.

You, too, will have your own way into this challenge – and that's good, because we are going to need everyone's insights and creativity and particular ways of viewing things to overcome global warming.

This book is intended to be a spiritual journey with the Risen LORD[1] to prepare us for the long road to overcoming global warming. It is also meant to be spiritual sustenance along that journey. My deepest hope is that this book becomes a source of encouragement for you as we follow the Risen LORD while He works to overcome global warming through us.

We often hear that we should lose ourselves in a great cause. I say no. Absolutely not. We find our true selves in *the* cause greater than ourselves – the LORD's will. But we can't set out focused on finding ourselves, because the focus is still on us. Rather, our self must become lost in the greatest cause there is: loving the LORD with all of our heart, soul, and mind, with everything that we are. And from that flows everything else.

But, as we will see, it is also true that loving the LORD and therefore doing the LORD's will ("If you love Me, keep My commandments," Jn 14:15, NKJV) today includes all of us playing our part to overcome global warming.

Intrigued? Then welcome to the journey. My hope is that this book will be a good spiritual travel guide as we walk this road together with the Risen LORD.

A key task will be to come to know the One we walk with more fully. The more deeply we know Him, the more deeply we will love Him. The more we love Him, the more our deepest desire will be to do His will – which in the twenty-first century includes overcoming global warming. And in this way, without even trying, we will find our true selves.

Let me offer a final comment for those of you who are not Christians. I hope you will let me introduce you to Christ in these pages. Though this might sound threatening, don't be afraid. As the angel of the LORD said to the shepherds when announcing Jesus' birth, "Fear not: for, behold, I bring you good tidings of great joy, which shall be to all people" (Lk 2:10, KJV). Because God loves you, everything will be all right.

INTRODUCTION

THE KEY TO OVERCOMING GLOBAL WARMING

Born in 1885, Betsie contracted pernicious anemia in early childhood and was sickly throughout her life. The disease prevented her from having children, so at a young age she decided not to marry. After finishing secondary school, she began working as the bookkeeper in her father's watch repair shop. Soon it was clear that her younger sister was better at bookkeeping; Betsie's real talents lay in hospitality and making their house a home, so she became a full-time homemaker.

During the Nazi occupation, Betsie and her family lived in Amsterdam. Because they were devout Christians, Betsie and her family helped whoever was in need, including Jews, whom they respected as God's chosen people. By 1942 they were working with the Dutch underground and also hiding Jews in their home. On February 28, 1944, Betsy and her family were betrayed by one of their fellow citizens and arrested by the Gestapo. They would provide no information about the Jews they were hiding or their underground activities or contacts. As a result they were sent to prison and then to Ravensbrück concentration camp.[2]

When Betsie died at Ravensbrück in December 1944 at age 59, no one in the world at that time would have taken notice. She was not highly educated, had no accomplishments of merit, no spouse or children, and was not rich or of noble birth. Her death was a profoundly obscure event. She simply would have been a Nazi statistic if her sister, who barely survived, had not lived to tell her story. Her sister was scheduled to be executed within a week, but was released due to a clerical error.

A few days before her death, Betsie's true self perhaps shone the brightest. She was being taken once again to the sick ward at Ravensbrück. In spite of experiencing all of the horrors of the camp, Betsie whispered to her sister, "There is no pit so deep that God's love is not deeper still."[3]

Betsie's testimony was kept alive through the extraordinary – some would say miraculous – survival of her sister, Corrie, who at age 53 "began a world-wide ministry which took her into more than 60 countries in the next 33 years."[4]

Nearly 30 years after Betsie Ten Boom's death, their story was told in the book and film, *The Hiding Place*. It was Betsie's conviction about the depth of God's love and the healing power of forgiveness that inspired Corrie to begin her ministry. As Betsie told Corrie a few days before she died, "We must tell people what we have learned here…they will listen to us, Corrie, because we have been here."[5]

Some may find it strange that a book on global warming would begin this way. But this book is not just about global warming. More deeply, it is about the Risen LORD's presence with us to help us with whatever challenges and opportunities we face. And I am hard-pressed to think of a story from the last 100 years that reflects the Risen LORD's presence more deeply than that of Betsie and Corrie in the Ravensbrück concentration camp. There is no doubt in my mind that the Risen LORD was present with them during their time at Ravensbrück, that His love filled Betsie's heart and inspired her to say her now famous words: "There is no pit so deep that God's love is not deeper still."

Corrie recounts how the sick ward, where Betsie had to be taken numerous times, was even more dismal than the rest of the concentration camp. But Betsie saw it differently. "To her it was simply a setting in which to talk about Jesus – as indeed was everyplace else. Wherever she was, at work, in the food line, in the dormitory, Betsie spoke to those around her about His nearness and His yearning to come into their lives."[6]

What was true then is still true today. Spiritually the Risen LORD walks with each of us. He is, always ready for us to turn to Him for strength and guidance. As it says in Rev 3:20: "Behold, I stand at the door and knock. If anyone hears My voice and opens the door, I will come in to him and dine with him, and he with Me" (NKJV).

The ever-present Risen LORD is our key to overcoming global warming. The only way we Christians are going to make our contribution is by situating ourselves deeply in Christ and in the reality of the Risen LORD's presence and Lordship. The wonderful thing is this is what our lives need to be about anyway. And to remain true to His Lordship we must all play our part to overcome this great challenge of the twenty-first century, because He is the One leading the way.

PURPOSE AND OVERVIEW

This book has a threefold purpose: (1) to get you started or help you continue on your journey of overcoming global warming with the Risen LORD; (2) to help you understand the terrain over which we will travel, i.e., both the challenges and the opportunities; and (3) to provide encouragement and spiritual sustenance along the way, especially by reminding you that we are not alone; the LORD is with us.

To get started we need to understand why Christians should care. This is a major goal of Part 1. My purpose is not to convince you of the reality of global warming or that we are causing it. Many other excellent resources do that.[7] Rather, the focus is on *how* we are causing it and that all of us contribute to it. As such, we all have a role to play in overcoming global warming.

But, frankly, why should we? There are plenty of worthwhile ways we can spend our time. The rest of Part 1 covers the hard reality of global warming's consequences, especially for the poor whom Jesus referred to as "the least of these." We will also explore why Christians in particular should care enough to do something about it. We must understand the current state of the world in order to be able to follow our Risen LORD as he leads the way in changing it.

Overcoming global warming is not a sprint. There is no quick fix. Rather it is a marathon, something we will be doing for the rest of our lives. It is going to require stamina – spiritual staying power. That's why it is comforting to begin to understand that it is the Risen LORD who is leading the way. He is calling each one of us to play our part. To stay true we need to plant ourselves deeply in Christ by reminding ourselves of what He has done for us. We must partake of the spiritual food He gives to sustain us and the teachings He provides to guide us. Because at times we will fail along the way, we also need to be reminded of the examples of the apostles who failed as much or more than they succeeded, but who are never failed by the Lord nor abandoned by God – just as He will never fail nor abandon us.

In Part 3, I share how it is possible for us to overcome global warming in ways that make our lives better and provide the poor with opportunities to create better lives for themselves. This challenge creates opportunities. This challenge can be met.

A BRIEF TOUR OF THE BOOK

To give you a sense of where we are going, here is a brief "travel guide" for your journey through this book.

Chapters 1 & 2, *Climate Change: The Basics* and *How We Are Causing Global Warming*, will describe what global warming is and how we are causing it, and that this abnormal warming is due primarily to the burning of fossil fuels, with deforestation an important secondary cause.

Chapters 3 & 4, *Consequences for the United States – Increased Floods, Droughts, and Wildfires* and *Coastal Consequences and Health Concerns in the United States*, will highlight some of what the United States will experience.

Chapters 5 & 6, *Consequences for the Poor in Developing Countries: An Introduction*, and *Seven Major Impacts on the Poor in Developing Countries*, will show how billions of the world's poor will be seriously impacted by increased hunger, malnutrition, and water scarcity; sea level rise and intensified coastal and inland flooding; and increased vulnerability to health threats like malaria. Consequently, millions will be at risk of becoming climate refugees and billions are at increased risk of violent conflicts.

Chapter 7, *The Next Great Cause of Freedom*, explains how the consequences of global warming will seriously impede the economic and democratic dimensions of freedom, especially for the poor. However, the solutions to global warming can enhance the economic freedom for the poor and provide chances to create a better life. As a result, overcoming global warming is the next great cause of freedom and an incredible opportunity for us to be engaged in such a cause.

Chapter 8, *Christ the Creator and Sustainer*, explores the New Testament's astounding claim that Jesus Christ is the Creator and Sustainer of all things. Therefore our efforts to overcome global warming are in concert with His creating and sustaining activities.

Chapter 9, *Christ the True Imago Dei*, shares that while we were made in the image of God to freely reflect His will, sin has impeded our capacity to do so. But Jesus of Nazareth was – and the Risen LORD is – the true image of God, the true *imago dei*. By reflecting Him and following His lead in overcoming global warming, we become more free, beautiful, and glorious.

Chapters 10 & 11 are focused on *Christ Our Savior*. They remind us that the Father sent the Son to set us free from our bondage to sin so that we can love Him back, and that His grace empowers us to be able to do His will, which includes following the Risen LORD in overcoming global warming. They compare this great moral challenge of global warming to civil rights, and conclude with the reassurance that with the LORD's help we shall overcome.

Chapters 12, 13, & 14 are focused on *Christ Our LORD*. They recount how the disciples misunderstood Jesus' messiahship, how they abandoned him in his time of greatest need, and how after the crucifixion, still in hiding in fear for their lives, it was the Risen LORD's *presence* that gave them eyes to see who He really was; allowing them to confess Him to be "My LORD and my God" (Jn 20:28). His presence provided them the courage to go forth boldly proclaiming the Gospel and following Him as LORD, thereby experiencing "life in His name" (Jn 20:31). And it is His love, grace, and presence that will sustain us as we live out His teachings about the five great loves.

Chapter 15, *The Spiritual Goal of Overcoming Global Warming: Transforming the Future*, anticipates the challenges and opportunities described in chapters 16-21. It asserts that the primary spiritual goal is to become Christ's agents of transformation as we strive with Him to work with others in overcoming global warming in this great cause of freedom.

Chapters 16 through 19 describe how it is possible to overcome both the causes and consequences of global warming. We can do so at a modest cost or even net economic benefit, but the scale and speed of the changes necessary offer a significant challenge. Overcoming global warming is therefore a tremendous opportunity for Christians to follow the Risen LORD in leading the way. These chapters provide a game plan and concrete examples of how the causes can be successfully addressed in ways that create jobs, enhance the lives of the poor, and bring about a cleaner, safer world.

Chapters 20 & 21 explain how the ways we already assist the poor in developing countries – via sustainable economic progress; addressing hunger, malnutrition, water scarcity, and health issues; the fostering of democracy and human rights; and the empowerment of local communities – are all part of the poor being able to cope with or adapt to the consequences.

These chapters also describe how more targeted adaptation efforts such as floating gardens, flood-resistant housing, rainwater harvesting and drought-resistant crops, are also necessary to cope with global warming's consequences. With sufficient funding and commitment, it will indeed be possible to help the poor in developing countries adapt. This will again provide Christians with tremendous opportunities to make a difference.

Chapter 22, *Walking into the Future with the Risen LORD*, outlines in practical terms what it means for us to be the Risen LORD's agents of transformation in this great cause of freedom. It also provides concluding reflections on how Christ the Creator, Sustainer, true image, Savior, and Risen LORD is with us as we grow in the five great loves by playing our part in following Him in overcoming global warming.

As this "travel guide" suggests, there will be some tough times while you journey through this book. In reading chapter 6 you will discover some hard truths about the impacts of global warming – that it is or will become:

> a "threat multiplier" from the perspective of national security;
> a "poverty intensifier" that could wipe out many of the gains Christians and others have supported and plunge those climbing out of poverty back into it; and
> the greatest driver of species extinction in this century.

Along the way you will meet individuals whose real-life situations illustrate what others might encounter in the future. People such as Doug deSilvey from Gulfport, Mississippi, who lost his entire family during Hurricane Katrina; or two young boys from Honduras, Olbin and Edgardo Casares, whose father was killed and home destroyed by Hurricane Mitch. Rejected by their mother's family, they took to living on the streets.

But you will also discover that overcoming global warming is a tremendous opportunity not only to blunt the impacts just highlighted, but also to

☑ transform the world's economies into clean energy economies;
☑ create millions of sustainable jobs;
☑ get rid of a tremendous amount of deadly and debilitating pollution; and

☑ have sustainable economic progress become the norm in developing countries, given that two-thirds of efforts to reduce global warming pollution need to occur there.

You will also be introduced to those who are leading the way in overcoming global warming, such as Bill Keith, a former roofer from Indiana who invented a solar attic fan and took out a second mortgage to start his company, SunRise Solar. And Lee Scott, who as CEO of Walmart led the company to adopt the following aspirational goals: (1) to be supplied by 100% renewable energy; (2) create zero waste; and (3) sell products that are produced sustainably. And Alec Loorz, who began to give presentations on global warming when he was 12, and has educated over 10,000 students.

THREE THINGS TO KEEP IN MIND ON YOUR JOURNEY

As you make your way through the book and begin your journey of playing your part to overcome global warming with the Risen LORD, there are three things I hope you will keep in mind: (1) the time is now; (2) Christians must answer the call; and (3) you are never alone.

1) THE TIME IS NOW

This decade, 2010-2020, is the most crucial one to overcoming global warming. If we don't seriously begin the transformation to a climate-friendly economy and way of life now, we shall *not* overcome.

2) CHRISTIANS MUST ANSWER THE CALL

Many Christians must join this journey. If, God forbid, Christians fail to answer this call to follow the Risen LORD in overcoming global warming, it will be hard to overcome – especially when one considers the practical reality of our sheer numbers. All things are possible for the LORD, of course. But for some inexplicable reason the LORD has decided that He will achieve His will in society through human beings – especially Christians, who are the Body of Christ. We must answer the call.

3) YOU ARE NOT ALONE

If there is one thing I hope you take away from this book it is that you are never alone, whether you are dealing with this challenge or any other challenge or problem you face. Even when you are physically alone, even when you feel alone in a vast crowd of people, even during times of this journey of overcoming global warming when no one is with you and there may even be some against you, you are never alone.

As the Risen LORD says in the very last words of Matthew's Gospel: "**And remember, I am with you always, to the end of the age**" (28:20, NRSV).

One of the striking things about the Risen LORD's last appearance in Matthew's Gospel is recorded several verses before: "When they saw him, they worshipped him; but some doubted" (28:17). If we are honest, for many of us this captures something about our own experience with the Risen LORD. At times, yes, we worship Him, we acknowledge Him. But in our finitude, as we hurry and stumble through our day, we forget that He is there. And this forgetfulness can lead to doubts that He really *is* there, that He really *is* with us, that He really *does* walk beside us. Why would He? Why would He bother, frankly? I mean, honestly, I'm really not that important in the vast scheme of things. I'm sure He has much better things to do than to spiritually stand with me in the grocery line.

This weariness of faith, this doubt, combines with our finitude to create a form of spiritually pernicious egocentricity: to think that God shares our limitations and that His love is so small. If we let it, such corrosive doubt can spiritually stop us in our tracks. Thankfully, God's ways are not our ways.

Here is where we can see the mercy for us in the Risen LORD's reply to doubting Thomas: "**Stop doubting and believe**" (Jn 20:27). For with such belief comes a spiritual strengthening. As the Risen LORD further says to Thomas, "Because you have seen me, you have believed; blessed are those who have not seen and yet have believed" (Jn 20:29).

Literally, the Risen LORD does not stand before us; we don't physically see Him. But we are literally blessed when through our faith we allow the reality of the Risen LORD's spiritual presence to strengthen us in our journey. To put it succinctly: the LORD is with you.

WHAT IT'S REALLY ALL ABOUT

This has already been hinted at, but before you turn to the rest of the book I wanted to lift your eyes to the spiritual horizon of our journey. As

surprising as it may sound, overcoming global warming is really about … freedom … and love … and beauty … and glory.

That's because our response to global warming is part of the larger story of human freedom and of God's wonderful and terrible gift of freedom – which is the key to our destiny.

In a time before time God, Father, Son, and Holy Spirit, embarked upon the creation of the universe. And in the vastness of the universe He created a small planet called the Earth, and from the earth of the Earth he formed Adam, an "earth-man" (Adam from *adamah*, Gen 2:7). He breathed the breath of life, the life-force or *nephesh*, into this earth-man and he became a living being. Along with the other creatures, this finite creature of dust became infused with the gift of *nephesh* (Gen 2:7; Ps 104:29-30; Eccl 3:19). But to this living dust He gave the terrible and wonderful gift of freedom that brought both sin and its miseries, but also brings the capacity to love and live life to the glory of God, to freely do the LORD's will. The LORD risked our freedom to allow us to freely love Him, to be in freedom what he created us to be: beings who love Him back.

The other earthly creatures glorify God by living the lives he intended for them. They do so naturally by being themselves. Our ability to glorify God is limited only by our inability to freely choose to do so. Sin has clipped our wings, but Christ has restored our capacity to fly. As it says in Galatians 5:1, "For freedom Christ has set us free" (NRSV).

And Christ has set us free to allow us to fulfill our destiny – loving God back. And as we do so, through overcoming global warming and the vast multitude of other ways we have been given, as we become whom God created us to be – *nephesh*-infused earth-men and earth-women who freely love Him in ways only we can as distinct individuals with our particular personalities, gifts, and abilities – as we do so we become what He created us to be, spiritually beautiful and glorious.

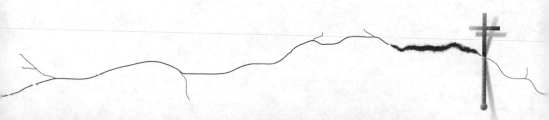

PART 1
A CHALLENGE FOR US ALL

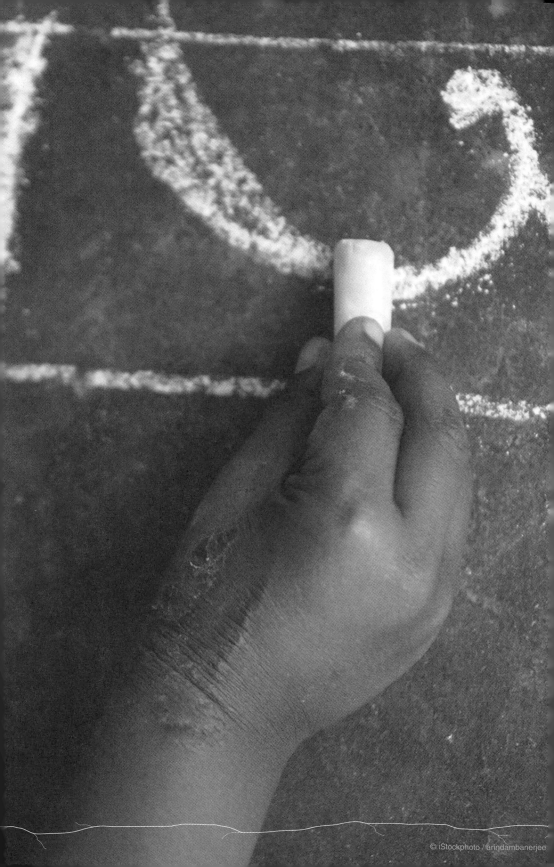

1

CLIMATE CHANGE: THE BASICS

This chapter will cover the following:
1) the difference between climate and weather
2) what makes up the climate and why it is foundational to well-being
3) how increasing the global temperature is like giving the planet a fever
4) evidence that the Earth is warming and the climate changing
5) the main reason why there is abnormal warming, which is changing the climate
   a. the greenhouse effect
   b. humanity is causing an abnormal warming.

THE DIFFERENCE BETWEEN CLIMATE AND WEATHER

We need to begin by understanding the difference between climate and weather. Let's say you were moving from North Carolina to New Hampshire. You might ask someone who has lived in New Hampshire a long time, "What's the weather like around here?" The thing is, you don't really mean weather. You mean climate. Your real question is, "What's the climate like around here?" That's because climate is the average weather of an area over time, while weather is what happens on a particular day or days in terms of temperature and precipitation. Weather is what we want to know when we ask what the weather forecast will be for today or over the next several days.

Under normal circumstances, climate is relatively stable and predictable. Weather, on the other hand, is transitory and unpredictable. That's why we have weather forecasters who make their living trying to predict the weather.

You don't need a weather forecaster to predict what the climate should be for where you live. So, returning to our example, if you were to move from North Carolina to New Hampshire and ask what the *climate* is like, you might hear, "Our summers are short and our winters are long and cold. You're going to want to get a good set of snow tires. My cousin John can set you up. It rains a good deal in the spring, and the fall colors are just breathtaking."

Global warming is going to make this "climate forecast" all wrong. If we continue to pollute at our current levels, scientific projections suggest that global warming will bring about a situation where "residents of New Hampshire would experience a summer climate more like what occurs today in North Carolina." The winter season will be cut in half, with hot summer temperatures arriving three weeks earlier and staying three weeks later.[8] The big city to the south, Boston, will experience over 60 days of 90°F-plus temperatures (compared to fewer than 10 in 1990) and 24 days of 100°F-plus temperatures.[9]

WHAT MAKES UP THE CLIMATE AND WHY IT IS FOUNDATIONAL TO WELL-BEING

A stable climate has been foundational to the health and progress of human civilizations. It has helped determine where we settle and invest in the creation of human societies.

Key determining characteristics of a particular area's climate are average temperature and precipitation patterns over thirty years or more. These combine with physical characteristics to compose the climate for a particular place.[10] In general, the climate, soil, landform (topography), and latitude of a specific location determine what type of plant life will grow there, and that determines the animal life. And the interrelationships

between plant and animal life and their surrounding environment make up the ecosystem of an area.[11] Thus, climate is a key defining characteristic of what lives in a particular place, i.e., what type of ecosystem exists on the landscape. Since the advent of settled agriculture over 10,000 years ago,[12] whether humans chose to settle and live in a certain place was based largely on climate.

By raising the average temperature, we are also changing the weather patterns as they relate to precipitation. And by changing temperature and precipitation, we are changing the climate.

Thus, when we start messing around with temperature and precipitation as we are with global warming, we are messing around with a lot. A stable climate such as the world has enjoyed for the past 10,000 years has been essential to human advancement[13] – a stability that is needed even more in a world with nearly 7 billion people that is projected to grow to over 9 billion by the half-century mark.

GLOBAL WARMING: CHANGING THE CLIMATE

Under normal conditions a healthy human body keeps its temperature within a certain range, a process called homeostasis. Normal temperature is 98.6°F. However, the body's normal temperature varies about 1°F in a 24-hour period, with the coolest temperature around 4 a.m. and the hottest between 4 and 6 p.m. If it rises above this normal rhythm and range, then the body has a fever. Even a few degrees difference can be accompanied by significant discomfort. For an adult, a temperature of 105°F constitutes a medical emergency.[14]

For the past 10,000 years, the Earth's climate and temperature have created a planetary equilibrium, analogous to the homeostasis of the human body. This equilibrium has been quite suitable for the growth and progress of human civilizations. There have been natural warm and cold periods during this time, the same way the temperature of our bodies naturally varies in a 24-hour period. But since the start of the Industrial Revolution, around

1750, many of our activities are abnormally warming the planet outside of these natural warming and cooling periods.

In effect, we are giving the planet a fever. We are causing the Earth to warm. This, in turn, is changing the climate. We are taking a normal situation of climate equilibrium or stability, one that enhances the well-being of humanity and God's other creatures, and changing it to one of disequilibrium and instability. As we explore more fully in chapter 9, this is the opposite of God's first commands to us in Genesis, namely to have dominion or rule as He would rule and to tend and keep the garden (Gen 1:26-28 and 2:15).

THE EARTH IS WARMING, THE CLIMATE CHANGING

Is the Earth warming? Yes. How do we know? By looking at temperature trends and by observing how God's creation is responding.[15]

We have reliable temperature records starting in 1850 from around the world taken with thermometers and other instruments on land and in the oceans. These instrumental records show a definite warming trend. From 1850 to around 1910, the Earth's average temperature was relatively constant. Then it began to rise until around 1940, when it declined for a few decades and then started to rise again in the 1970s.[16] The rate of warming is now accelerating. In the last 50 years the rate has nearly doubled that of the previous 100 years.[17] The last two decades (1990-2009) have been the warmest in the instrumental record.[18]

Historical temperature trends can also be determined by looking at what are termed "proxy" data, such as evidence from tree rings and coral reefs. Based on such evidence from the Northern Hemisphere (which are much more complete than from the Southern Hemisphere), there is a 90% or greater probability that "the temperatures during the second half of the 20th century were higher than for any other 50-year period in the last 500 years." And there is a 65% or greater probability that this period is the warmest in the last 1,300 years.[19]

God's creation is responding to the warming. A recent analysis of nearly 30,000 observational data series from 75 scientific studies found that approximately 90% of the changes studied in God's creation were consistent with the warming trend.[20]

Biologically, this includes the earlier timing of spring events such as leaf-unfolding, bird migration and egg-laying, shifts in the ranges of plants and animals, earlier fish migrations in rivers, and changes in the oceans, such as shifts in ranges for algae, plankton, and fish, as well as impacts on coral reefs.[21]

Major physical changes consistent with warming have also occurred. After 2,000 years of stability, sea level rose eight inches in the twentieth century and the rate is accelerating; sea level is now rising twice as fast in this century as it was in the twentieth.[22]

Precipitation patterns have also changed. "Warming accelerates land surface drying and increases the potential incidence and severity of droughts, which has been observed in many places worldwide." Warming also increases the water-holding capacity of the atmosphere "by about 7% for every 1°C rise in temperature" which leads to more heavy downpours when the rains do come. As would be expected, "Widespread increases in heavy precipitation events have been observed, even in places where total amounts have decreased." Finally, rain is falling rather than snow in many areas, "leading to increased rains but reduced snowpacks, and consequently diminished water resources in summer, when they are most needed."[23]

Other physical changes consistent with warming include the "melting of Arctic sea ice, the retreat of mountain glaciers on every continent, reductions in the extent of snow cover … and increased melting of the Greenland and Antarctic ice sheets."[24] Finally, "Heat waves have become longer and more extreme" and "Cold snaps have become shorter and milder."[25]

Many of these signs of global warming from God's creation are happening faster, sometimes much faster, than predicted.[26]

To return to our analogy of the human body, the current warming is like a fever that is producing unhealthy responses in the form of a changing climate. While climate has a range of natural variability, because of global warming we are quickly moving outside that range.

THE MAIN REASON WHY THERE IS ABNORMAL WARMING, WHICH IS CHANGING THE CLIMATE

To understand what we are doing to warm the Earth, we must first understand something called the greenhouse effect.

THE GREENHOUSE EFFECT: PUSHING A GOOD THING TOO FAR

As will be discussed more fully in chapter 8, it is through God the Son that the Triune God creates and sustains all things. Christ the Creator created a world that He declared good not once, not twice, but seven times in the first chapter of Genesis. Psalm 104 also describes the wonderful balance that Christ's creation displays, with everything working together to bring glory to God. Indeed, it is precisely when the things of creation work as they should that they glorify the Creator.

Psalm 19 proclaims:

[1] The heavens declare the glory of God;
   the skies proclaim the work of his hands.

[2] Day after day they pour forth speech;
   night after night they display knowledge.

[3] There is no speech or language
   where their voice is not heard.

⁴ Their voice goes out into all the earth,
   their words to the ends of the world.
   In the heavens he has pitched a tent for the sun,

⁵ which is like a bridegroom coming forth from his pavilion,
   like a champion rejoicing to run his course.

⁶ It rises at one end of the heavens
   and makes its circuit to the other;
   nothing is hidden from its heat.

When the sun and the sky work together as they should, they communicate in a manner that everyone can understand. Indeed, as Paul will later say, their witness and that of the rest of creation to the Creator means that we are "without excuse" in knowing who God is (Rom 1:20).

One of the ways Christ the Creator made the sky (or atmosphere) and the sun to work together is through a process known as the *greenhouse effect*. Without this effect, the average surface temperature of the Earth would be about 50°-60°F colder.[27] In other words, the greenhouse effect makes the Earth habitable.

Here's how it works. The Sun's rays warm the surface of the Earth. The warmed Earth then reflects some of this energy back towards space. But some of this reflected energy is in turn captured by heat-trapping gases in the atmosphere.

While the physical process is actually different from a greenhouse, the effect is the same. The same way the glass of a greenhouse or rolled up windows of a parked car keep things warmer inside than outside, naturally occurring heat-trapping gases keep the Earth about 50°-60°F warmer than it otherwise would be. These gases act like the glass of the greenhouse and trap heat. As such, they are called *greenhouse gases* (GHGs). The naturally occurring GHGs are water vapor, carbon dioxide (CO_2), methane (CH_4), nitrous oxide (N_2O), and ozone (O_3). Man-made gases called chlorofluorocarbons or CFCs are also potent heat-trapping gases.[28]

The main problem is that human activities are adding significant amounts of greenhouse gases to the atmosphere. This increases the naturally occurring greenhouse effect and throws out of balance the relationship God the Son created between the Sun and the sky and the Earth and the creatures of the Earth. These relationships are no longer working as they should, and this disharmony is diminishing creation's glory to God.

HUMANITY IS CAUSING AN ABNORMAL WARMING

As the evidence of warming and climate change started to mount over the last several decades, the world's leading climate scientists began to create major reports to help policymakers and the public understand the causes and consequences. The National Academies of Science and governmental scientific agencies from many nations around the world and numerous prestigious scientific associations have issued reports and statements.[29]

What are the scientists telling us in all of these reports? In a nutshell: **we are the primary cause of global warming since 1950, and our actions will only increase this trend unless changes are made**. As a major report from the US National Academy of Sciences (NAS), issued in May 2010, says, "Climate change is occurring, is caused largely by human activities, and poses significant risks for — and in many cases is already affecting — a broad range of human and natural systems."[30] The report regards both global warming and that human activities are the primary cause as "settled facts."[31] It goes on to state, "The ultimate magnitude of climate change and the severity of its impacts depend strongly on the actions that human societies take to respond to these risks."[32]

Internationally, the major reports of the Nobel prize-winning Intergovernmental Panel on Climate Change (IPCC), considered the world's most authoritative body on the subject, have provided the scientific basis upon which the world's governments have collectively made their decisions

about climate change. The IPCC's summary reports are approved line by line in open meetings by representatives from nearly all the governments of the world (except some of the smallest ones).[33]

While to some it is still news that the actions of humanity are warming the planet, and to others this scientific conclusion still is not something they are willing to accept, the IPCC concluded in 1995 that we were doing so. In their now famous words, the IPCC in its Second Assessment Report said, "the balance of evidence suggests that there is a discernible human influence on global climate."[34]

In 1995 Sir John Houghton oversaw the process and meetings where this conclusion was accepted. He is one of the world's leading climate scientists; he also happens to be a devout evangelical Christian.[35] As he chaired the key meetings, countries like Saudi Arabia, buttressed by the American Petroleum Institute and American coal companies, tried to weaken the report's conclusions. Others, influenced by some of the more radical environmental organizations, tried to exaggerate the threat. During this crucial turning point in humanity's willingness to face up to the truth of this challenge – where the major players knew the stakes but the rest of the world was oblivious – it was Sir John's personal Christian commitment to the truth that guided him, and it was the presence of the LORD and the prayers of Christian colleagues that sustained him.[36]

Scientific findings since 1995 have confirmed and strengthened the conclusions of the IPCC. The 2007 IPCC report involved over 3,750 scientific experts from more than 130 countries.[37] It concluded that "Warming of the climate system is unequivocal,"[38] and that there is "*very high confidence* [at least a 9 out of 10 chance] that the global average net effect of human activities since 1750 has been one of warming … its rate of increase during the industrial era is *very likely* [90% or greater probability] to have been unprecedented in more than 10,000 years."[39] Furthermore, "Most of the observed increase in global average temperatures since the mid-20th century is *very likely* [90% or greater probability] due to the observed increase in anthropogenic [human caused] greenhouse gas concentrations."[40]

In other words: (1) human activities since 1750 have been warming the planet; (2) the rate or speed of the warming is faster than it has been in at least 10,000 years; and (3) most of the observed warming since 1950 is due to our actions. Just as is the case with the temperature of the human body, while some warming and cooling is natural, we have been creating *abnormal* warming. And most of the warming since 1950 is abnormal warming caused by us.

Furthermore, if we keep doing things like we have been, there is a good possibility that the *rate* of change during this century will exceed anything that has been seen for hundreds of thousands of years or more.[41]

To sum up:
1. Climate is not the same as the weather, which changes moment by moment. Rather, climate is the average weather – or temperature and precipitation patterns – for a particular area over several decades or more.
2. The Earth has enjoyed a stable climate over the last 10,000 years that has helped provide the foundation for the flourishing of human civilizations.
3. Both the temperature record and observations from God's creation demonstrate that the Earth is warming. The second half of the twentieth century was probably warmer than any period in the last 1,300 years.
4. Human activities are responsible for most of the warming since 1950. If we keep going as we have been, the rate of change will be something not seen in the last 50 million years.
5. By adding heat-trapping greenhouse gases to the atmosphere we are warming the Earth; we are enhancing an otherwise naturally occurring and beneficial process called the greenhouse effect.

Global Warming and the Risen LORD

2

HOW WE ARE CAUSING GLOBAL WARMING

If humanity is warming the Earth, how precisely are we doing so? There are two primary ways that share common causes: greenhouse gases and particles called black carbon. Greenhouse gases are by far the biggest contributor and most well understood, creating at least 84% of the problem. We have only recently begun to understand the role of black carbon. Let's discuss each in turn.

GREENHOUSE GASES

According to the Intergovernmental Panel on Climate Change (IPCC), the principal greenhouse gases that human activities are adding to the atmosphere are:

► carbon dioxide or CO_2

► methane or CH_4

► nitrous oxide or N_2O

► halocarbon man-made gases or F-gases.[42]

GREENHOUSE GAS EMISSIONS WORLDWIDE

The chart below utilizes figures from the IPCC to represent worldwide greenhouse gas emissions from 1970 to 2004.

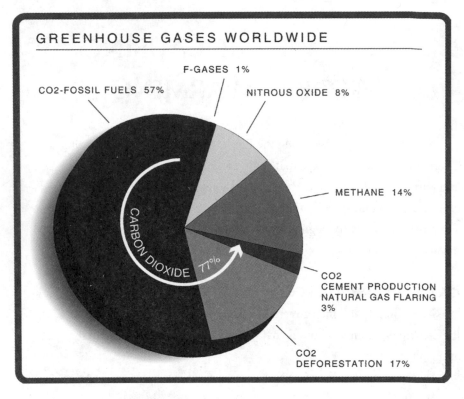

At 57% of the total, the largest contributor is the burning of fossil fuels, which we use to power our homes, businesses, industries, cars, trucks, and planes. Next at 17% are CO_2 emissions from deforestation primarily in developing countries, to provide the world with lumber and firewood and to clear the land for agricultural products, a good deal of which are exported. (Trees and other plants naturally store carbon as they grow. When they are burned or decay they release this stored carbon into the atmosphere.) Combined, the burning of fossil fuels and deforestation contribute nearly three-fourths of human-caused greenhouse gas emissions.

Global Warming and the Risen LORD

We can also look at where emissions are coming from regardless of gas type by seeing how much each major sector of the world's economy is contributing:

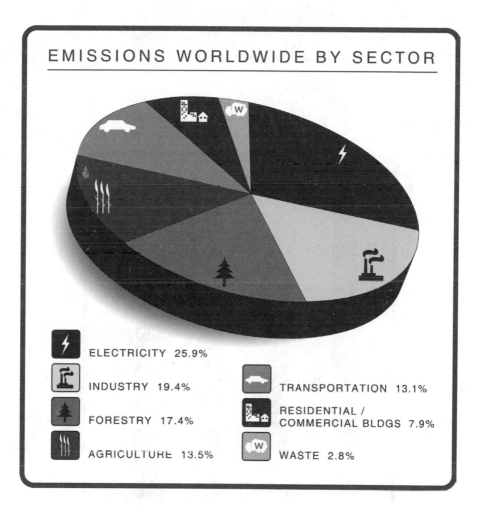

EMISSIONS WORLDWIDE BY SECTOR

ELECTRICITY 25.9%

INDUSTRY 19.4%

TRANSPORTATION 13.1%

FORESTRY 17.4%

RESIDENTIAL / COMMERCIAL BLDGS 7.9%

AGRICULTURE 13.5%

WASTE 2.8%

GREENHOUSE GAS EMISSIONS FROM THE US

In the United States, greenhouse gas emissions are dominated by carbon dioxide (CO_2) even more than is the case worldwide. The greenhouse gas contributions from the United States are as follows: 99% of the CO_2 comes from the burning of fossil fuels (with oil comprising 42% of these energy-related CO_2 emissions, coal at 36.5%, and natural gas at 21%.)[43]

If you break down the United States carbon dioxide emissions by end use:

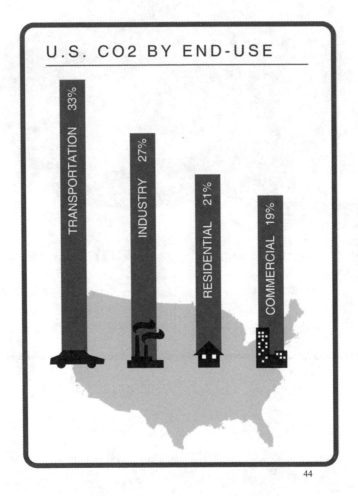

U.S. CO2 BY END-USE

TRANSPORTATION 33%
INDUSTRY 27%
RESIDENTIAL 21%
COMMERCIAL 19%

[44]

It is important to note that emissions from electric utilities contribute a significant amount to the last three. Nearly half of the CO_2 emissions from industry come from electricity; for residential it's 72% and for commercial it's 79%.[45]

All told, 41% of US greenhouse gas emissions are generated by electric power utilities with 82% of this generated by coal and 15% by natural gas.[46] Coal produces over half of our electricity.

Methane, at 10% of greenhouse gases in the United States, comes in roughly equal amounts from the energy sector, agriculture, and waste management, while nitrous oxide (4.3%) mainly comes from agriculture, with smaller contributions from industry and waste management.[47]

Thus, the two biggest sources of total US greenhouse gas emissions are (1) coal-burning utilities that power our homes, businesses, and industries, contributing 34%,[48] and (2) 33% from vehicles that burn gasoline and diesel derived from oil.

PAST AND FUTURE EMISSIONS

An important thing to understand about the principal greenhouse gases is that they trap heat for a long time, some much longer than others. The lifetime for atmospheric methane is estimated to be about 12 years, whereas that for nitrous oxide is around 110 years.[49] One of the F-gases (CF_4 or tetrafluoromethane) could be trapping heat for over 50,000 years.[50]

Carbon dioxide is a bit of a special case when trying to estimate how long it traps heat in the atmosphere. Unlike the other heat-trapping gases, whose lifetimes are determined by the fact that they break down in the atmosphere, CO_2 is removed from the atmosphere in its existing form through a natural process.[51] As the IPCC states: "While more than half of the CO_2 emitted is currently removed from the atmosphere within a century, some fraction (about 20%) of emitted CO_2 remains in the atmosphere for many millennia." The IPCC concludes, "A lifetime for CO_2 cannot be defined."[52] The EPA gives a range of 50-200 years,[53] while one prominent scientist suggests, "the likely upper limit may be 50 or 60 years."[54] A commonly accepted lifetime is 100 years,[55] which is what we will use in this book – recognizing that some of what we put up there today could be trapping heat more than 1,000 years from now.

Therefore, *looking forward* we should assume that what we emit today will be trapping heat for the remainder of this century, which is one of the reasons for urgency. *Looking back*, much of what was emitted, say, in 1950 is still causing warming today.

This latter point on past emissions is important when considering why the world should work together to overcome global warming. Up until 2005 a majority of CO_2 emissions from energy had come from the developed countries.[56] Whether we start by looking at emissions in 1850 through this past century or start in 1900, the result is the same: the United States has been the world's largest emitter of CO_2, contributing nearly 30%. From 1850 to 2002 the developed countries contributed 76%, the developing countries 24%.[57] Thus, during this time period the US contributed more CO_2 from the burning of fossil fuels than all of the developing countries combined.

In effect, past emissions from developed countries have created a temperature platform upon which humanity is building towards highly dangerous temperature levels that will lead to thresholds better left uncrossed. Those who have done the least to create our current situation, the poor in developing countries, are the most vulnerable to the impacts of global warming.

Now before you start to feel like I'm laying a big guilt trip on those of us in the United States, I'm not. I don't think that's helpful. We did not set out to burn these fossil fuels to create global warming, something the world has realized as a major problem only relatively recently. We utilized the energy contained in coal, oil, and natural gas to create electricity and to power our vehicles – in short, to power our economy and create a better life. And with our wealth derived from burning fossil fuels we have been able to help others around the world.

Global Warming and the Risen LORD

The simple fact is **we didn't know** we were helping to create this problem. **But now we do**. Now we know the facts. A fair-minded person, when presented with such facts, would say, "I'll help to clean up the mess I inadvertently made." And Christians especially, when presented with a case like this, will recognize that we have an even greater moral calling to ensure that "the least of these" (Mt 25) are helped, not hurt, by our actions.

There is a "silver lining," so to speak. The wealth and benefits we in the United States have received from the burning of fossil fuels can now be used to overcome the causes and consequences of global warming. The benefits derived from creating the problem can now be used to "fertilize" the seeds of the solutions – and there are plenty of solutions to be nurtured in both the United States and in developing countries, as will be seen in chapters 16-19.

Such assistance to developing countries will be needed for all our sakes, because in this century it is the developing countries who will be the biggest emitters. In fact, they already are. As of 2005, the developing countries surpassed the developed ones in terms of total energy-related CO_2 emissions, a trend that is projected to continue and grow.[58] As will be seen in chapter 16, the projected contribution of developing countries means that approximately two-thirds of all greenhouse gas reductions will need to occur in these countries in order to reach our global reduction goals.

Remember that greenhouse gases have long lifetimes in the atmosphere. The developing countries are building upon a temperature platform created largely by the developed countries. If the world keeps going on a "business-as-usual" trajectory, it is our twentieth and twenty-first century emissions combined with their twenty-first century emissions that will have us exceeding dangerous tipping points. Simply put, we are in this together. We will sink or swim together.

The good news is that overcoming global warming will help shift the world into a clean energy future. Those who think that developing countries must develop just like we did are trapped by the thinking of the past. The future belongs to clean energy. And as Christians, we must work to ensure that the poor in developing countries benefit from this transition; this is not so much about sinking or swimming together, it's about freedom. Overcoming global warming is the next great cause of freedom.

EMISSIONS INCREASING RAPIDLY

One of the most important aspects of this challenge is how fast greenhouse gas emissions have risen – the *rate* of increase. As we have seen, carbon dioxide is the biggest contributor. It has increased by over 35% since 1750 (when the Industrial Revolution began) and is now higher than it has been in at least 800,000 years.[59]

Historically, it has taken about 5,000 years to increase CO_2 emissions by 80 parts per million (ppm). During the last three decades of the twentieth century the world added 50ppm[60] and from 1970 to 2004 global greenhouse gas emissions grew by 70%, with CO_2's growth rate at 80%.[61]

Between 2000 and 2007 the rate of growth in energy-related CO_2 emissions continued its ascent at 3.5% per year (compared to 1.1% in the 1990s). This has produced more CO_2 in the atmosphere than the worst case scenario projected by the IPCC in the late 1990s.[62] The Great Recession slowed the rate of growth for 2008 to 2.2%, which was still higher than most of the IPCC emissions scenarios.[63]

The earth's soils, vegetation, and oceans store CO_2 naturally. During the last 50 years, they have soaked up 45% of our emissions.[64] However, there is "increasing evidence of a decline in the efficiency of CO_2 sinks in oceans and on land in absorbing anthropogenic emissions." Again, this is happening faster than expected.[65] We are increasing our CO_2 emissions at an unprecedented rate. At the same time, a vital natural process that had been soaking up nearly half of our emissions is starting to be overtaxed and is losing its ability to continue to do so. In effect, when it comes to natural absorption of CO_2 via the oceans, we have nearly eliminated this natural "margin for error," further throwing things out of balance.

In summary, over half of greenhouse gas emissions worldwide come from CO_2 produced by burning fossil fuels (over 80% in the United States), while over 17% comes from deforestation primarily in developing countries. Historically, developed countries dominated the nineteenth and twentieth centuries in terms of emissions (the United States contributing more than all developing countries combined). However, the developing countries will dominate this century. Emissions are increasing even faster than the worst case projections of the IPCC, while nature's ability to store CO_2 may be close to its limit.

BLACK CARBON

Many discussions of global warming proceed as if greenhouse gases constitute 100% of the problem. While these gases are responsible for the vast majority of global warming pollution, they are not the whole story. Another global warming pollutant is called "black carbon."

For our purposes a basic definition of black carbon is airborne soot particles that cause global warming.[66] Such soot comes from the burning of both (1) fossil fuels and (2) plant materials or animal waste used as fuel (also called biomass). The blackest or darkest of black carbon soot particles, such as those from the burning of diesel fuel, cause the most global warming and do so wherever they are emitted. One gram of such black carbon, which, for example, would be emitted from the burning of "about half a gallon of diesel fuel in a mid-1990s engine," adds the same amount of energy to the atmosphere as the energy used by a large barbecue grill.[67] And "black carbon adds 2-3 orders of magnitude more energy to the climate system than an equivalent mass of CO_2."[68]

Recent research has concluded, "Emissions of black carbon are the second strongest contribution to current global warming, after carbon dioxide emissions."[69] A leading scholar in the field estimates black carbon's contribution at 16%.[70] A 2008 report by the US government suggests that by 2050 it could comprise 20%.[71] Such studies and reports are good science, and there is no doubt that black carbon is a warming agent. However, the scientific community has not yet settled how significant a role black carbon and other related cooling particles play in global warming.[72] Thus, the suggested size of its role – 16% – should be considered a rough estimate.

While black carbon is a very potent global warming pollutant, thankfully it is also a temporary one. It is normally flushed out of the atmosphere in 1 to 2 weeks.[73]

The combination of high potency but short lifetime makes black carbon an ideal candidate for a pollution-reduction strategy designed to get more bang for the buck in the short term as a *supplement to* a realistic but aggressive greenhouse gas reduction effort. An all-out effort on black carbon in the next several decades *combined with* such a greenhouse gas program could potentially help us avoid some of the more severe global warming impacts in this century.[74]

However, if we fail to act soon on black carbon, initial studies indicate that we would need to achieve reductions in greenhouse gases "one or two decades sooner" than we would otherwise if we hope to have a reasonable chance of overcoming global warming (i.e., keeping the temperature 2°C below preindustrial levels).[75]

While black carbon's contribution needs further research and analysis to form a scientific consensus about its exact contribution to global warming, there is no doubt that it is making an important contribution.

TWO IMPORTANT WAYS THE BLACK CARBON PROBLEM IS DIFFERENT

1) WARMING AND COOLING PARTICLES

Black carbon particles are not the only type of particles that are released from the burning of fossil fuels and biomass. One of the major, complicating factors of the black carbon discussion is the fact that there are particles that warm (black carbon) and there are particles that cool. Indeed, the cooling particles play a major role in actually dampening global warming. According to several prominent scientific experts, approximately

half of the global warming that could have happened since 1750 has been dampened or masked by such cooling particles.[76]

About two-thirds of this cooling effect comes from sulfate particles, with over 70% of the sulfate contribution generated from sulfur dioxide (SO_2) pollution,[77] which is a major health hazard and component of acid rain.

Worldwide, more than half of SO_2 emissions come from coal, with oil the second largest source.[78] Global emissions peaked in 1980 and have been in decline since then primarily due to air pollution laws.[79] Until the 1950s the United States and Europe dominated SO_2 emissions. But by 2000 Asia had surpassed them. In the United States, even as our economy grew by 126%, SO_2 was reduced by a remarkable 56% from 1980 to 2007.[80] This was due to the 1970, 1977, and 1990 Clean Air Acts, with the latter especially effective, employing the first cap-and-trade systems to significantly reduce costs.

So while we have rightly been reducing sulfur dioxide emissions for health and environmental reasons, we have not reduced the black carbon or the CO_2. We are reducing the cooling particles – which have dampened or masked approximately 50% of global warming – while doing nothing about the warming particles. This is a deadly recipe. If this continues there is no way we could overcome global warming.[81]

The burning of plant materials or animal waste for fuel results in a mix of cooling and warming particles that can have a range of effects locally. Depending upon the mix the result could be anywhere between slightly warming to slightly cooling.[82] However, particles can be transported by atmospheric currents and come to rest on snow or ice in places like the Arctic and the glaciers of the Tibetan Plateau (which includes the Himalayas). The cooling effects of the particle mix are essentially neutralized or turned into warming because the black carbon particles in the mix are darker than the snow and ice on which they land. This reduces the ability of the ice and snow to reflect back the sun's rays and simultaneously absorbs the heat – a global warming one-two punch.[83]

2) REGIONAL CONSEQUENCES

Unlike greenhouse gases, black carbon and other associated pollutants can have climate impacts at a regional level greater than their global contribution. These regional impacts are estimated to affect about 3 billion people.[84] They include three important consequences:

1. The melting of snowpacks and glaciers in the Himalayan region, where black carbon may be just as powerful a factor as greenhouse gases.[85] As we will see more fully in chapter 6, this melting is not trivial matter, since these glaciers and snowpack feed rivers that provide water to 40% of the world's population.

2. 20%-50% of the heating in affected regions felt at the surface and in the lower atmosphere comes from black carbon.[86]

3. Through complex interactions in the atmosphere, black carbon and associated pollutants can also lead to regional drought. A key study concluded, "Precipitation trends over many regions of the tropics during the last 50 years have been negative, particularly over Africa, South Asia and northern China."[87] This includes "a decrease in monsoon precipitation over India and Southeast Asia by about 5-7 percent."[88] Such changes have had serious consequences for the poor. One study found that annual rice output for nine states in India between 1985-1998 was reduced by "the total annual consumption of 72 million people."[89]

NET WARMING FROM PARTICLES BY REGION

Looking by region at the net atmospheric effect of both the cooling and warming effects of particles (i.e., subtracting the cooling from the warming), one finds net global warming contributions as displayed in the graph on the following page.[90]

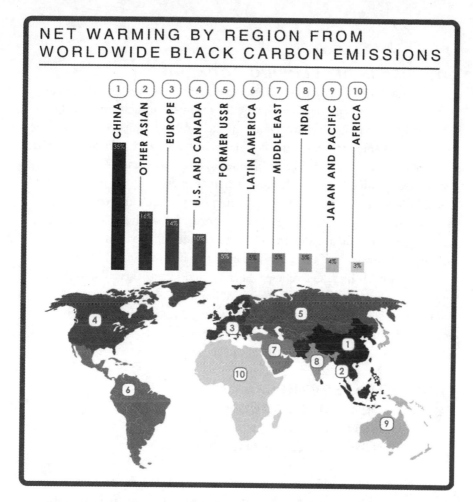

NET WARMING BY REGION FROM WORLDWIDE BLACK CARBON EMISSIONS

1. CHINA — 35%
2. OTHER ASIAN — 16%
3. EUROPE — 14%
4. U.S. AND CANADA — 10%
5. FORMER USSR — 5%
6. LATIN AMERICA — 5%
7. MIDDLE EAST — 5%
8. INDIA — 5%
9. JAPAN AND PACIFIC — 4%
10. AFRICA — 3%

SOURCES IN THE UNITED STATES

Looking at black carbon emissions from their sources in the United States is rather straightforward. Eighty-nine percent of black carbon emissions come from transportation sources that use diesel fuel, with 45% coming from so-called "road sources," i.e., trucks and cars, while 44% come from "off-road" sources, i.e., tractors and other off-road diesel vehicles. The remaining emissions come from industry and other coal-burning sources.[91] It is clear that reducing black carbon emissions from both on- and off-road diesel vehicles will significantly reduce global warming pollution.

SOURCES IN DEVELOPING COUNTRIES

Black carbon emissions from developing countries are complicated by the fact that, except for China where these emissions come primarily from coal-burning power plants, much of the soot emissions come from: (1) the "open burning" of trees and other plants in forests or savannahs, whether due to naturally occurring fires or fires set to clear land, and (2) the "contained burning" of wood for fuel, whether in cookstoves, furnaces, etc. Depending on the cooling/warming particle ratio of each emissions mix, "open burning" is either cooling or on the border between cooling and warming, while "contained burning" is slightly cooling to slightly warming.[92] However, if these particle emissions travel via atmospheric currents to places like the Arctic or the Tibetan Plateau, the effect becomes one of warming. In this regard, India's 5% contribution is closer to 7%.[93] Some scientists studying black carbon have concluded that black carbon primarily from India and China "is as important as greenhouse gases in the observed retreat of over two thirds of the Himalayan glaciers."[94]

THREE STRATEGIC RESPONSES TO REDUCE BLACK CARBON

While all of this is in one sense very complicated, in another sense it is clear what the world can do to reduce black carbon particles. There are three strategic responses.[95]

First, we should concentrate on reducing black carbon diesel emissions anywhere around the world, but especially in the United States and Europe (due to emissions transport to the Arctic).

Second, we should reduce soot particles that are warming the Tibetan Plateau. This includes those that might be primarily cooling elsewhere but become warming when they land on the snow and ice. The major focus should be to reduce black carbon emissions from Asia. With China, this means tackling coal emissions (which are also a major source of greenhouse gases, a global warming two-fer). As for India and other Asian nations, black carbon comes primarily from burning firewood and other biomass in inefficient cookstoves.[96]

The contribution of better cookstoves to overcoming the causes of global warming could be dramatic. One study suggests that implementing a program using "black carbon-free" stoves in South Asia could reduce black

Global Warming and the Risen LORD

carbon emissions 70%-80%, while in East Asia reductions would be 20%-40%. The health impacts would also be substantial, as over 400,000 premature deaths in these areas are attributed to indoor air pollution from such cooking.[97] As we will explore in chapter 19, field studies in India found that certain efficient cookstoves reduced emissions of both black carbon and greenhouse gases by 50% to 70%.

Third, as the science becomes clearer we should target particle emissions that are helping to create the regional "hot spots," leading to drought, among other things.

WHY BLACK CARBON CANNOT BE IGNORED

- It constitutes up to 16% of the problem.
- It can create significant regional consequences, including impacting the 40% of the world's population who depend on water from the Himalayan glaciers.
- Because of its short lifetime, reducing black carbon can have an almost immediate positive impact on overcoming global warming.
- Reducing black carbon via efficient cookstoves for the poor will bring health and other important benefits.
- Reducing only the cooling particles (such as sulfates) in the fight against air pollution will unmask up to half of current warming; we simply can't ignore the warming particles.

CONCLUSION

Climate change is an incredibly complex issue, and yet the bottom line on the causes of global warming is pretty straightforward. Without the burning of fossil fuels at the levels we have used them there wouldn't be anthropogenic (or human-caused) global warming. This abnormal warming is changing the climate and leading to consequences described in the next four chapters. At the dawn of the industrial revolution, when our fossil fuel bonfire began, we didn't intentionally set out to cause global warming. We were trying to make a better life. We didn't know about global warming and its causes. But now we do.

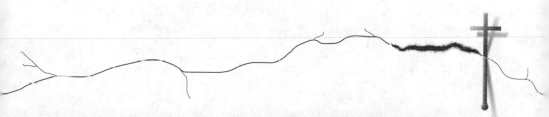

3

CONSEQUENCES FOR THE UNITED STATES – *INCREASED FLOODS, DROUGHTS, AND WILDFIRES*

Love each other as I have loved you. Greater love has no one than this, that he lay down his life for his friends. John 15:12-13

Twenty-one year old Joshua Landtroop was described by his friends as "the best dad in the world" to his sons, Nicholas, two, and Tristan, one. On May 1, 2010, Joshua left his job at the local Olive Garden in the Nashville area. Floodwaters had unexpectedly engulfed the area and Joshua was determined to make sure his boys were safe. He was last seen alive when the fast moving currents swept Joshua off his feet. A friend found his body later that day.[98]

Joshua was one of 29 deaths from storms and flash floods around Nashville, Tennessee, in early May 2010. Scientists call storms like this a "heavy precipitation event." It created 500-year flooding for many of the tributaries of the Cumberland River, which in turn flooded Nashville. The Associated Press reported, "The flash floods caught the city off guard, and thousands of residents and tourists were forced to flee."[99] Nashville was unprepared in part because for over 70 years a system of levees and reservoirs had been created to prevent such massive flooding. As Rodney Scott, a scientist with the US Geological Survey, put it, "That a regulated river like the Cumberland could have such high flooding is unusual and is a testament to the severity of this event."[100]

At the height of the flooding, the stage of the Grand Ole Opry, the epicenter of country music, was under two feet of water. The nearby Gaylord Opryland Hotel complex had 10 feet of water. Two elderly German tourists staying at Opryland, Gerdi and Kurt Bauerle, recalled, "We had just finished eating and suddenly they said 'Go! Go! Go!'"[101]

Over 2,000 homes were also destroyed, including that of Lisa Blackmon, 45, who didn't have flood insurance and recently lost her job at a trucking company. "I know God doesn't give us more than we can take," said Lisa, "but I'm at my breaking point."[102] Another victim, Woody Hall, remarked, "You know, you see it on TV all the time, but you never expect to live it."[103]

As many of us will discover in the coming decades, global warming will increase the number and severity of heavy precipitation events like the one that struck Nashville in 2010 because global warming intensifies natural disasters.

This chapter highlights the consequences of global warming in our country, the United States. Because the impacts will be so extensive, those we love will feel the effects of global warming. Jesus commands us to love each other as Christ loves us. In our day, loving our loved ones includes playing our part in overcoming global warming.

The following discussion is illustrative – certainly not exhaustive. Literally, only God knows what all the impacts will be. But we don't need to know everything, just enough to understand why we should follow the Risen LORD as He leads the way in overcoming global warming.

THOSE WE LOVE AND INCREASED FLOODS, DROUGHTS, AND WILDFIRES

Global warming makes droughts drier, storms stronger, and floods fiercer.[104] Such intensification of floods, droughts, and wildfires happens because global warming pushes weather patterns to their extremes. This has already begun in the United States and will get worse.

HEAVY PRECIPITATION AND FUTURE FLOODING

In many parts of the United States, rainstorms are becoming less frequent but more intense. For example, over much of the East an extreme storm that would have occurred once in 20 years is now expected about every eight years.[105] Scientists are convinced that such intensification is due to human-induced global warming. As a Bush Administration study released in June 2008 stated: "It is well established…that the global warming of the past 50 years is due primarily to human-induced increases in heat-trapping gases. …The increase in heavy precipitation events is associated with an increase in water vapor, and the latter has been attributed to human-induced warming."[106]

These increasingly intense rainstorms could lead to more floods like the Nashville flood of 2010 or to the type of floods that hit the Midwest in the summers of 1993 and 2008, both described as 500-year floods. Floods of this magnitude are expected only once every 500 years – not twice in 15 years.[107]

Up until the 2008 flood, the 1993 Midwest flood was the largest flood to ever hit the United States. It covered nine states and 400,000 square miles and caused $15 billion in damage, 50 deaths, and the evacuation of tens of thousands. Ten thousand homes were destroyed; 75 towns were under water; and 15 million acres of farmland were inundated. The flooding was due to rainfall amounts across the Midwest that were 200%-350% higher than normal. Many areas experienced rain for 20-plus days. Rainfall amounts for the summer across the Midwest exceeded 75-to 300-year frequencies. Some areas were flooded for nearly 200 days.[108]

Among the evacuees was Mike Johnson, an out-of-work machine operator, and his wife Roberta, their three children, two dogs and two cats.

They were awakened at 3:30a.m. and told they must leave their house in the St. Louis area. When he returned in a boat several days later he found 6 feet of water in his living room. Their temporary abode was their Bronco SUV. Mike slept on the roof to make room for his family in the vehicle.[109]

While the 1993 flood was devastating, it was exceeded by the flood of 2008, when precipitation levels in the Midwest were higher and covered a much larger area. Due to the extreme downpours, extensive flash flooding occurred. Over 1,100 daily precipitation records were broken across the area. In Dorchester, Iowa, 2-day and 4-day rain totals exceeded events that would be expected once in 1,000 years.

As for damages, more than 40,000 people had to be evacuated across the Midwest. Thirty percent of the corn and soybean crop were impacted. Total crop losses exceeded $8 billion. Iowa was hardest hit, with 83 of its 99 counties declared disaster areas.[110]

One of the most devastated cities was Cedar Rapids, Iowa. The city began 2008 in high hopes. They had named 2008 "The Year of the River" and planned outdoor concerts and a water skiing show in the hopes of revitalizing downtown. Unfortunately 2008 became the year of the river for another reason when the Cedar River crested at 19 feet above flood level – 11 feet above the earlier record.[111]

In February 2009, eight months after the flood, the *Wall Street Journal* reported on how Cedar Rapids was coping. Almost 10 square miles of neighborhoods were still largely abandoned. Five thousand four hundred apartments and houses and 300 municipal buildings were damaged. Estimated damages were around $2 billion. About 1,200 families were still living with relatives or in temporary situations, including City Council member Jerry McGrane, who was living in a FEMA trailer.[112]

Linda Seger, 64, a part-time substitute cook at a local school, and her husband, Gary, 62, a welder, were still living in temporary housing. They had depleted their savings to repair their home. The federal government contributed funds but stopped when the city declared their home was in an area planned for a levee. The city won't buy their home for at least five years. "'It's a state of suspension that we're in,' said Mrs. Seger. 'What is in our checkbooks is all we have.'"[113]

While there have always been natural disasters like floods, global warming has made and will make floods fiercer. The fact the Midwest has had two floods within the span of 15 years that were considered 500-year events is disturbing. Unfortunately it may be a sign of things to come.

DROUGHTS

In the United States, harsher droughts "may become increasingly likely"[114] due to both a lack of rainfall and reduced snowpack resulting from warmer temperatures. The American West and places already plagued by drought will become drier.[115]

Global warming may already be intensifying droughts in other areas like the Southeast. There are some indications that the severe drought of 1998-2002, the so-called "Turn of the Century" drought, which occurred in the American Southeast and other parts of the United States and the world, may have been intensified by global warming.[116] Significant impacts began in Florida, as fires began to burn out of control around the 1998 Memorial Day weekend and continued through July. Thick clouds of smoke covered much of the state, blocking out the sun in some areas. Over 120,000 Florida residents were forced to evacuate and over 300 homes and businesses were damaged or destroyed.

One of the homes was that of Hugh and Geraldine Conklin. When they were able to go back to see what had happened to their house, they found it destroyed. "I cried a lot last night … I'll probably cry some more," said Geraldine.

The cost to fight the fires was over $100 million. Another $300 million in timber was torched.[117] But that was just the beginning, for this drought was to last for four more years with billions in additional damages, including $10 billion in 2002 alone.[118]

At its peak in 2002 the drought in the Southeast was as bad as other major droughts in the instrumental record. North and South Carolina experienced their driest summer months ever recorded up until that time.[119]

Normally, such major droughts typically occur only every 40 years. However, the next major drought to hit the Southeast began only four years later, in 2006; 2007 was the driest year on record. One study of Tennessee found only two other years with such extreme drought in the past 400 years.[120]

These droughts heightened conflicts over water that had been building for years between Georgia, Florida, Tennessee, and Alabama. Population increases and poor planning created the conditions for the conflict and the droughts pushed the situation into a crisis. Previously, "water wars" were thought to happen only out west in the United States. This isn't the case anymore.

The American West has also suffered and will continue to suffer from an intensification of drought. A severe drought began in 1999 and has continued through 2008. In August 2002 it affected nearly 90% of the West (Rocky Mountains westward), making it the second most extensive and one of the longest in a century. The IPCC labeled it an "extreme climate event" that is "consistent with physically based expectations arising from climate change."[121] There's a good chance human-induced global warming has intensified this drought.

In December 2008, the Bush Administration issued its last major report on climate change. It warned that the West and Southwest could experience a "mega drought" of historic proportions with "a permanent drying by the mid-21st century" and "may reach this level of aridity much earlier." They concluded that "the strategies to deal with declining water resources in the region will take many years to develop...if hardships are to be minimized, it is time to begin planning to deal with the[se] changes," and there is cause for "immediate concern."[122]

Diminishment of mountain snowpack from global warming is another significant cause of water problems in the American West. Melting snowpack accounts for 75% of the flow of streams and rivers of the West.[123] One study found that in the second half of the twentieth century global warming was responsible for up to 60% of the climate-affected changes in river flow and snowpack.[124] A recent report concluded that global warming has already diminished and will continue to diminish mountain snowpack.[125]

The impacts to the Colorado River, which carved out the Grand Canyon, illustrate the serious consequences for Western water.[126] It is the main source of water for seven states and such major cities as Los Angeles,

San Diego, Los Vegas, Phoenix, Albuquerque, and Denver. In 2007 it provided water to approximately 25 million and is projected to serve 38 million by 2020.[127] Even without global warming's impact the river is overtaxed, and the region's population increases will only heighten the problem. But add in global warming and a bad situation becomes even worse, unless major adaptive measures are taken.[128] Runoff decreases due to global warming could reach 20%-30%.[129] By 2025, when you factor all of this in, the current water allocation agreement may be met only 60%-75% of the time.[130]

As soon as the 2020s, 41% of Southern California's supply will be vulnerable to global warming.[131] There is an excellent chance that the present conflicts over water in the West will only get worse.

WILDFIRES AND TREE INFESTATIONS

Wildfires are another consequence of global warming pushing weather patterns to their extremes. And an independent but related problem is the explosion of the pine beetle population. Wildfires have and will continue to increase due to drought intensification. Insects like the pine beetle are able to live longer and increase its range due to warmer temperatures, thus enhancing its capacity to kill trees.[132]

Let's focus first on the pine beetle. Since 1990 it has killed millions of trees in North America, especially in the West. The fact that the pine beetle kills trees is not unusual. But, according to the US Forest Service, "several of the current outbreaks, occurring simultaneously across western North America, are the largest and most severe in recorded history."[133] A key factor in all of these outbreaks is global warming, because cold temperatures help keep the beetle in check. The warmer climate has allowed the pine beetle to avoid cold-induced death, move into new territory, and breed up to three times faster. Many of the trees are weakened due to drought and are thus ripe for the kill.[134]

To try to save some of the infested trees, many others are being cut down to reduce the competition for scarce water. For landowners who have had to cut most of the forest surrounding their homes, it has been painful. "I've literally had people in my office crying," said Gary Ellingson, a forestry expert.[135] Others are preemptively cutting to avoid the safety hazard of the dying trees falling on people, roads, or buildings. A large number of campgrounds have been closed. "'We know they are going to fall,' said Clint Kyhl, a forest official. 'And they are going to fall in the next 10 to 15 years. There are campgrounds, thousands of miles of road, picnic areas, power lines and trails. How do we keep the facilities open for people to use?'"[136]

These problems, as well as the visual blight of the dead trees, could have a significant impact on tourism. In Grand County, Colorado, where beetles and drought have already killed much of the forest, four million tourists had been coming annually to enjoy the beauty of God's creation. "'What happens,' said Ray Jennings, director of emergency management for Grand County, 'if this becomes an ugly place to be?'"[137] To make matters worse, these dead, beetle-infested trees are also kindling for wildfires.

Since the mid-80s, there has been a dramatic increase in major wildfires in the West due in large measure to global warming. Earlier melting of snowpack increases the fire season, and increased temperatures, drought, and beetle infestation contribute to a widespread rise in tree mortality rates.[138] Snowmelt in the western US is occurring one to four weeks earlier with five times as many fires in early snowmelt years. The wildfire season has increased by 64%. The number of fires themselves has increased by nearly four times, and they are over six times as large, lasting nearly 30 days longer.[139] As of January 2008, six of the ten worst wildfire seasons have happened since 2000.[140] In 2006, the worst season on record, there were 84,500 wildfires with 9.4 million acres burned.[141] These wildfires are now costing taxpayers $1.7 billion a year to fight.[142] Unfortunately, with global warming there will be no end in sight. A recent report from the National Academy of Sciences suggests that "warming of 1°C (relative to 1950-2003) is expected to produce increases in median area burned by about 200-400%" in the American West; portions of western Colorado could see an increase of 656%.[143] Given that 1°C of warming is quite likely, Westerners can expect a tremendous increase in wildfires. The only thing to stop it in some areas will be when all the trees are burned.

Global Warming and the Risen LORD

Wildfires have significant additional economic and personal costs. In 2002 wildfires in Colorado cost tens of millions in tourism losses.[144] The story of the Hakanson family of Northern California illustrates a human dimension of these tragedies. "'We basically all ran for our lives,' said Steve Hakanson, 58, as a fire quickly spread to their neighborhood. 'I'm looking in my rearview mirror, and all I could see was red,' said Steve's wife, Linda, 47, as they sped away from their home of 20 years. When they were allowed to briefly return, they found their home burned to the ground. Besides their home, they lost two vehicles and irreplaceable treasures such as the china Linda had since high school and her grandmother's handwritten recipe for peach pie. 'It's just a lifetime of everything lost,' Steve said. 'There's just some things that can't be replaced.'"[145]

4

COASTAL CONSEQUENCES and HEALTH CONCERNS IN THE UNITED STATES

IF THOSE WE LOVE LIVE ON OR NEAR A COAST: SEA LEVEL RISE, STORM SURGE, AND HURRICANE INTENSIFICATION

Chances are excellent that either you or someone you love lives on or near a coast. That's because over 50% of the US population lives in a coastal area, even though the coasts are only 17% of the US land mass.[146] An additional 25 million people are projected to move in over the next 25 years.[147] Unfortunately, the coasts will be some of our areas most affected by the projected impacts of climate change. They will face sea level rise, stronger storm surges and flooding, increased coastal erosion, saltwater intrusion, ocean acidification and the death of coral reefs, and the possibility of more intense hurricanes. These impacts in combination with population increases on the coasts, water pollution, unsustainable fishing practices, and harmful development practices (e.g., the destruction of natural coastal defenses such as mangroves and wetlands) will all contribute to economic harm and a diminishment of the quality of life. For our purposes we'll only highlight sea level rise, storm surges, and hurricane intensification.

SEA LEVEL RISE

If there is one consequence of global warming that has penetrated public consciousness it is sea level rise. Global warming has raised and will continue to raise sea level. This is a result of the water of the oceans expanding when they are heated combined with the melting of ice caps and glaciers. The rate of sea level rise has nearly doubled since 1993 when compared to the previous 100 years.[148] The IPCC conservatively projects it will rise anywhere from 7 to 23 inches this century. More recent studies project three to six feet.[149] To give a sense of what this means for America's coasts, for much of Florida a one foot rise would send water 100 feet inland.[150] Currently 23,000 square miles along the Atlantic and Gulf Coasts are two to three feet above the tides, mainly in Louisiana, Florida, Texas, North Carolina, and the Chesapeake Bay area.[151]

Looking at the coast of the Gulf of Mexico, even without global warming this region would experience sea level rise due to subsidence, or sinking, in many areas. Putting both together, sea level rise is conservatively projected to be one to seven feet (with, for example, 1.5 feet out of 7 due to subsidence). Assuming a two to four foot rise, an area stretching from Houston to Mobile will be inundated in the next 50 to 100 years, and "27% of the major roads, 9% of the rail lines, and 72% of the ports" are at or below four feet in elevation and are at risk, with only a portion of such structures currently protected in some way.[152]

STORM SURGES

A key impact related to sea level rise is storm surges. Because of sea level rise, they "will build upon a higher base, and hence reach farther inland."[153] An example of the costs of storm surges is the Chesapeake Bay, where storm surges from Hurricane Isabel caused $84 million in damages to the western shore.[154]

The Gulf of Mexico is probably the most vulnerable area in America to storm surges. Right now approximately 60,000 miles of coastal highways are exposed to storm flooding.[155] Add in projected sea level rise,

and anything currently at or below 30 feet is subject to direct storm surges.[156] Scientists, modeling a likely storm surge of 23 feet, found that 64% of the Interstates, nearly half of the rail lines, 29 airports, and virtually all of the ports would be subject to flooding.[157] (Hurricane Katrina's surges were up to 28 feet.[158])

In light of this, a major report by the National Academy of Sciences asks a vital question:

What if critical evacuation routes were themselves to become submerged by rising seas and storm surge? …Many seaside communities count on coastal highways for evacuation in a major storm. …Highways in low-lying areas that provide a vital lifeline could themselves become compromised by encroaching seas and storm surge. Some communities could be cut off in a severe storm or would be forced to evacuate well in advance of the storm's known trajectory to avoid that risk.[159]

For those trying to flee the flooding associated with Hurricane Katrina, this was not an academic exercise.

HURRICANE INTENSIFICATION: KATRINA AS HARBINGER?

One of the more troubling aspects of global warming is evidence that climate change will probably intensify hurricanes in the future, both in terms of wind speed and the amount of rainfall.[160] Such intensification could turn a Category 3 storm into a Category 4, for example. A recent study suggests that Category 4s and 5s could nearly double.[161] Such jumps from one Category to another can belie the increase in destructiveness, since destructiveness increases exponentially. For example, a Category 3 hurricane has 50 times the destructive potential as a Category 1 hurricane; a Category 4 has 250 times the destructive potential, and a Category 5 has 500 times the destructive potential. So intensifying a hurricane from a 3 to a 4 doesn't make it a bit more destructive, it makes it a lot more destructive.[162] If there is a silver lining it is that global warming probably won't increase the frequency or absolute number of hurricanes and they may even decline slightly.[163] But because intensity will increase, overall destructiveness is projected to rise by around 30%.[164]

Scientists cannot tell us at this time whether global warming intensified a particular hurricane. However, the consequences of Hurricane Katrina could be a harbinger of things to come, in terms of the possibility of stronger hurricanes and what happens when we are not fully prepared for major disasters.

Katrina has been the most expensive disaster in US history at around $125 billion, with over 1,800 lives lost.[165] After initially passing over Florida as a Category 1 hurricane, Hurricane Katrina strengthened into a Category 5 in the warm waters of the Gulf. It made landfall on August 29, 2005, as a Category 3 hurricane, with the center of the storm near the Louisiana-Mississippi border. As it moved inland, Katrina also spawned a total of 43 tornados in Georgia, Alabama, Mississippi, and the Florida Keys.[166]

MISSISSIPPI AND THE deSILVEYs

In Mississippi, 238 people died from Hurricane Katrina. The storm surge "penetrated at least six miles inland in many portions of coastal Mississippi and up to 12 miles inland along bays and rivers."[167] According to Mississippi's Governor, Haley Barbour, "An estimated 35,000 homes outside of the flood plain without flood insurance were destroyed by Hurricane Katrina's storm surge on the Mississippi Gulf Coast."[168] A report by the National Hurricane Center states, "The storm surge of Katrina struck the Mississippi coastline with such ferocity that entire coastal communities were obliterated, some left with little more than the foundations upon which homes, businesses, government facilities, and other historical buildings once stood."[169]

The deSilvey family of Gulfport was one the families devastated by Katrina. Tragically, Douglas P. deSilvey lost his entire family: his wife, Linda Allen deSilvey, 57, daughter, Donna K. deSilvey, 35, mother-in-law, Nadine Allen Gifford, 79, and father-in-law, Edward "Ted" Gifford, also 79.[170]

For 20 years whenever there was a severe storm Mr. deSilvey and his family went to his in-law's house where they had always been safe. At one point during the storm he went to a back bedroom because he heard glass breaking. He looked out a window and saw that the water was rising "so fast it was unbelievable." He was about to try to get his family out, but suddenly the roof caved in. He found himself submerged and his lungs filling up with water. Here is his story in his own words:

> I just kept on asking Jesus not to let me go like this. I had to get my family out … Losing a family – I don't think there's any words for it. Kinda makes you wonder what life's all about. Many, many questions I have. Not received answers for yet. I've got the house full of their belongings that I don't know what to do about. As a father and a dad and a husband you always plan for the future of everybody and it's just the opposite now. I have nobody to plan for, or work on retirement for, or to save to buy a house for. It's just me.[171]

NEW ORLEANS

Hurricane Katrina is most known for the destruction of New Orleans, mainly due to the failure of the levees designed to protect the city. "About 80% of the city of New Orleans flooded, to varying depths up to about 20 ft."[172] Before Katrina New Orleans had a population of about half a million. Six months after Katrina the population was about 155,000.[173]

Besides the failure of the levees, two other examples – what happened at the Convention Center and at local hospitals – highlight the failure of the government to have adequate defenses, plans, and resources in place to protect all of its citizens from such disasters, especially vulnerable populations like the poor and the ill.

THE CONVENTION CENTER

"'It was chaos. There was ... nobody in charge ... nobody giving even water. The children, you should see them, they're all just in tears. There are sick people. We saw... people who are dying in front of you."[174] That was CNN Producer Kim Segal's description of what she encountered when they reported from the Convention Center.

The searing ordeal at the Convention Center began Monday, August 29, the day the levees failed and the city flooded. It did not end until Saturday September 3, when the last of over 16,000 were finally evacuated.[175]

At 6:10a.m. Central Time Monday August 29, Hurricane Katrina made landfall in Louisiana. That afternoon the levees started to be breached, resulting in the flooding of 80% of the city. Government officials told citizens who could not evacuate by car (mainly the poor and working families) to go to the Superdome and the Convention Center. But while the Superdome was stocked with basic provisions for evacuees, the Convention Center was not. It was merely intended to be a gathering point from which citizens would be evacuated. As Capt. Pfeiffer, the operations officer of the New Orleans Police Department put it, "It was supposed to be a bus stop where they dropped people off for transportation. The problem was, the transportation never came."[176]

The Doby family found themselves at the Convention Center: Leon, 26, Khaylin, 3, and Leah, 1. On Monday Leon swam away from his home headed for the Superdome with his daughters in a crate, tied to his waist. But when he arrived he was told they were at capacity and to go to the Convention Center, 10 blocks away. He spent the next four days there, until he gave up on being evacuated and struck out on his own. (Eventually he was picked up by someone who drove him to Dallas.) For four days at the

Convention Center all he had for himself and his two daughters "was a sandwich and two bottles of water that a stranger had given him." At one point when he was searching for a restroom for his daughters he saw bodies prostate on the ground in front of the doors. He thought they might be dead. Instead of taking his daughters in, he gave them permission to soil themselves.[177]

The Cash family, Linda 26, and her sons Clarence, 6, and Cyrin, 2, also found themselves at the Convention Center Monday expecting to be evacuated. "'Soon as I got there,' Cash recalled, 'I saw fighting. I saw people throwing chairs. People pulling guns out, right in front of little children.'" That evening, Mrs. Cash found a young boy struggling to breathe. Ms. Cash was able to locate a police officer, a rarity during this ordeal. "'The officer came over, his gun drawn.' Cash pointed to the young boy. 'The officer checked the boy,' Cash remembered, 'then turned to us and said there was nothing he could do.'" The officer left. The boy was dead.[178]

Amazingly, for three of the most chaotic days during this time, from Tuesday evening until Thursday morning, there were 250 National Guard troops staying in part of the Convention Center. Although they were fully armed, their mission was not to help the evacuees but to clear roads. Instead of assisting, they used their trucks to barricade themselves in one of the exhibit halls. "Almost as soon as they arrived, Guard commanders became concerned enough about the safety of their troops that they ordered more weapons and ammunition."[179]

"There was way too many of them and way too few of us," said Master Sgt. Anderson, 37. "Since we couldn't help them, it was best to avoid them." As for sharing provisions, another Sgt. stated, "We didn't have any for them. We had to feed our own people." She felt pain at the knowledge that teenage girls were wandering around the center, alone, knowing they were possible prey. "There were prisoners, mobsters, gangs," she said.[180]

The Convention Center was not stocked with provisions for disaster victims but it did have alcohol. Gang members "broke into the locked alcohol storage areas and suddenly had 50 cases of hard liquor and 200 cases of beer."[181]

On Wednesday a few buses arrived, but with more than 16,000 evacuees, only a small number could get on board. When other buses arrived they left without taking any passengers (probably because no one was in charge and there was no security).

At 11a.m. on Thursday morning, the New Orleans Police finally showed up in force. The SWAT team arrived, entered the Convention Center and called out the name of the wife of a fellow officer and that of another female relative. This officer had asked the SWAT team Captain for help in rescuing his wife. "'He knew they were there and was hearing nightmarish stories,' said one of the SWAT team. The two women rushed to the team and they 'put the women in the middle of the team, then backed out the door.'"[182]

An added dimension to this particular part of the story is that the two women were white and the vast majority of those at the Convention Center were African-American. "Once it became clear that the SWAT team had come with the single goal of rescuing two white women, anger exploded. 'Racists!' one man cried out. 'Some people were upset we weren't rescuing them' one of the SWAT team members later remarked. 'It's hard to leave people behind like that, but we were aiding an officer.'"[183]

Friday morning, National Guardsmen from Arkansas arrived to help the evacuees. Many had just returned from Iraq, and came in full body armor and helmets, and their assault rifles.[184]

One of the first orders of business was to deal with those most vulnerable: pregnant women, women with small infants, those in urgent need of medical care, and the frail and elderly. Commanders worried that others might rush the medivac helicopters that were being brought in, but encountered no serious problems. Food lines were set up and the 16,000 citizens, who had arrived days before, finally receive help.[185]

On Saturday, the sixth day of the Convention Center ordeal, the soldiers again lined up the 16,000 evacuees and in "an eerily quiet process" these American citizens boarded buses and left.[186]

THE HOSPITALS

The situation for the area's 11 hospitals that were flooded was also quite shocking. Having discharged all the patients they could before the storm, these hospitals were left with 1,749 patients. The staff, their families, and those who came to hospitals as a known place of refuge made up another 7,600 in the hospitals.[187]

The floodwaters knocked out the electricity and the backup generators ran out of fuel and became useless. The plumbing stopped working, so there was a lack of running water and the toilets filled up. Doctors and nurses worked around the clock and made rounds with flashlights. They climbed many flights of stairs to get to critically ill patients located on the upper floors. Emergency surgery was done by flashlight with little or no anesthesia. The temperature inside the hospitals rose to upwards of 115 degrees. By Wednesday many of the hospitals were running out of food and rationed to one meal a day.

Doctors at Charity Hospital grew so desperate they called the Associated Press to plead for help. "'We have been trying to call the mayor's office, we have been trying to call the governor's office … we have tried to use any inside pressure we can. We are turning to you. Please help us,' said Dr. Norman McSwain, chief of trauma surgery. 'We need coordinated help from the government.'"[188]

Dr. Sanjay Gupta, CNN's medical correspondent, was at Charity Hospital. He reported that because the hospital's morgue in the basement was flooded, corpses were being put in the stairwells. Summing up the entire situation, he said, "It's gruesome. I guess that is the best word for it … It is one of the most unbelievable situations I've seen as a doctor, certainly as a journalist as well."[189]

Attempts to evacuate hospital patients were hampered by the chaos, lawlessness, and fear. "'My people are in harm's way,' said Richard Zuschlag, the CEO of a local ambulance service. 'They are scared. Our command station about an hour ago had the generator stolen off the back of it. We've had an ambulance turned over.'"[190] On Thursday Zuschlag reported that at one of the hospitals one of his helicopter pilots "got scared because 100 people were on the helipad and some of them had guns. He was frightened and would not land."[191]

Despite all of this, some hospitals were evacuated by Thursday, the remainder by Friday.

Some may say it wasn't Hurricane Katrina – a hurricane that may or may not have been intensified by global warming – that destroyed New Orleans. Rather it was the failure of levees, which should have been able to protect the city from a Category 3 storm like Katrina, if they had been built properly.

For me this shows how Katrina should be a warning to us. Since global warming is a natural disaster intensifier, its consequences increase the chances of finding society's weaknesses. Global warming will find our failed levees, so to speak. What happened to New Orleans was the most anticipated disaster in US history, in the richest nation the world has ever seen, but still many of the precautionary measures of experts were not implemented. It was literally a disaster waiting to happen that did happen.

How many such preventable disasters will the impacts of global warming expose in the future? Global warming's capacity to intensify natural disasters will increase the risk that a bad but still manageable problem becomes a catastrophe, either because of poorly designed defenses and disaster plans or because the strength of the storm or wildfire or drought was not anticipated. As one government report dryly sums it up, "North American urban centers with assumed high adaptive capacity remain vulnerable to extreme events."[192]

THOSE WE LOVE AND GLOBAL WARMING'S CONSEQUENCES FOR THEIR HEALTH

Another obvious consequence of global warming is that it is going to get hotter. There will be more heat waves. Climate-intensified heat waves alone will kill more people, especially vulnerable populations like the elderly, very young children, and poor urban populations who lack air conditioning. Areas that are currently not prepared for such heat, such as the Northeast, will be hit especially hard. One study projected if a heat wave like the one to hit Europe in the Summer of 2003 – where over 70,000 people died[193] – were to occur in the United States, heat-related deaths would be 11% greater in Detroit, 29% greater in St. Louis, and an astounding 155% greater in New York City, when compared to the hottest summers between 1961 and 1995.[194] Another prediction suggests that by the end of this century, if emissions are not reduced, the number of heat wave days in Los Angeles is projected to double. The number of heat days as well as heat wave-related deaths in Chicago is likely to quadruple by 2050.[195]

When excessive heat is combined with smog, the result will be more deaths from heart and lung illnesses. By 2050, Red Alert Days for smog are estimated to increase by 68% in the 50 largest cities in the East and Midwest.[196] Forest fires, increased in the West and Southeast due to growing drought, will also make life harder and possibly deadly for those with pulmonary problems.

Rising temperatures will enable the expansion of vectors (e.g., rats, mosquitoes) into new areas. This will lead to the introduction or intensification of vector-borne diseases, such as West Nile Virus and Rocky Mountain spotted fever.[197]

Food and water-borne pathogens could also increase, especially after heavy downpours lead to sewage overflows.[198] The possibility of contaminated water after heavy rains is especially acute in the 770 cities that have "combined sewer systems" where storm water and sewage are carried in the same pipes, including New York, Philadelphia, Washington DC, Chicago, and Milwaukee.[199]

Some bothersome health consequences involve pollen and poison ivy. Between 1900 and 1999 pollen associated with allergies doubled due to the increasing concentrations of CO_2 in the atmosphere, and is projected to double again by 2050.[200] One study found that poison ivy doubles its growth rate and significantly increases its toxicity in an environment where CO_2 is twice the pre-industrial level.[201]

Christians should note, "Climate change is very likely to accentuate the disparities already evident in the American health care system. Many of the expected health effects are likely to fall disproportionately on the poor, the elderly, the disabled, and the uninsured."[202]

CONCLUSION:
KEEPING OUR LOVED ONES OUT OF HARM'S WAY

Global warming has and will continue to be a natural disaster intensifier. Collectively, our actions are contributing to this. Yet as Christians we also contribute billions of dollars each year to help victims of such natural disasters and spend billions insuring ourselves from them. Our left hand doesn't know what our right hand is doing.

Let us be reminded again of Jesus' command: "Love each other as I have loved you. Greater love has no one than this, that he lay down his life for his friends" (Jn 15:12-13).

Are those you love, your Christian friends, your family, or your neighbors in harm's way because of global warming?

Does the Risen LORD want more cases of someone sleeping on the roof of their Bronco because of flooding, like Mike Johnson? Does He want to see a family run for their lives as drought-stoked fires burn their home to the ground, like the Hakansons? Will He want to watch someone lose his entire family, like Doug deSilvey or swim through flood waters with his little girls in a crate only to end up in some place like the New Orleans Convention Center, like Leon Doby?

Of course not. The Risen LORD was with Mike on the roof of his Bronco. He was with Doug when he lost his family. He swam beside Leon

and his little girls in the Katrina floodwaters. And he beckons us to be with them, too.

To love someone is to do things that make their lives better. But global warming is going to make the lives of many people here in the United States worse - even those you've been called to love. It may not require us to literally sacrifice our life like Joshua Landtroop did to save his two boys; but that is the depth of the love we are to have for those God has placed in our lives. We are to love them with a sacrificial love.

Jesus told us to remain in His love by obeying His commands and emphasized that we love one another as He loves us. Today, to remain in His love, to love one another, is to do our part and follow the Risen LORD in overcoming global warming.

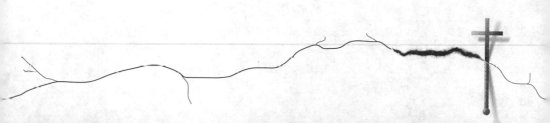

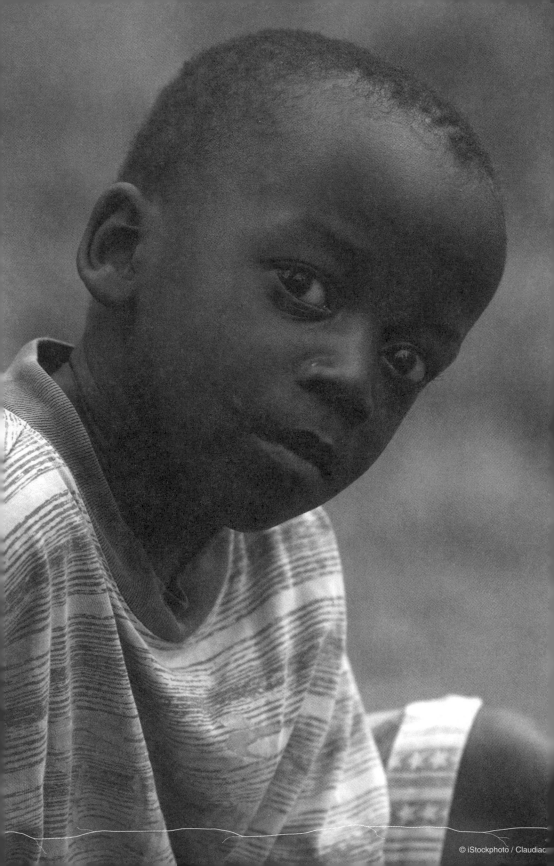

5

CONSEQUENCES FOR
THE POOR
IN DEVELOPING COUNTRIES:
AN INTRODUCTION

Seventy-year-old Tulsi Khara from India feels that abnormal climate conditions made her family's land disappear:

"We are not educated people, but I can sense something grave is happening around us. I couldn't believe my eyes — the land that I had tilled for years, that fed me and my family for generations, has vanished. We have lost our livelihood. All our belongings and cattle were swept away by cyclones. ...Displacement and death are everywhere here." [203]

Because of what Jesus says in Matthew 25, I believe the Spirit of the Risen LORD walks with people like Tulsi, or "the least of these" as Jesus describes them. He also walks in Bangladesh with young girls like Bithi. Her parents, two sisters, and Bithi had to flee their home because of unusually heavy flooding in 2007. "We do not sleep in the night, fearing that our little children may fall into the nearby deep floodwater," says Bithi's mom. [204]

The Risen LORD is spiritually present beside them. Whatever happens to them happens to Him. As such, if we love the LORD, we must love the poor. You can choose to not love the LORD but you cannot choose to love and follow the LORD and to not love the poor.

This chapter and the next focus on the impacts of global warming on the poor around the world and the way global warming will intensify their plight in the coming years. It is not a happy picture. Our human tendency is to avoid pain and deny problems. It's easier to turn away and remain ignorant of what global warming will do to the poor. Denial is not an option for someone who confesses to know the Truth, the Truth who has set us free (Jn 14:6; 8:32).

But, we do not take this journey alone! The LORD is with us and He is with the poor around the world. Since the LORD is with Tulsi, Bithi, and the billions who will be impacted by global warming, we cannot turn away from what it is doing and will do to them.

Remembering the Gospel stories of Jesus feeding the multitudes will be helpful before we begin our journey into the impacts of global warming on the poor around the world.

Before Jesus fed the 4,000 he looked out upon them and said to the disciples, "I have compassion for these people; they have already been with me three days and have nothing to eat. If I send them home hungry, they will collapse on the way, because some of them have come a long distance" (Mk 8:2-3).[205] In the accounts of the feeding of the 5,000 the disciples urge Jesus to send the crowd away to buy their own food. But Jesus says, "They do not need to go away. You give them something to eat" (Mt 14:16).

It is a gloriously shocking reply. The disciples are overwhelmed by Jesus' command. They say to Jesus, "That would take eight months of a man's wages! Are we to go and spend that much on bread and give it to them to eat?" (Mk 6:37). Philip adds even this "would not buy enough bread for each one to have a bite!" (Jn 6:7).

Just imagine the disciples' thinking: *We didn't ask these people to come out here. This wasn't planned. It just happened. We don't have the resources to care for all of these people. Lord, be reasonable!*

It's the same with global warming. It's not something we planned. It's an unforeseen negative consequence of a process (i.e., the burning of fossil fuels) that has brought a great deal of good. We may feel

overwhelmed by the resulting needs and feel the LORD is being unreasonable when He tells us, "You give them something to eat!"

Yet it's amazing what can happen when the LORD is with us. As we roll up our sleeves and begin to solve these problems we may find that the results can be miraculous.

But first we must see the need – and the needy, like Tulsi and Bithi and their families. We cannot pass by on the other side like the priest and Levite did in the parable of the Good Samaritan. We must have compassion for them and be willing to follow the LORD's lead in overcoming global warming – even when the need feels overwhelming.

In these chapters focused on the poor worldwide we will review some facts and figures. These big statistics are abstractions. But to the Risen LORD nothing is an abstraction. He continuously sustains all things. He walks beside every person on the planet. He lies in the ditch with the destitute. These abstractions are individuals He loves and died for.

Mother Theresa once said, "If I look at the mass I will never act. If I look at the one, I will."[206] She reminds us that while big numbers may overwhelm us, we are willing to help *individuals* in distress. What I am suggesting is that as we grapple with the magnitude, we keep our eyes on the Risen One. If we look at the One, we will have the spiritual capacity to act, to love the least of these.

We also need to understand why we in the United States have a role to play in overcoming global warming. Some may say, "Global warming? That's not my issue." The evidence says otherwise. Everyone in the United States contributes to the problem. Our country has the capacity – and I would add, the responsibility – to lead the world in overcoming global warming. We all have a responsibility as citizens in this democracy to be a part of overcoming global warming.

Some may not like that they are part of the problem. (Count me with you.) Some may deny that it's their problem. But that doesn't mean that it isn't.

Of course, it isn't just a problem that individuals bear alone. Far from it. It's a problem we all share and we must overcome it together. However, it's even more important to understand the spiritual fact that the

Risen LORD is the One who is leading the fight to overcome global warming. To walk with Him today is to do our part in overcoming global warming.

RISK & VULNERABILITY

Changes brought about by global warming will affect us all, including the potential for a few positive changes for some, at least in the first half of this century. However, the major negative impacts will not be evenly distributed. Some places, such as sub-Saharan Africa, low-lying deltas, or communities dependent upon glacial melt and snowpack for their water, will be hit harder than others.

Whether climate impacts for a certain place are mild or harsh, one thing will be constant: the poor will have less capacity than others to prepare for the impacts and deal with the consequences. The reason is simple and straightforward: they are poor. This by itself makes them more vulnerable.

During Hurricane Katrina in the richest nation on earth with the most anticipated natural disaster in our history, it was the poor who were literally left behind and suffered the most. When global warming intensifies natural disasters in this century, all will be at risk but the poor will be the most vulnerable.

Simply put, the vulnerable among us will be more vulnerable, no matter what the risk. If you are poor, you are vulnerable to global warming, no matter where you live. If you live in a poor country, and in an area that will be hit hard by global warming, then you are potentially the most vulnerable of all.[207] You have done nothing to create this new vulnerability you must face. Others have put you at risk. Others have made you more vulnerable.

Even though the United States and other developed countries are the most responsible for creating this new vulnerability, it was not done intentionally. But now that we know, we cannot pass by on the other side. Empowered by Christ's grace and with the Risen LORD walking beside us, we are to care for the least of these by helping to overcome global warming.

CURRENT SITUATION FOR THE POOR WORLDWIDE

The impacts of global warming in poor countries began in the last century, but the consequences will be much more severe and apparent in this century. Such effects won't happen in isolation, even though we mainly talk about them one at a time. Instead, the impacts will interact with each other and with the other situations people are facing at the time.

To give a better sense of their capacity to cope with the consequences, I will briefly sketch the current situation of the poor worldwide. How vulnerable does their current situation make them?

In 2005 the global population was 6.5 billion and is projected to be over 9 billion by 2050.[208] Currently 3.6 billion, or over half the world's population, lives on less than $2 a day, with 1 billion living on less than $1 a day. Three-quarters of this 1 billion, the extremely poor, depend directly on agriculture – which is important, given the vulnerabilities of agriculture in poor countries.[209]

Some good news is that "the share of the population living in developing countries on less than $1 a day has fallen from 29 percent in 1990 to 18 percent in 2004" meaning 135 million fewer individuals were living in extreme poverty. This is mainly due to the economic development of China and India.[210] This positive development, therefore, is not evenly spread out worldwide. It is also unevenly spread within China and India themselves. For example, in rural India in 2004 one-half of the children are still underweight, roughly the same percentage as it was in 1992.[211]

Worldwide, more than 80% of the population lives in a country where the income gap is increasing. A developing country's GDP (gross domestic product) may be growing, but today it takes over three times as much growth as it took 30 years ago to achieve the same amount of poverty reduction, according to one estimate.[212]

More good news is the fact that for the last four decades per capita food production has gone up. And until recently, the percentage of those who are undernourished had been going down. However, around 2005 it began moving up again. In 2003-05 those undernourished worldwide had dropped to 16% of the population, down from 20% in 1990-92. However, in 2007 it had grown to 17%, with the total number of undernourished at 923 million, more than 80 million higher than in 1990-92.[213] Troubling, also, is the prospect that some of the per capita increase that has occurred will not be maintained due to unsustainable practices (see below).

In some cases whole regions have been left behind. For example, the economic divide has widened between sub-Saharan Africa, where global warming will hit hardest, and the rest of the world. "A child born in sub-Saharan Africa is 20 times more likely to die before age 5 than a child born in an industrial country, and this disparity is higher than it was a decade ago."[214] Per capita food production has declined. By 2015 sub-Saharan Africa will account for almost one-third of world poverty, up from one-fifth in 1990.[215]

In other cases, certain groups remain mired in poverty and therefore more vulnerable to the impacts of global warming. In Guatemala, for example, "malnutrition among indigenous people (or the Mayans) is twice as high as for non-indigenous people."[216] When Hurricane Stan came ashore in 2005 it killed 1,600, mainly Mayans. It also created major impacts on agriculture, which hit the indigenous the hardest, since the majority are subsistence farmers or agricultural laborers who work for others.[217]

The lives of the poor are more vulnerable to the condition of the natural resources in their area. They "live off the land" – or the sea, as the case may be – more than do those with more financial resources. Three-quarters of those living in extreme poverty –on less than a $1 a day – depend directly on agriculture and hunting. As for the sea, it supplies "more than half of the protein and essential nutrients in the diets of 400 million poor people living in tropical coastal areas."[218] Much of this abundance is connected with coral reefs, which are at serious risk due to global warming and other factors.

For the poor to have a chance for a better life and not slip further into poverty, such natural resources must be kept healthy and not be pushed beyond their limits. They must have their "Sabbath rest," as God intended. If not, the Bible warns of consequences.[219]

But the evidence indicates that in the second half of the twentieth century we pushed the rest of creation too hard. Approximately 60% of the benefits that the rest of God's Creation provides to humanity are being used unsustainably. For example, at least 25% of commercial fish stocks are overharvested. Up to 25% of freshwater globally "exceeds long-term accessible supplies and is now met either through engineered water transfers or overdraft of groundwater supplies" and "15–35% of irrigation withdrawals exceed supply rates and are therefore unsustainable."[220]

Currently "1.1 billion people (17% of the global population) lack access to water resources,"[221] including two-thirds of Asia and 42% of sub-Saharan Africa.[222] In addition, "more than 2.6 billion lack access to improved sanitation."[223] Lack of water and sanitation result in 1.7 million deaths each year, as well as various diseases in half the urban populations of Africa, Asia, Latin America, and the Caribbean.[224]

Other health deficits involve the cases of deadly or debilitating diseases. For example, approximately 34 million of the world's poor have AIDS, with three million deaths in 2004.[225] Each year an estimated 300-500 million worldwide become ill with malaria.[226] Approximately 700,000 to 2.7 million people die of malaria each year, the majority of them young children.[227] One estimate has a child dying every 30 seconds.[228] In areas with malaria the disease kills an estimated 25% of children before their fifth birthday.[229]

The world's poor disproportionately feel the impacts of natural disasters. But in keeping with global warming, natural disasters have been increasing recently "from around 200 annually in the period 1987–97 to about double that in the first seven years of the 21st century."[230] In 2000-2004, one in nineteen of those living in a poor country was significantly impacted by a natural disaster and someone in a poor country was 79 times more likely to suffer consequences from a disaster than someone from a rich one.[231]

Finally, another problem that many of the world's poor have to contend with is social instability and societal failure. Many of the world's poor live in a state that is close to failure or highly unstable. In 2007 all but three of the 20 states ranked as "critical" in terms of their capacity to fail are poor.[232]

All of these problems can lead to tragic consequences for children. They are still developing physically and intellectually, and are therefore more vulnerable than adults. "Around 28 percent of all children in developing countries are estimated to be underweight or stunted."[233] One study followed children in Zimbabwe after a drought and found that it permanently stunted their growth. Such stunting, especially before the age of three, leads to poorer school performance, lower scores on cognitive tests, and a failure to acquire both physical and intellectual skills at normal rates.[234] This same study found that drought leading to malnutrition "results in a loss of lifetime earnings of 7-12 percent and that such estimates are likely to be the lower bounds of the true losses."[235] These consequences from malnutrition can stay with children for the rest of their lives and the physical consequences can even be passed on to the next generation. For example, women who were stunted as children are at greater risk of having complications giving birth, delivering babies with a lower birth weight, and dying in childbirth or having their baby die.[236]

And so as the poor face a future with significant global warming impacts as soon as 2020, they do so with major social, economic, and natural deficits that makes them much more vulnerable.

FIVE CAVEATS TO KEEP IN MIND

Five caveats are in order before considering the scientific estimates presented in the next chapter.

First, chapter 6's discussion of seven major impacts for the poor worldwide is not intended to be exhaustive. (It may be exhausting, but it's not exhaustive!) I do not cover all of the various types of impacts the poor will face. As an example, I won't cover how global warming will influence the incidence of the disease leishmaniasis.[237]

Second, the scientific estimates in chapter 6 do not consider the impacts of extreme weather events of modest duration, e.g., flash floods and wind damage from strong storms, or an intense heat wave of several days. Such impacts are currently too hard for scientists to model,[238] yet their impacts will be significant. For example, recent evidence suggests heavy downpours that produce flash flooding will occur about twice as frequently as had been projected. Flash floods can bring major cities to a halt, increase water-borne diseases, contribute to landslides, and destroy crops.[239]

Third, chapter 6 does not include impacts that would be associated with various types of abrupt climate change (e.g., the melting of the Greenland ice sheet). Such abrupt changes would increase harmful impacts significantly.

Fourth, unless otherwise noted, as a general rule I provide mid-range scientific projections on impacts from the world's leading experts (e.g., the Intergovernmental Panel on Climate Change, or IPCC). In other words, you are getting solid, even conservative, numbers from sound sources.

Fifth, by presenting – out of necessity – a series of impacts that are primarily talked about one by one, there is a distortion. These impacts won't be happening in isolation in some neutral laboratory. A whole variety of climate impacts could be occurring simultaneously in one place. They will interact in ways no one can predict. And if that place is a poor country wracked by AIDS and violent conflicts, the synergistic whole will be worse than the sum of the parts.

While it is important to understand that billions will be affected, we must never lose sight of the fact that young girls like Bithi and her family make up the billions. These are not just numbers. They are all known and loved by the Risen LORD Himself, who walks besides each of them as He does us. And so, with the Risen LORD beside us, let us explore the potential consequences of global warming for the poor worldwide.

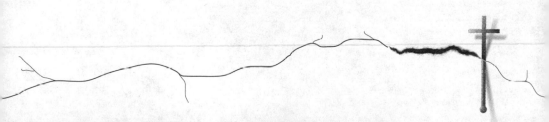

6

SEVEN MAJOR IMPACTS
ON THE POOR
IN DEVELOPING COUNTRIES

Many of us over the years have given money to help the poor deal with natural disasters such as floods and droughts or to help the poor lift themselves out of poverty through development. Unfortunately, global warming is a poverty intensifier that in a myriad of ways will push those climbing out of poverty back down into it, wiping out painstaking gains – with deadly consequences for some. The most vulnerable among us have little margin for error, and global warming will help erase that margin.

In this chapter we will consider seven major impacts on the poor in this century. It will be important to keep in mind, however, that for the poor global warming is not some problem in the distant future. Significant impacts for the poor are projected to happen in about a decade. For example, in Africa from 90 to 220 million will face water scarcity by 2020, and crop yields could be reduced by 50% in some areas.[240]

The seven major concerns we will explore are as follows:
1) Hunger and Malnutrition
2) Thirst and Water Scarcity
3) Floods
4) Disease or Health Impacts
5) Extinction of God's Other Creatures

6) Refugees

7) Violent Conflicts, Oil Dependence, and Terrorism

1) HUNGER & MALNUTRITION

"I was hungry and you gave me nothing to eat."
Matthew 25:42

A significant impact of global warming will be the disruption of rain-fed agriculture, which comprises 80% of agricultural land worldwide.[241] This type of agriculture requires predictable rainfall during the right season, in moderate amounts over time. Any disruption has serious consequences.[242]

Global warming will also damage or destroy crops due to intensification of strong winds, heavy downpours, and floods. The crop production of Mohammad Iyar Ali of Bangladesh was "damaged three times because of sudden and untimely floods" followed by a devastating drought. Given that he only has one and a half acres from which to feed his eight-member family, these new climate conditions threaten their survival and "life is becoming tougher day by day."[243]

In the case of drought, during this century global warming is predicted to increase the amount of land in extreme drought from 1% to 3% today to 30% by 2090 if no significant actions are taken to reduce global warming pollution.[244] In addition, climate change will impact agriculture by shifting the times growing seasons begin and end and increasing agricultural pests. According to the IPCC 40-170 million additional people annually could be at risk of hunger and malnutrition.[245] Another major report suggests that it could be as much as 550 million annually.[246]

These numbers only reflect those who could suffer from hunger and malnutrition due to agricultural production losses. This does not include how many more could go hungry due to the other consequences of global warming.

One of the major sources of freshwater in the world is from glacial and snow melt. Glaciers serve like huge water savings accounts, storing water in the wet winters to be released in the dry summers. The IPCC conservatively estimates that more than one-sixth of the world's population depends on water from glacial and snow melt.[247] Currently that is over 1 billion people and growing.

Most impacted people live in Asia. In the Tibetan Plateau, which includes the Himalayas, there are over 45,000 glaciers that contain the largest concentration of freshwater in the world outside the polar ice caps.[248] These glaciers feed rivers that provide water to 40% of the world's population. Unfortunately, global warming could reduce these glaciers significantly this century.[249]

The glacial loss has already begun due to global warming from greenhouse gases as well as the regional and localized warming effects of black carbon.[250] China's glaciers have declined 21% since the 1950s. As much as two-thirds of the glaciers in the Himalayas are retreating at a rate that is the quickest in the world.

Unfortunately, for many areas 90% of the enhanced water flows from rapid glacial melts are

2) THIRST AND WATER SCARCITY

"I thirst."
John 19:28, KJV

Fresh water is necessary for human survival and essential for our well-being. In many respects, most of the impacts of climate change are related to water in one way or another – how it comes, whether it comes, how it is stored, and how much is available in a given area for human use.

happening at the same time as the summer monsoon season and is simply flowing out into the sea.[251] As such, in the near term glacial melt will increase flooding and landslides in deforested mountainous areas. And because a good number of these regions were already using water unsustainably for irrigation, industrial uses, and to provide water to burgeoning populations, the decrease in water flow will make existing water scarcity worse. Significantly more area will convert to desert. Agricultural production that relies heavily on irrigation will eventually experience important declines in areas that can ill afford such losses. For example, due to water loss and increasing temperatures, by 2050 China's grain production could be reduced by up to 10% if successful adaptation measures are not found and implemented. Near the end of the century production of wheat and rice output could decline as much as 37%.[252]

Major rivers that could be significantly impacted include: the Indus, found mainly in Pakistan; the Ganges and the Brahmaputra, found mainly in northern India and Bangladesh; and the two major rivers of China, the Yangtze, the longest river in Asia, and the Huang He, or Yellow River. Others that will see more moderate impacts include the Mekong,[253] which runs through China, Burma, Thailand, Laos, Cambodia, and Vietnam. These rivers "provide household water, food, fisheries, power, jobs and are at the heart of cultural traditions." [254]

In Asia alone, over 500 million people in China, over 170 million in Pakistan and Northern India, and over 500 million in India and Bangladesh could face water shortages.[256]

Global warming will make floods fiercer and droughts dryer. In certain places already struggling with water availability, global warming will result in too much water coming too fast at the wrong time (e.g., during the wet season instead of the dry season when the crops need it). This will result in more flooding. It's as if you were thirsty and instead of offering you a glass of water someone shot you in the face with a hose. Unless you captured that water somehow, it would be essentially useless to you.

In other places, global warming will lead to less water being available – just what the poor don't need. It's as if you were walking in the desert and found a small hole in your canteen and half your water had leaked out. What you started out with wasn't enough, but it was all you had – and now you have even less. Worldwide, roughly 1-2 billion people already in a water-stressed situation could see a further reduction in water availability by 2085.[256]

In addition to a lack of rain, water stress will also come from the loss of glaciers and snowpack around the world. One billion people worldwide rely on melt from glaciers and snowpack to provide their water. Glacial melt provides water to seven of Asia's largest rivers, including 70% of the summer flow of the Ganges River, which provides water to around 500 million people. Almost one-quarter of China's population receives most of their water from glacial melt. In Latin America 50 million people rely on water from the Andean glaciers.[257] When taken together, the lack of rain and the loss of glacial melt and snowpack will bring thirst and worse to billions in this century.

MAJOR IMPACT

Worldwide, roughly 1-2 billion people already in a water-stressed situation could see a further reduction in water availability.

3) FLOODS

Global warming has and will continue to increase flooding. This is due to the increase and intensification of downpours, strong storms, the intensification of hurricanes/cyclones, and to sea level rise.

Floods are among the most dangerous and destructive disasters humanity faces and global warming will significantly increase both inland and coastal flooding. It is currently too difficult to predict accurately the amount of additional inland flooding. However, worldwide such flooding has increased from "about 50 floods per year in the mid-1980s to more than 200 today."[258] We know that global warming's intensification of future flooding will be significant. And we know that more and more poor people are moving into inland areas that will be flooded. Over 1 billion of the world's poor already live on riverbanks and areas highly susceptible to landslides.[259]

As for the coasts, currently 10% of the world's population – over 600 million – lives in areas that are less than 10 meters above sea level.[260] The IPCC conservatively estimates that 100 million additional people will be impacted by coastal flooding due to sea level rise and storm surge from global warming.[261] The National Academy of Sciences suggests that up to 300 million would be impacted by one meter (3.28 feet) and up to 200 million by 50 centimeters of sea level rise alone. These numbers could be significantly reduced if adequate adaptation measures are taken.[262] Sea level will rise because (1) warmer temperatures will warm the water in the oceans, and warmer water takes up more space than colder water; and (2) melting glaciers will increase the amount of water in the oceans.

With flooding comes death from drowning and the destruction of crops and homes. Whole communities and cities can be devastated. (This is true even in a rich country like the United States, as seen in New Orleans.) Health impacts include diarrheal diseases such as cholera. Poor children are especially vulnerable. In central Mexico illnesses in children under five increased by 41% as a result of flooding.[263] Furthermore, floods can also push people into poverty. In Ecuador after the 1997-98 floods, those impoverished increased by 10%.[264]

In keeping with global warming, floods have been on the increase, doubling in the 1990s.[265]

The intensity of flooding events also appears to be increasing. The 2010 Pakistani floods are the worst ever recorded in that country and one of the world's worst natural disasters to date. At its height more than one-fifth of the nation, or 62,000 square miles, was flooded.[266] (It would be as if all of Illinois or Georgia or New York State was covered with water.)[267] The flooding affected more than 20 million people; 6.5 million faced urgent needs for shelter, food, and medicine; 3.5 million children were put at risk

BANGLADESH:
LIVES SAVED BUT OPPORTUNITIES LOST

As one of the world's lowest nations and one of its poorest, Bangladesh is at high risk from the impacts of global warming, especially flooding. However, Bangladesh is also an example of how even in one of the poorest countries, vulnerability to global warming can be reduced through measures designed to adapt to the consequences.

This particular case involves how the death toll from cyclones (the name for hurricanes in the Pacific) has been dramatically reduced.

In the mid-1990s the government of Bangladesh, working with relief and development organizations, created a cyclone preparedness program.[272]

Comparing the death tolls before and after the program's implementation demonstrates its tremendous success. "Cyclones in 1970 and 1991 were responsible for killing around 500,000 and 140,000 people respectively. The November 2007 cyclone, Sidr, which was of greater intensity but occurred after the cyclone preparedness program was implemented, killed fewer than 4,000 people."[273] Even though 4,000 deaths are still unacceptably high, saving up to half a million lives is astounding. With the likelihood of increased flooding in Bangladesh due to global warming-intensified storm surges from cyclones,[274] adaptation programs will be essential to saving many lives in the future.

Bangladesh will also face flooding from glacial melt and extreme rainfall – events that can occur rather quickly like cyclones. Additional flooding will come from sea level rise, which will occur more gradually.

Reducing direct loss of life from flooding has to be the first prior-

for diarrhea-related illnesses. Hundreds of thousands of homes were destroyed.[272]

After visiting the devastated areas during the flooding, United Nations Secretary-General Ban Ki-moon said he had never seen anything like it, that it was worse than the 2004 Indian Ocean tsunami and 2005 Pakistani earthquake combined. "People are marooned on tiny islands with the floodwaters all around them. They are drinking dirty water. They are

Global Warming and the Risen LORD

ity. However there will be other serious impacts. The 1998 flood, deemed "the flood of the century," flooded 70% of the country (compared to 20%-25% in normal years).[275] It was the result of snowmelt in India and Nepal, a 20% increase in Bangladesh's major rivers from heavy rainfall, and "elevated tides in the Bay of Bengal from the monsoon."[278]

While thankfully only about 1,000 direct deaths occurred, 30 million were displaced, 20 million were made homeless, and there were several hundred thousand cases of diarrhea.[279] The floods prevented the replanting of crops, and while large-scale food aid prevented starvation, "the proportion of children suffering malnutrition doubled after the flood." [280] Fifteen months later the situation was still quite dire. Poor households had sold off many of their assets and had borrowed an average of 150% of their monthly expenditure (twice the pre-flood level). Even so, "40 percent of the children with poor nutritional status at the time of the flood had still not regained even the poor level of nutrition they had prior to the flood." [281]

When all of the flooding threats from global warming are considered, and assuming a one meter rise in sea level, up to 60% of Bangladesh's population could be impacted – over 70 million people.[282] Furthermore, a one meter (or 3.28 feet) sea level rise would inundate approximately 18% of the country, including much of the richest farmland in Bangladesh. This would have devastating consequences by itself, ranging from loss of 28%-57% of GDP,[283] not to mention what such losses could mean for hunger and malnutrition. Protection measures and eventual resettlement of the most threatened would cost over $20 billion according to several studies[284] – costs that are much too high for Bangladesh to bear on its own.

While we celebrate the reduction in loss of life from the implementation of successful adaptation initiatives like the cyclone preparedness program, we cannot be sanguine about the loss of opportunities for a better life. The Risen LORD isn't simply working in the world to have people survive, but to provide them the chance, the freedom, to be who the Father intends them to be.

Chapters 20 and 21 will explore adaptation's possibilities more fully.

---

living in the mud and ruins of their lives. Many have lost family and friends. Many more are afraid their children and loved ones will not survive in these conditions."[280] One small-scale farmer worried about the long-term consequences. "It'll take three to four years before we can grow anything on our land again."[281] An early estimate of the cost to rebuild one of the harder hit provinces was $2.5 billion.[282]

Another example is the 2000 floods in Mozambique, one of the poorest countries in the world where approximately 40% of the population live on less than $1 a day and another 40% live on less than $2 a day.[283] These floods resulted in two million people being displaced, with 350,000 jobs lost impacting the livelihoods of up to 1.5 million people.[284] As one older gentleman recounted:

> When the waters came in the middle of that night I was forced to flee my house and take refuge in a tree, where I was stranded with nothing to eat for four days … we tried to make platforms in the trees for the children to sit on. Some people died, and some of the children fell into the water, only to be swept away. The adults could not swim after them – they had to sit and watch them float away.[285]

The recent intensification of flooding has caught short even those accustomed to it. As one man in the Nadia District of West Bengal, India, put it in 2007:

> We had never seen such floods before. Lots of houses were destroyed, lots of people died, our agricultural land was submerged; crops stored in houses were lost. Many livestock were lost too. We were just not prepared to face such big flooding. So we didn't have any savings of money or food.

There are also psychological after-effects. One study from Great Britain found that three-fourths of those impacted by flooding experienced anxiety and depression, which lasted for two or more years.[286] Given the vulnerabilities of those in poor countries, the psychological impacts of increased flooding could be devastating. Just imagine yourself in Intsar Husain's shoes in northwestern Bangladesh in 2007: "There are more floods now and the river banks are being washed away faster. There's nowhere to go. My land is in the river. I have nothing now."[287]

MAJOR IMPACT

Not counting inland flooding, the IPCC conservatively estimates that climate change could result in 100 million more people being impacted from coastal flooding alone due to sea level rise and storm surge. More recent scientific findings suggest this number could be substantially higher.[288]

4) DISEASE OR HEALTH IMPACTS

"And people brought to him all who were
ill with various diseases ... and he healed them."
Matthew 4:24

Arriving at precise projections of the number of people at greater risk of harmful health concerns because of global warming has proven difficult. However, we know that global warming has and will continue to produce a variety of direct and indirect impacts to human health, from an increase of heat stress and allergens to worsening effects of air pollution. Global warming could have implications for yellow fever, encephalitis, meningococcal meningitis, West Nile virus, and Lyme disease among others.

The major concerns for developing countries are diarrheal diseases, malaria, and dengue fever. Such diseases hit poor children particularly hard; for example, 90% of worldwide deaths from diarrheal diseases and malaria are children five and under, mainly from poor countries.[289]

Diarrheal diseases, which currently kill 2.2 million,[290] will increase due to changes caused by heat increases, heavy downpours, floods, and drought. One study in Peru found that for every 1.8°F rise in temperature, hospitalizations for children under 10 with diarrheal disease increased by 8%.[291] Floods and heavy downpours can overwhelm sanitation systems and stir up rivers, increasing cholera, cryptosporidium, E. coli, and typhoid.[292] A study in Bangladesh found that after major flooding, diarrhea was the most common illness and cause of death for all age groups.[293] But droughts can also increase the incidences of diarrhea. Low river flows and pooling of water leads to increasing concentrations of pathogens in water supplies.[294]

Malaria is the most important vector-borne disease in the world today. As mentioned earlier, each year an estimated 300-500 million worldwide become ill and approximately 700,000 to 2.7 million die from malaria, the majority of them young children. According to one estimate, a child dies from malaria every 30 seconds. In areas with malaria, the disease is responsible for one-quarter of all deaths of children who die before their fifth birthday.

The malaria parasite is transmitted through the bite of a certain type of mosquito.[295] Temperature is a major determinant of malaria transmission. Warmer weather speeds up the parasite's lifecycle as well as the biting activity of the mosquito. More bites transmit a greater number of parasites per bite. Colder weather slows down the parasite's lifecycle until it shuts it off completely. As the temperature falls, the mosquitoes stop biting.

Because global warming has and will continue to cause changes in temperature and precipitation, it has and will continue to increase the geographical expansion and lengthen the season of malaria. This will introduce malaria into new areas where people have little or no immunity. It will also intensify its transmission in areas already struggling with the disease. On a positive note, a beneficial side effect of the decrease in precipitation could be the decrease of malaria at that time due to reductions in breeding grounds for the mosquito.

It is difficult to determine how many additional cases of malaria will result from global warming, since we don't know how successful any future eradication responses might be, whether an effective vaccine could be made available in areas with little or no immunity,[296] or the number of people who will move in and out of transmission areas. Recognizing these limitations, scientists have made projections as to how many additional individuals could be put at risk. Worldwide, estimates range from over 200 million to over 400 million by 2080.[297] There is a chance, however, that worldwide those at risk could go down by 200 million or more due to drought and decreased precipitation.[298] When reviewing all of the studies, the IPCC concluded that overall the increases of those at risk would outweigh the decreases.[299]

For Christians we must also pay attention to *who* will be made more vulnerable. This means paying particular attention to the projected increases in malaria for developing countries in areas where transmission is currently low because temperature levels naturally keep it that way, but where rising temperatures caused by global warming will change the equation for the worse. Unfortunately, worldwide these areas make up the "vast majority of additional population at risk" according to a key study.[300] In such countries an additional 90 to over 200 million could be made more vulnerable by 2080.[301] As this study on malaria and climate change puts it, "It is unlikely that these countries will have the structural or economic capacity to cope with any increases in malaria that climate change will bring."[302] Accordingly their citizens are not just at risk, they are vulnerable. Consider Mbiwo Constantine Kusebahasa, a Ugandan farmer living at the foothills of the Rwenzori Mountains:

> When I was young, this area was very cold. Now the area
> is much warmer. Before the 1970's, we did not know
> what malaria was. The mosquitoes that spread malaria are
> thriving due to the higher temperatures. At present, there
> are many cases of malaria in the Kasese area.[303]

These countries are ground zero for the malaria/climate change nexus. Unless outside funding and assistance helps them reduce their climate-enhanced risk, their vulnerability will increase.

Malaria will continue to have major implications for those in poverty. Poor individuals and countries are less able to cope with the disease, and the impacts of malaria help to keep them poor. As a major study puts it: "Where malaria prospers most, human societies have prospered least."[304] There is a five-fold difference in GDP between countries with malaria and those without. Factoring out other causative factors, it is estimated that malaria reduces GDP per capita by more than half.[305]

Dengue fever has been described as "the world's most important vector-borne viral disease."[306] While there is a more severe form of the disease that can be fatal, the common type causes high fever, severe headache, backache, joint pains, nausea and vomiting, eye pain, and rash. The symptoms last for about seven days. It is also commonly known as "break bone fever" because of the severe pain patients feel in their bones. The disease is primarily an urban phenomenon, and about one-third of the world lives in areas where the climate is suitable for transmission.[307] When an epidemic hits, it can infect 70%-80% of a local population.[308] Scientists project that global warming will increase those at risk of dengue fever by 900 million by 2055 and 1.4 billion by 2085.[309]

MAJOR IMPACTS

- **Diarrheal diseases are the #1 cause of death for children age 5 and under in poor countries. Global warming will significantly increase the number of children in the poorest countries who are vulnerable.**
- **In poor countries with little capacity to cope, an additional 90-200 million could be more vulnerable to malaria by 2080.**
- **Worldwide, an additional 900 million by 2055 and 1.4 billion by 2085 will be at increased risk from dengue fever.**

Global Warming and the Risen LORD

5) EXTINCTION OF GOD'S OTHER CREATURES

"Look at the birds of the air;
they do not sow or reap."
Matthew 6:26

In communion with and obedience to the Father, Christ the Creator has graciously provided from the abundance of His creation all that we need to live how God intends, including the use of plants and animals for food and other necessities. It has been estimated that over the years humanity has utilized 7,000 of the 270,000 known plants, although today just 30 crops provide approximately 90% of the world's food. Although humanity has utilized several hundred different types of animals over the years, currently 14 domesticated animals account for 90% of worldwide livestock production (yet many indigenous communities today continue to use 200 or more different types of species for food).[310]

However, humanity is currently overexploiting natural resources. Approximately 60% of what we derive from God's creation is being used unsustainably. We are pushing creation beyond its limits. This includes wild fish, game, and plants, for which the poor around the world depend upon significantly, especially when times are hard. It is a part of their safety net when the harvest fails, their "margin for error" in terms of survival. One study has found that such resources provide 22% of household income for the rural poor in poor countries.[311] A study of the rural drylands of Kenya during the 1996 drought found that 94% of the households depended upon wild resources both for food and to make products to sell or trade, a finding consistent with other studies.[312]

One of the major impacts from global warming will be that more of God's creatures will vanish from the Earth. Approximately 20%-30% of God's creatures are likely to be at increased risk of extinction this century, making global warming the single biggest threat to biodiversity today.[313]

In some cases this will mean that only a number of species disappear from a particular ecosystem that was helping to sustain a local human population, especially the poor. However, depending upon the species, such losses can have significant consequences for human well-being. Pollinators, for example, are essential for "three quarters of the world's principal crops."[314] Their "economic value worldwide has been estimated at $30-60 billion."[315] The loss of even one species can have important economic impacts. A study in Costa Rica found that "forest based pollinators increased coffee yields by 20% within 1 kilometer of the forest (as well as increasing the quality of the coffee)." They "increased the income of a 1,100-hectare farm by $60,000 a year."[316] What happens if global warming causes this pollinator to go extinct, or to disappear from local areas?

In other areas there will be a vast reduction of species because global warming will have pushed ecosystems beyond their tipping points.

What could this mean for the poor? Let's say in a certain area global warming contributes to the failure of crops due to drought intensification. The rains come as violent windstorms at the wrong time and there is an increase in pests. And then it also contributes to the fraying of the safety net because the poor can no longer harvest wild resources. What will happen to them then?

While in this chapter we are highlighting the consequences of global warming on the poor, it is also right for us to take a moment to underscore the profound consequences global warming represents to God's other creatures.

Being created in the image of God we are called to rule God's other creatures as He would. Christ the Creator and Sustainer provides sustenance and homes for His other creatures. But through global warming, destructive land practices, and other human activities, we are doing the opposite. Christ the Creator has given them life, but we are taking it away. Through His death on the cross, Jesus reconciled all things (Col 1:20), yet global warming works against such reconciliation in a myriad of ways, including the

Global Warming and the Risen LORD

extinction of other creatures. To be *imago Christi*, to follow the Lifegiver, is to maintain and even enhance the conditions for life for God's other creatures. Thus, being good stewards is one more reason to follow our Risen LORD in overcoming global warming.

MAJOR IMPACT

Approximately 20%-30% of God's creatures are likely to be at increased risk of extinction in this century, making global warming the single biggest threat to biodiversity today.[317]

6) REFUGEES[318]

"I was a stranger, and ye took me not in."
Matthew 25:43, KJV

As will be touched upon in chapter 9, the Holy Family themselves were refugees, fleeing to Egypt to escape the persecution of King Herod. The Risen LORD experienced what it was like to be a refugee in His earthly life, before His resurrection.

In 2008 the UN estimated that there "were 11.4 million refugees outside their countries and 26 million others displaced internally by conflict or persecution at the end of 2007."[319] But these estimates don't include refugees who basically have had no other choice but to leave their homes and communities because of poverty, natural disasters, and/or environmental degradation. Although no official government tallies are kept of such numbers, estimates by the Red Cross and others suggest refugees fleeing for these reasons are roughly equivalent to those forced to leave due to violence and persecution.[320]

Global warming will add to these numbers. Although difficult to estimate, climate impacts could create, roughly, 200 million refugees by 2050.[321] According to the CIA's National Intelligence Council, this would represent "a ten-fold increase over today's entire documented refugee and internally displaced populations."[322]

As former Army Captain and national security expert Andrew Sloan recently stated as he testified before the Senate:

> Who among us would stand by and watch our loved ones slowly wither away and die from starvation? Who would not look to relocate if the areas where you lived contained less and less drinking water, year after year? Or if the land you lived on was flooded so often that you and your family were almost permanently living in water, unable to find food and increasingly susceptible to diseases such as malaria, dengue fever or cholera? … The question that we must be asking is not just where will people go, but how are the people already living there going to react?[323]

As is widely recognized, refugees are highly vulnerable – especially women and children. Both are at greater risk of rape, molestation, and human trafficking. A study by World Vision on refugee camps in Africa discovered that child sexual abuse ranged from 50% of children in one camp to 87% in another.[324] Children are also in danger of being misplaced from their parents and loved ones, becoming orphans, or being forced to become soldiers. Refugee camps are often crowded and lack basic sanitation. Children encounter a variety of diseases and health care is either scarce or financially out of reach. Malnutrition is high, especially in camps for so-called "internally displaced persons" – refugees who stay within their own country. Education is inconsistent or nonexistent. Child labor is common. As one boy in a refugee camp in Africa put it:

Global Warming and the Risen LORD

We eat once a day. How can you learn well? This is
why we have to go and look for work so that we too can
have something to eat. People who support us
sometimes insult us before giving us food. How can you
eat in such a condition?[325]

Darfur, an area in western Sudan about the size of Texas, has seen
genocide in recent years. Refugees from here are fleeing their homes due to
political strife exacerbated by poverty, environmental degradation, and
possible climate impacts.

Over 80% of those living in Darfur are farmers who produce rain-
fed crops. Severe and prolonged droughts in the 1970s and 1980s forced
farmers to cultivate marginal land utilized by nomads/pastoralists for their
herds. The droughts also attributed to desertification. Out of desperation,
residents resorted to unsustainable land management practices, primarily
deforestation which leads to soil erosion. This further reduced land suitable
for farming. The scarcity of resources combined with increased population
is exacerbating the long-standing strife between nomads/pastoralists and
farmers. They belong to different ethnic groups and these tensions have in
turn been exploited by Sudan's national government in its struggle with
rebel factions.[326]

Since 2003, when major fighting began in Darfur, 200,000 to
400,000 have been killed and over 2.5 million have fled their homes. These
refugees are now living in camps inside Darfur and in neighboring Chad and
the Central African Republic. [327]

In 2007 the UN estimated that 7,000–10,000 refugee children from
Darfur were kidnapped from encampments in neighboring Chad. By June
2008 the situation deteriorated to the point where adults were selling boys as
young as nine years old. The numbers of children kidnapped had grown
significantly.[328] Commenting on the situation, one refugee leader said, "We
fled because of the war and we don't want to have any rebel activities inside

the camp… We want at least the children left to us to be educated. To have a start to life instead of being fighters."[329]

As for the vulnerability of Darfur women, rape has been used as a weapon to terrorize and cause people to flee their homes. But even when women get to refugee camps they are not safe, sometimes becoming rape victims as they venture out to collect firewood. The scarcity of firewood due to environmental degradation is so pronounced a humanitarian group called "Darfur Cookstoves" is providing fuel-efficient stoves "that can reduce the need to venture outside the camps — reducing the risk of rape and other violence."[330]

There is the possibility that global warming has contributed to the current situation in Darfur by intensifying the droughts that have been the main environmental driver of the conflict.[331] But whether it has or not, the situation in Darfur helps illustrate what can happen when longer and more intense droughts ravage an area already struggling with multiple problems. Such situations are bad enough – they don't need global warming making them worse in the future. Unfortunately, as mentioned earlier, if no significant actions are taken to reduce global warming pollution, climate change is predicted to increase the amount of land in extreme drought from 1%-3% today to 30% by 2090.[332]

MAJOR IMPACT

Climate impacts could create, roughly, 200 million refugees by 2050.

7) VIOLENT CONFLICTS, OIL DEPENDENCE, AND TERRORISM

"Blessed are the peacemakers,
for they will be called children of God."
Matthew 5:9, TNIV

Global warming will not automatically lead to conflicts and violence. However, as we have seen with Darfur, the impacts of global warming could help create conditions that lead to or exacerbate them. One group's speculative estimate suggests that as many as 2.7 billion people in 46 countries could be at high risk for violent conflicts when global warming is added to the mix.[333]

Because of this potential, the national security community in the United States has begun to see the dangers of global warming to peace and prosperity for our nation and the world.[334]

In the spring of 2007, 11 retired generals and admirals in conjunction with a national security think tank, where they served as senior advisors, issued a remarkable report, *National Security and Climate Change*.[335] One of the generals, Gordon Sullivan, former chief of staff of the US Army, captured the seriousness of the situation regarding our national security this way: "The Cold War was a specter, but climate change is inevitable."[336]

The generals and admirals introduced a new and important concept. They termed global warming a "threat multiplier." The following quotation captures their conclusion that global warming and its impacts will multiply threats around the world:

> Climate change has the potential to result in multiple chronic conditions, occurring globally within the same time frame. Economic and environmental conditions in already fragile areas will further erode as food production declines, diseases increase, clean

water becomes increasingly scarce, and large populations move in search of resources. Weakened and failing governments, with an already thin margin for survival, foster the conditions for internal conflicts, extremism, and movement toward increased authoritarianism and radical ideologies.[337]

Recently one of the generals, Chuck Wald, former deputy commander of the US European Command (which includes 91 countries in Europe, Eurasia, the Middle East, and sub-Saharan Africa) testified on behalf of the group before the Senate Foreign Relations Committee. General Wald believes global warming helped cause the conflict in Darfur, and uses the situation to illustrate what is meant by the term "threat multiplier." He states, "The Darfur region was already fragile and replete with threats – but those threats were multiplied by the stresses induced by climate change."[338]

Thus, global warming is a threat multiplier in at least two respects: (1) global warming interacting with already unstable situations could foster violent conflicts occurring simultaneously around the world; and (2) in unstable areas, the stresses created by global warming could tip concerns into threats, multiplying the threats to peace and at times serving as the spark that turns threats into violence.

The admirals and generals point out, "Struggles that appear to be tribal, sectarian, or nationalist in nature are often triggered by reduced water supplies or reductions in agricultural productivity."[339] Thus, global warming's intensification effect may not be immediately recognized or appreciated as a cause in conflict.

Concerning **oil security**, the report highlights that we currently import more oil from sub-Saharan Africa – the world's poorest region and the one that will be hit hardest by global warming – than we do from the Mideast.[340] In 2006, the year Africa claimed the top spot; we imported 22% from African countries.[341] But the United States isn't the only country that wants Africa's oil. China has begun to increase its African oil imports as well, receiving about a third of its oil from Africa in 2006.[342]

"If Nigeria's access to fresh water is reduced or additional stresses on food production – which could be a result of projected changes in rainfall patterns – millions of people would likely be displaced. If the Niger delta were to be flooded from sea level rise, or if major storms damaged oil-drilling capacity, the region would lose its primary source of income. Again, millions of people could be displaced. There really is no controlled place in Nigeria for displaced people to go, no organically controlled capacity for an organized departure, and an extremely limited capacity to create alternative living situations. And the movements would be occurring in a country with a population of 160 million people that is split geographically between Muslims and Christians. These stresses would add dramatically to the existing confusion and desperation, and place even more pressure on the Nigerian government. It makes the possibility of conflict very real." [343]

General Chuck Wald
before the Senate Foreign Relations Committee

If there is currently one country in Africa where we don't want to see potentially destabilizing global warming impacts it is Nigeria, Africa's most populous country with 146 million in 2008.[344] As of July 2008, it was one of America's top five oil suppliers, and our largest supplier in Africa.[345] Yet much of Nigeria's oil production is vulnerable. Flooding and sea level rise threaten approximately 259 oil-producing fields in the Niger Delta.[346]

Oil, which began flowing in the late 1950s in Nigeria, has had a corrupting influence on the country. The rich and powerful have become more so, but most of the country is still mired in poverty. In many respects oil has become a curse. The Niger Delta has suffered an oil spill the size of the Exxon Valdez spill every year for the last five decades.[347] As a *National Geographic* article put it, "It stains the hands of politicians and generals, who siphon off its profits. It taints the ambitions of the young, who will try anything to scoop up a share of the liquid riches—fire a gun, sabotage a pipeline, kidnap a foreigner."[348]

THE INTENSIFICATION
OF SMALL CONFLICTS

Besides being a "threat multiplier" that could to lead to large violent conflicts, global warming will also help create the conditions for small, obscure conflicts, ones so small that the outside world does not notice. Studies have shown that "When rainfall is significantly below normal, the likelihood of conflict outbreak is significantly elevated in the subsequent year."[349] Another finding has been that in years of low rainfall in the Horn of Africa, "human deaths and livestock losses (from starvation, theft or killing) increased dramatically during the end of the dry season." [350]

The story of what happened in 2005 in a remote village in northern Kenya called Sambarwawa, serves as an example. At one point the drought had driven 10,000 nomadic pastoralists with 200,000 animals to this tiny village in search of water. Many had come over 300 miles only to find there wasn't enough water to go around. Some resorted to violence. Ethnic tensions and divisions surfaced as people began to form into ethnic groups. In a little over a month nine people were killed. Over 3,000 animals – a family's only asset – were stolen. [351]

One young herder who was injured while tending his family's livestock was Arkan Athan Hussein. His friend, Abdi Maalim, was killed. "Six armed people emerged from nowhere … one of them raised his AK-47 and shot Abdi in the chest and shoulder. As I fled, they shot at me." Arkan's father reported that in 40 years of traveling to Sambarwawa this was the first time people had been murdered. [352]

Before oil, Nigeria was self-sufficient in food production and agricultural products like palm oil and cacao beans made up most of its exports. Now, oil accounts for 95% of revenue from exports and 80% of government revenues.[353] The country currently imports more food than it exports.[354] Even though the national government receives 55%-60% of oil revenues, up to 70% of the funds are stolen or wasted according to a 2003 report from the government's own anticorruption agency.[355] Very little of the rest trickles down to help with local community services and infrastructure. The region that produces 80% of the oil revenue has the lowest life expectancy in the country.[356] "One local government spent 2 percent of its share on its crumbling primary school and 30 percent on its own salaries and offices. Another local chairman claimed to have spent huge sums on projects, including a fish pond with neither fish nor water."[357] Noting the corruption, an analyst for the oil companies remarked, "So you have communities with no hospitals, roads or schools, and that creates a lot of anger against the government and the operating companies there."[358] Because of Nigeria's chronic poverty, despite its oil wealth, and the resentment and opposition fueled by inequity, it ranks in the top 20 of countries vulnerable to becoming "failed states."[359] Comments by General Wald on a recent trip to Lagos, a city of 17 million people, illustrate the situation. "The best way to describe our drive from the airport to the hotel is that it reminded me of a 'Mad Max' movie … It was just short of anarchy." Massive numbers of people filled the roads. Huge piles of trash were everywhere. Fires dotted the roadside with massive fires in the distance.[360] From General Wald's perspective, when you add global warming's impacts into Nigeria's combustible mix, the chances for serious conflict increase significantly (see quote on page 121).

CARE, a major relief and development group, agrees with Gen. Wald about Nigeria. In a recent analysis identifying areas of the world where global warming could heighten the potential for conflict, Nigeria is identified as a conflict hotspot.[361]

Besides Nigeria, Chad, Sudan, Kenya, and the Democratic Republic of Congo (formerly Zaire) are other African countries that produce oil and are potential conflict hotspots due to global warming.[362] All are currently listed as "critical" or "in danger" by the Failed States Index. Sudan and Chad are already rife with conflict, some of it potentially climate-induced and climate-intensified. In 2008 they were ranked #2 and #4, respectively. The Democratic Republic of the Congo was ranked # 6, while Kenya was ranked #26 (despite having significant violent ethnic conflicts related to disputed elections in December 2007).[363]

Disruptions of African oil could have serious consequences for US energy and economic security. It would force us to rely even more on countries that don't share our values such as despotic regimes in the Middle East that imperil religious freedom and the security of Israel, Chavez's Venezuela that fosters instability in South America, and a newly resurgent Russia. If such disruptions were to occur at financially troubled times, the economic consequences for the United States, indeed the world, could prove quite harmful. Given that we have fought wars over oil, disruptions also have national security implications.

But when asked why US citizens should care about Africa, General Wald begins not with threats to our economic or national security. Surprisingly, he believes we *should* care because we *do* care. "We have a humanitarian character; it's one of our great strengths, and we shouldn't deny it." In other words, we should care about Africa in order to remain true to our deepest selves.

General Wald goes on to say, "Some may be tempted to avert their eyes, but I would hope we instead see the very real human suffering taking place there. We should be moved by it, challenged by it." Wald believes that such considerations should be given weight in the realistic environment of national security discussions, "because part of our security depends on remaining true to our values."[364] In other words, when we stray too far from who we are, we become weak in character, which erodes our security.

According to the generals and admirals, global warming will also foster the conditions for **terrorism**. Admiral Joseph Lopez, the former top NATO commander in Bosnia, believes global warming "will provide the conditions that will extend the war on terror ... more poverty, more forced migrations, higher unemployment. Those conditions are ripe for extremists and terrorists."[365] The report cites Lebanon's experience with Hezbollah as an example where extremist groups fill the vacuum when governments fail to deliver basic services and maintain the rule of law.[366]

Another recent example is the 2010 Pakistani floods. In areas where aid was slow to reach the poor, militant groups were quick to fill the void.[367] "Within days, groups including Falah-e-Insaniyat, the charitable arm of Lashkar-e-Taiba (LET)—which carried out the 2008 Mumbai terrorist attack—were serving meals to victims and providing them with clothing, medicine and even money." As one person put it, "For us they're angels."[368]

For the militants, the flooding is simply another recruitment opportunity. A recent study by an institute based in Pakistan's capital, Islamabad, found that the areas with "the worst food insecurity were also home to the worst militancy." According to the head of the institute, Abid Suleri, "It's a class conflict exploited by [the militants] who say, 'If you are living in misery, it's better to at least kill the infidels.'"[369] Also, while about 60,000 troops whose normal assignment is to engage the Taliban were dealing with flood relief, the Taliban used the chaos created by the floods as an opportunity to stage attacks.[370]

Finally, while climate impacts could lead to violent conflicts, such conflicts increase vulnerability and diminish the capacity to adapt. One study found that "Not only does environmental stress make countries prone to conflict, but conflict exacerbates vulnerability and reduces the adaptive capacity of a country and its population to deal with environmental stress."[371]

MAJOR IMPACTS

- Global warming will increasingly act as a threat multiplier for instability in some of the most volatile regions of the world, and has the potential to result in multiple chronic conditions occurring globally at the same time.
- Global warming poses a threat to our energy security, our economic security, and our national security, given that Africa, our #1 supplier of oil, will be hit hard by climate impacts, and that the consequences of climate change will create or intensify conditions that could lead to terrorism.
- One group's estimate, although quite speculative, suggests that as many as 2.7 billion in 46 countries could be at high risk for violent conflicts.

CONCLUSION

As we conclude this review of major ways global warming will impact the poor in developing countries worldwide in this century, let me summarize some of what we have covered thus far:

- ➢ Up to 170 million could face hunger and malnutrition.
- ➢ 1-2 billion already in water-scarce situations could see a further reduction in water availability.
- ➢ Conservatively, 100 million could face coastal flooding, and millions more will be impacted by inland flooding.
- ➢ 90-200 million could become more vulnerable to malaria, 1.4 billion could become at increased risk of dengue fever, and the number of children vulnerable to diarrheal diseases – the number one killer of children – will increase significantly.
- ➢ 20%-30% of God's other creatures will likely be at increased risk of extinction by the end of this century.
- ➢ Approximately 200 million could become refugees by 2050.
- ➢ Billions could be at increased risk for violent conflicts, including in areas sensitive to energy security and the growth of terrorism.

As we will explore in more detail in Parts 2 and 3, walking with the Risen LORD includes helping people to be less vulnerable and helping those in poverty climb out of it. Chapters 5-6 have demonstrated that the impacts of global warming will do the opposite. Global warming will

(1) make the poor and vulnerable *more* vulnerable, and

(2) be one of the great poverty intensifiers of this century.

We can't stop all of the projected impacts from happening. But, as we will explore in Part 3, by aggressively addressing both the causes and the consequences we can reduce the number affected and also help to lessen the impacts for those who will face the hunger, thirst, disease, displacement, and violence from global warming's wake.

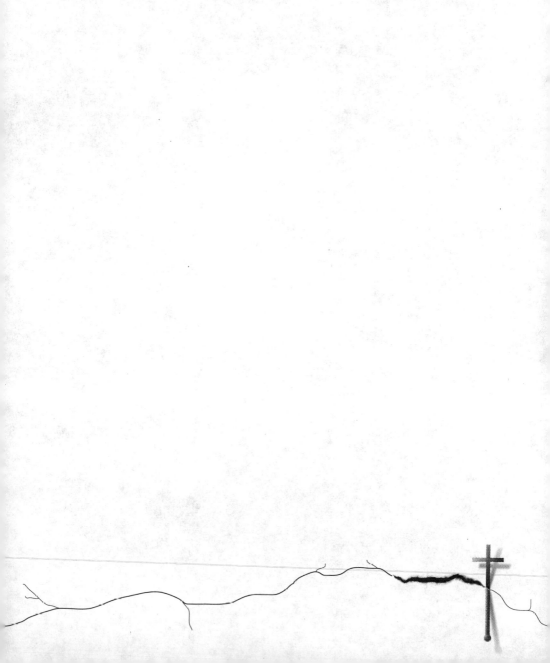

7

THE NEXT GREAT CAUSE
OF FREEDOM

> **They have a passion for liberty that is almost unconquerable, since they are convinced that God alone is their leader and master.**
>
> Josephus, a first century historian writing about early Christians[372]

We live in a broken world. Our time is filled with imposed restrictions on the opportunities we should have to become whom the LORD created us to be. Made in the image of God, we have all been created to be free of restrictions such as under-nutrition, prejudice, the lack of religious freedom, or the denial of education. The list could go on. But today global warming must be added to near the top. The impacts of global warming are now a part of these false, harmful limitations. All Christians have a role to play in following our Risen LORD as He leads us in overcoming global warming.

The United States of America came into being because of the belief that we are created equal. (For Christians, this belief is founded in the understanding that we are created in the image of God.) While it took us two centuries and a Civil War, a women's suffrage movement, and a civil rights movement to begin to live up to the fullness of this belief, we now recognize that all adult citizens must have a say in how our country is governed.

For those who fought to found our country, this crystallized in the phrase "no taxation without representation." Before the American Revolution our Founding Fathers worked hard to create a better life for themselves and their children. But those in power across the ocean, through taxation without representation, took what our founders had earned. Without having any say in the matter, our founding fathers had their opportunities limited. This denial of freedom constituted tyranny, and to overcome such tyranny they were willing to give their lives.

For the poor in developing countries today, the tyranny of global warming is the equivalent of taxation without representation. Through a process in which they have no say, by decisions made by those far, far away, are profound limitations placed upon their freedom to create a better life for themselves and their loved ones.

Americans hold freedom especially dear. We are the beneficiaries of the blood of patriots who gave their lives on the altar of freedom, from the American Revolution through the Civil War and up to our own time. We are a freedom-loving people who fight to maintain and expand our freedom because we know how precious and costly it is.

For freedom-loving people like us, global warming is a worldwide scourge, similar to how communism was in the twentieth century. Global warming is a freedom denier, a freedom destroyer, not only in terms of denying opportunities for individuals, but potentially for the cause of freedom in entire countries.

A recent study found poor countries lacking a literate population are most vulnerable to climate impacts. Why? "A literate population will be better able to lobby for political and civil rights, which in turn will allow it to demand accountable and effective government. Where such rights exist, governments are more likely to become accountable for reducing the impact of successive high mortality disasters, and are thus more likely to address vulnerability."[373] The history of our own freedom proves the point. If our

Global Warming and the Risen LORD

Founding Fathers had not been literate there would have been no American Revolution, Declaration of Independence, Constitution, or Bill of Rights.

It is natural and right for people in poverty to want a better life for their loved ones and themselves. The reason is because the LORD created us to live in a world where our basic physical needs are met in order to enjoy our relationships with Him, one another, and the rest of His creation. This is the beautiful image presented to us in the first chapters of Genesis. Living in the Garden of Eden, Adam and Eve walked with the LORD in the cool of the evening (3:8-9). Only after the Fall did the production of food involve toil and difficulty (3:17-19) or the killing of animals (9:2-3). Only after the Fall did the destructive capacity of our sinfulness lead brother to kill brother and all types of injustice to enter the world, where righteousness fails by the fact that some don't have enough even to meet their basic needs.

But the LORD continued to show us that He desires for everyone's basic needs to be met. When He fed the Hebrews with manna in the wilderness, everyone was provided what they needed, no more, no less (Ex 16:18). When the Promised Land was partitioned, each of the 12 tribes and each family received their fair share (Num 34, Josh 13-21). When people fell on hard times and had to take out loans, the LORD instructed that the tools and resources needed to earn a living must not be taken as collateral, "because that would be taking a man's livelihood as security" (Dt 24:6). When crops were being reaped, the Israelites were not to go back and get what produce might have been missed. Rather they were to "Leave it for the alien, the fatherless, and the widow, so that the LORD your God may bless you in all the work of your hands" (Dt 24:19-22).

Every seventh year, the Sabbath year, the land was to be given a time of rest so that it would continue to remain productive. Whatever the land yielded on this year was for the "owner," (or more theologically accurate, the steward) but also for the servants, hired workers, refugees, and the livestock – even the wild animals (Lev 25:1-7). In the 50th year, the year of Jubilee, all agricultural land was to be returned to its original Israelite owners to ensure that no family lived in perpetual poverty (25:10-17). God reminded the Israelites, "The land must not be sold permanently, because the land is mine and you are but aliens and my tenants" (25:23). In addition, those who had sold themselves because of poverty must be set free (25:54).

The alien or refugee, the fatherless or orphan, the widow, the slave, were the most vulnerable, the ones without power. And precisely because they are powerless, the LORD has a special concern for their just treatment in a sinful world. Precisely because they are vulnerable the LORD wants to ensure that their basic needs are met.

Jesus himself began his life as a refugee where his very life hung by a thread. When Jesus launched his ministry, his first words were about the poor and freedom:

> The Spirit of the Lord is on me,
>> because he has anointed me
>> to preach good news to the poor.
> He has sent me to proclaim freedom for the prisoners
>> and recovery of sight for the blind,
> to release the oppressed,
> to proclaim the year of the Lord's favor (Lk 4:18-19; quoting Isa 61:1-2; 58:6).

This last phrase, "the year of the Lord's favor," alludes to the Jubilee.

Once Jesus finished the reading he said to them, "Today this scripture is fulfilled in your hearing" (v. 21). It was accomplished in His own person, in His words and deeds. He ministered to the poor; He healed the blind and those who were shunned, such as lepers and the woman with feminine bleeding. He set them free from their infirmities and their shame. He fed thousands and thousands of hungry people. Jesus' ministry was a manna-in-the-wilderness-type of ministry, a Jubilee-type ministry wherein he met the basic needs of the poor, the oppressed, the outcast, the forgotten.

The teachings of Scripture and Jesus' life and ministry confirm why it is natural and right for the poor to want better lives for themselves and their loved ones, lives of sufficiency and contentment. It is because God wants such lives for them as well, as He does for everyone.

But right when many of the world's poor around the world are lifting themselves up, global warming is going to push them back down.

In chapters 5-6 we have seen how global warming will hit the poor the hardest. And yet many times when we hear about natural disasters striking the poor, we tend to think of such impacts as only temporary setbacks from which individuals, families, communities, and whole countries simply snap back. On the news we see a poor family in distress because of a flood, pray for them, maybe give some money to a Christian relief organization, and go on with our day. We'll help to tide them over until normalcy returns. Tomorrow is another day, etc.

But we know from our own lives that wounds take time to heal and some things cannot be fixed. Particular opportunities once lost never come back and loved ones who die can never be replaced.

Whether we are talking about an individual, a family, a community, a country, a region, or a continent, coping with the consequences of global warming will take time, energy, and economic resources that could have been used in another way.

When disasters strike and hard times hit the poor, there are things they have to do in order to cope. They withdraw children from school and send them to work. They sell household goods, cut nutrition and then meals. They sell their animals and eat the seeds for next year's crop. At a certain point they may forsake their homes and become refugees. Such a terrible turn of events is what I call the **downward disaster spiral**.

Each of these decisions can have lifetime consequences, including ones that can prove fatal.

The simple fact is global warming-intensified natural disasters and its other impacts will rob people of opportunities to create a better life. It may even rob them even of one of their greatest motivations to create such a life – their loved ones.

While the loss of loved ones is incalculable, we have seen that some opportunities lost can be calculated, at least in approximate terms. For example, after Hurricane Mitch, the poor in Honduras lost over 30% of their assets,[374] pushing many who were climbing out of poverty back into it. As mentioned previously, drought-induced stunting can result in a 7%-12% loss in lifetime earnings. Under-nutrition also leads one to be more vulnerable to diseases such as malaria and diarrheal illnesses.[375] And when impacted poor families cope with diseases, in many cases they do not seek medical treatment – if it is even available – until it is too late.

Education, of course, is key to potential advancement. The impacts of global warming will create educational disruptions and hindrances for many. Children may be forced to work because of economic hardships rather than attend school. Their families may become refugees. In such circumstances education may become inconsistent or nonexistent as their families focus on survival. For example, during the flooding in Bihar, India, in 2007, considered at that time to be the worst in living memory, Sheila Devi's six daughters were unable to go to school. "They barely can get anything to eat. How can we even think of school?"[376]

After Hurricane Mitch hit Nicaragua, the number of children in affected areas who had to work more than doubled from 7.5% to 15.6%.[377] In Sri Lanka, two years after the 2004 tsunami, nearly 30% of children were still displaced and their education had yet to return to pre-tsunami levels.[378] And in the Ivory Coast, between 1985 and 1988, enrollment rates declined about 20% in areas impacted by drought.[379]

Education can also be thwarted through the premature selling of animals, which represent a family's long-term savings. Some keep them in reserve to sell them when the time comes for their children to attend secondary school.[380] If they are forced to sell them prematurely to avoid starvation, it can rob children of such opportunities to better their life.

A child taken out of school to work can have ripple effects for generations. For example, the greatest cause of inequality of opportunity in Latin America is the lack of the parents' education and the father's occupation.[381] For most, only by staying in school is there a good chance of breaking this cycle.

Hard times enhanced by global warming will also increase various types of exploitation of children, as they are put to work in hazardous conditions, forced to become child soldiers, and abused sexually or forced into prostitution – all, of course, with resulting lifetime impacts and lives squandered.

We have discussed the effects of global warming individually and in isolation from each other. However, this is not how they will present themselves. In many instances they will be happening simultaneously. Or as an area begins to recover from one impact another one strikes. Furthermore, chronic problems and new ones that are unrelated, or marginally related, to global warming also will be present.

A fairly recent interaction of drought, coffee prices, and natural disasters in Central America can serve as an illustration of what these synergistic impacts might look like in the future.

Hurricane Mitch in 1998 was followed in 2001-02 by a drought, which reduced agricultural yields over 18%. "Consequently, stocks declined and more resources had to be spent on food imports, while living conditions deteriorated for some 600,000 inhabitants of rural areas."[382] During this same time, due to a significant reduction in worldwide coffee prices, the agricultural sector suffered a sharp decline in income from coffee. This curtailed the revenues of nearly 300,000 coffee producers in Central America; 170,000 permanent jobs and about $140 million in wages were lost.[383] And so, in this instance, drought, disasters, and falling coffee prices combined into a potent brew for the poor.

Similarly, many of the world's poor will face simultaneous and numerous climate impacts during their lifetimes. A study of successive droughts in Ethiopia suggests what the consequences could be in such downward disaster spirals. The country weathered five major nationwide droughts since 1980, and numerous local droughts. More than half of all families experienced at least one major drought. Poverty increased by 14%, keeping at least 11 million from escaping it.[384]

Even in good times, many of the poor hedge their bets against disasters by planting crops such as sorghum that bring less of a return but are more resilient to drought. A study in Tanzania showed that due to such decisions, the poor earned 25% less than their richer counterparts who planted riskier but more profitable crops.[385] Thus, dryer droughts from global warming could lead to fewer profits for the poor as they become even more risk-averse. In so doing they reduce their chances of emerging from poverty. After being thwarted time and again, some may simply give up.

So, in this century, unless we work to change the situation, global warming will be a major factor in keeping the poor, poor. But it will also make others poor or return them to poverty.

While it is impossible to take full measure of such situations, one simple gauge to use is the reduction in GDP. Based on this very incomplete indicator, by the end of this century 145 million more people could be living on less than $2 a day due to global warming.[386]

In the future, how many poor children will drop out of school because of global warming? How many families who were emerging from poverty will be pushed back into it? How many opportunities will vanish? How many loved ones will be lost? How many dreams will be denied?

In the drought-stricken Vidarbha region of India there was a young farmer named Satish Bhuyar. In the summer of 2004 Satish was banking on a big crop to help pay off loans for his sister's wedding. In early June the monsoon rains arrived and Satish planted his crops. But the rains stopped and the saplings died. When the rains returned in late June Satish planted again, but again the rains stopped and his crops withered. Without knowing what else to do, even though there was no rain, Satish planted a third time. The rains didn't return. On July 13 Satish "walked out into his barren fields and swallowed pesticide," committing suicide.[387] Unfortunately, Satish is not alone. Tens of thousands of Indian farmers in this region have killed themselves since the drought began in the late 1990s.[388] In 2006, farmers from Vidarbha committed suicide at a rate of one every eight hours.[389]

The reasons why so many farmers are in such dire straits that they commit suicide are complex.[390] Furthermore, this region is naturally prone to drought. Whether or not global warming intensified this particular drought is an open but legitimate question. But we do know that global warming will intensify droughts in the future, and it will make the lives of those like Satish even more precarious. For many of those teetering in the balance, it may push them over the edge.

GLOBAL WARMING AND FREEDOM

As global warming works to keep the poor, poor, it could also help rob them of their chance to become free in the democratic sense of being able to petition and influence their government. Increased malnutrition, stunted children, maternal mortality, loss of educational opportunities, and increasing conflicts over scarce resources – these consequences and others of global warming could either erode the democratic dimension of freedom or strangle it in its cradle.

The point is this: the most vulnerable countries to global warming are poor countries that lack the freedoms we enjoy. Unless we work to overcome global warming, it will help keep them poor and will strengthen the possible stifling of the democratic dimension of freedom – one of the very things needed to make them less vulnerable.

To be on the right side of history, the right side of the cause of freedom, means overcoming the tyranny of global warming. This is the great cause of freedom in the twenty-first century.

As Christians we know that we don't simply have freedom for freedom's sake. We have it for God's sake. We have it for the sake of doing God's will. We have the gift of freedom so we can freely become the ever increasingly glorious images of Christ, as we love God, love our neighbors as ourselves, and care for the least of these as if they were the Risen LORD Himself.

As the Apostle Paul said to the Galatians, "It is for freedom that Christ has set us free ...You, my brothers and sisters, were called to be free. But do not use your freedom to indulge the sinful nature; rather, serve one another in love. The entire law is summed up in a single command: 'Love your neighbor as yourself'" (5:1, 13-14). Jesus died because of the abuse of our freedom, because of our sinfulness. But he also died to set us free from the tyranny of our own sinfulness so that we could love God back, including loving others as ourselves.

The terrible price Christ paid for our freedom is the ultimate demonstration of how costly and precious our freedom is. Christ has set us free to become fully and completely free; free from the bondage of sin so that we can live out, "Not my will, but Thine be done." The good news is that in doing His will are true freedom and glory and spiritual beauty and love all joined in a grace-empowered embrace. It is His will that we do our part in overcoming global warming by reducing future impacts, addressing the causes, and helping the poor to benefit from the clean energy future we will create.

To paraphrase Dr. Martin Luther King, to overcome global warming is to let freedom ring. Through our love we must let freedom from global warming ring in every African village and along the shores of Bangladesh and in the slums of La Paz and along the Indus River in Pakistan. We must let freedom from global warming ring in the lives of the poor and distressed around the world like the mother during the Niger famine of 2005 who said as she watched her daughter die, "God did not make us all equal – I mean, look at us all here. None of us has enough food." And we must not forget that we must let freedom from global warming ring in our own country as well.

As we work to overcome global warming, each of us here in the United States must let the freedom from the responsibility of global warming ring in our hearts and minds and in the daily living of our lives. Overcoming global warming is also something we must do together to be successful: as families, churches, communities, and the whole country – indeed, all the nations of the world.

In America's best moments we have been the harbingers of freedom around the world. So too can we be in overcoming global warming. We must rise to this occasion. And in so doing we will find that we are following the Risen LORD.

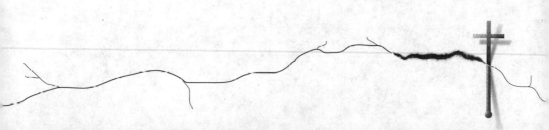

PART 2
WALKING WITH
THE RISEN LORD

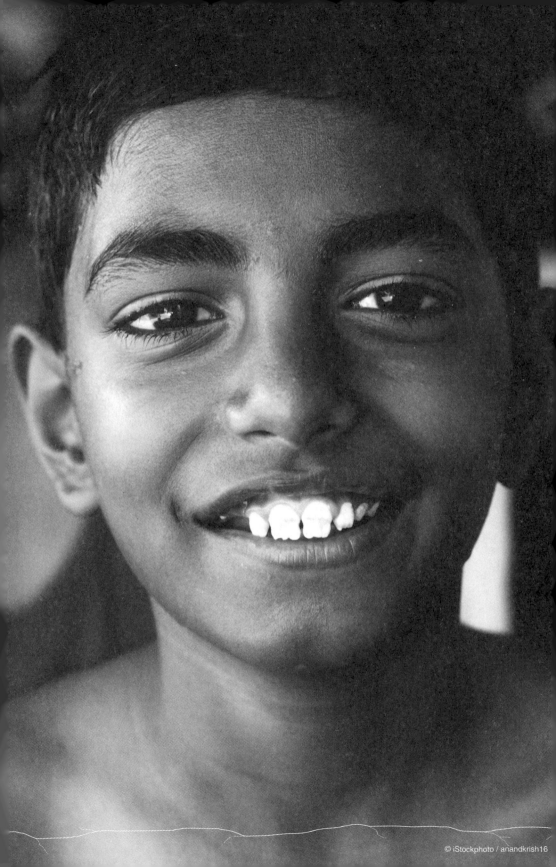

8

WALKING WITH CHRIST –
THE CREATOR AND SUSTAINER

I have to go to school.

That's what Galib Mahmud, a 12-year old from Dhaka, Bangladesh, said in response to a question about why he was in the flooded streets of his hometown, a city of over 10 million that was 40% submerged. While flooding in Bangladesh is normal, in the summer of 2004 the flooding was extreme, causing $6.7 billion in damages and over 2,000 deaths across the region.[391] Eighty percent of the country's crops were destroyed and 30 million people became refugees.[392] Regardless of the danger, Galib waded to school barefoot in his crisp white shirt "through waist-deep, filthy floodwaters, carrying a backpack on his head and dangling his black shoes in a string bag as he trekked across the flooded streets of Dhaka."[393]

When I think about Galib and other children like him, I ask, "Will something happen to him simply because he's trying to learn, to make a better life for himself?"

In this century Bangladesh will be at the epicenter of one of the most well-known effects of global warming: sea-level rise. As one of the world's poorest nations and one of its lowest, Bangladesh will be hit hard. Combine sea level rise with increased flooding from glacial melt and heavier monsoons, and up to 60% of the population could be impacted – over 70 million people.[394]

Yet Galib and others living in Dhaka contribute only one-eightieth the amount of global warming pollution as does someone from the United States.[395]

IN THE BEGINNING

In the beginning was the Word, and the Word was with God, and the Word was God. He was with God in the beginning. Through him all things were made; without him nothing was made that has been made. In him was life, and that life was the light of men. (Jn 1:1-4)

This passage begins the story of a love affair, of unrequited love in many respects. It is also a retelling of an older version of the story, and further reveals one of the central characters. That older version of the story began:

In the beginning God created the heavens and the earth …

This one starts with:

In the beginning was the Word …

It is probably no coincidence that these accounts start the same way. When early Christians began to fully comprehend who Jesus was, they must have asked themselves, "How do we understand how everything came into existence in light of the coming of the Son of God?"

The astonishing answer is "through Him" (Jn 1:3, 10). Christ created the Universe. Let that sink in a bit.

But another question tumbles out: "How is this possible?"

For that, John pushes us back, back before time and space, back before the beginning as it was described in Genesis. For there the "beginning" is when God created all of existence. John says, before that beginning "the Word was with God, and the Word was God." Not a demi-god. Not a "son" who is in some way less than God the Father. Plain and simple: the Word was God; the Word is God, the same as or coequal with God the Father. As Jesus will later say, "Anyone who has seen me has seen the Father" (Jn 14:9).

It is this same Word that "became flesh and made his dwelling among us" (Jn 1:14). Or more literally, in keeping with the nomadic history

of the Israelites, Christ "pitched his tent among us." God not only became a human being, he sought to join us, be one of us in the sense of belonging. "Hey, can I pitch my tent with you folks?"

The Creator came, but "the world did not recognize him." Even his own people "did not receive him."

"Can I pitch my tent with you?"

"No!"

Yet a few did receive him and "beheld his glory" – meaning they understood Him to be the Son of God through whom the Universe came into existence. And for those who accepted Him for who He was, "he gave them the right to become children of God" (v. 12).

Nearly 2,000 years later many people, including many Christians, do not understand that the Creator of the Universe is the same Jesus of Nazareth. Yet it is precisely because the Word, the preexistent Son of God, was with the Father and the Holy Spirit in the beginning, indeed, before "the beginning," it is precisely because the Word was God, that the entire Universe can come into existence "through Him."

Thus, it is the Second Person of the Trinity, the Word, the preexistent Son of God, who is the One through whom the Triune God creates the Universe. Simply put, Christ is the Creator of all things. Christ is the Creator.

So because the Word became flesh and modestly, humbly, quietly pitched his tent among ours, we can therefore say:

In the beginning Christ "created the heavens and the earth" (Gen 1).

Christ is the Creator – really? This seems so radical, so frankly unbelievable. Maybe it's a typo. Maybe an overzealous scribe somewhere along the way inserted this into the text of John's Gospel. Maybe this is too momentous a belief to hang on a few words in one text.

And yet the New Testament proclaims that Christ is the Creator in several other instances from the early Christian communities of the first century who wrote and preserved the New Testament.

The Apostle Paul puts it succinctly: "Yet for us there is but one God, the Father, from whom all things came and for whom we live; and there is but one Lord, Jesus Christ, **through whom all things came** and through whom we live" (1 Cor 8:6, emphasis added).

All things come from the Father but through the Son, Jesus Christ.

Hebrews 1:2 proclaims the same belief: "in these last days he has spoken to us by his Son, whom he appointed heir of all things, and **through whom he made the Universe**" (emphasis added).

CHRIST THE CREATOR REIGNS

But it is an incredible text in Colossians, specifically 1:15-20, that I want to introduce now, focusing in particular on verse 16.

> [15]*He is the image of the invisible God, the firstborn over all creation.* [16]*For by him all things were created: things in heaven and on earth, visible and invisible, whether thrones or powers or rulers or authorities; all things were created by him and for him.* [17]*He is before all things, and in him all things hold together.* [18]*And he is the head of the body, the church; he is the beginning and the firstborn from among the dead, so that in everything he might have the supremacy.* [19]*For God was pleased to have all his fullness dwell in him,* [20]*and through him to reconcile to himself all things, whether things on earth or things in heaven, by making peace through his blood, shed on the cross.*

In prison in the late 50s or early 60s Paul writes to the Colossians, whom he had never met, to correct ideas perpetrated by false teachers. These false teachings were causing trouble within the church. Paul comforts and strengthens them with the truth about who Christ actually is. He does so by quoting familiar teachings to the Colossians, especially our current text, 1:15-20. It is an early summary of the Christian faith, accepted by the Colossians, and quoted by Paul to help them understand how wrong the false teachers were and how they don't need to follow their false practices. In effect, Paul is saying, "Look, you've been led astray by these guys from the Christian teachings that you yourselves know as the truth. The church accepts these teachings and passes them on to you and me. Let me remind you of what you already have professed."

It's incredible and mysteriously wonderful that we don't know who was inspired by the Holy Spirit to compose this summary of the Christian faith that both Paul and the Colossians received. Who were the early Christians, unknown to us, who first believed and then composed these words, who were returned two millennia ago to the dust from whence they came, who thought these revolutionary thoughts? Did they sit and think these ideas together, knowing the ridicule and even persecution they would face? Who did they think they were, to think such thoughts? In the eyes of society they were probably nobodies, or possibly troublemakers. Who would think that a crucified peasant from a hick town in a puny, insignificant part of the Roman Empire could be the Creator of the Universe? It's delusional, and even socially and politically dangerous (especially when Caesar is supposed to be in charge). It's certainly spiritually dangerous, even blasphemous. These unknown, gloriously crazy folk are our spiritual ancestors who allowed the Holy Spirit to work through them so that nearly 2,000 years later we can better understand who Christ really is.

While we don't have the complete picture of the false teaching Paul is refuting, its proponents appear to have claimed some type of secret knowledge about angels and "the elemental spirits of the Universe" (2:8, 20). They claimed these supernatural powers were in fact the ones in charge and were to be appeased by various ritualistic acts (2:16, 20-23). They taught Christ's death provided for the forgiveness of sins, but humans must still appease these supernatural beings – or so this false teaching apparently went.

To refute these false ideas Paul quotes this early summary of Christian faith that the Colossians themselves had affirmed to remind them that Christ has "supremacy" in all things (1:18). Indeed, when we see how profoundly it proclaims Christ's supremacy, we see that Paul is being quite gentle and pastoral with the Colossians. Instead of responding, "Are you crazy!? The junk these guys are peddling is so contrary to what we profess I can't believe you're even asking me about it!" He instead reminds them this astounding early confession of the Church.

This text proclaimed to the Colossians, and now to us, that Christ is the Creator, and because He is the Creator He reigns. Christ has supremacy in all things, in all areas. He exists before all things, holds all things together, is the head of the Body (the Church) and He is the first one to be raised from the dead to show what our future resurrection bodies will be like. All of this makes Him the LORD of all that exists. Christ the Creator reigns.

Thus, there is no need for the Colossians to have some type of syncretism where Christ as redeemer is combined with beliefs and rituals appeasing other powers, because the whole cosmos is subject to Christ.

Are we any different from the Colossians in that we have confessed Christ to be our LORD and yet we appease all kinds of "principalities and powers?" We allow him to provide for the forgiveness of our sins, sure, but is He LORD of our lives? Will we allow Him to be LORD when it comes to overcoming global warming? (We will explore Christ's Lordship further in chapters 12-14.)

FOR HIM

There are many mind-blowing aspects to our Colossians text, but the next affirmation I want to lift up really can turn things upside down – or rather, right side up.

The end of verse 16 says, "All things were created by Him and for Him." We've already seen that Christ created all things. But we also see that all things were created *for* Christ. Hebrews 1:2 suggests the same thing when it proclaims that Christ is "heir of all things."

When is the last time you heard this idea, "All things were created *for Christ*"? When a person creates something in our society they own it. Because Christ is the Creator, He is therefore the owner.

However, we hear often the idea that it was created *for us*. There are even many Christian leaders who fall into this type of thinking. But it puts us in the place of the LORD, one of the greatest spiritual mistakes we can make. It puts us at the center, instead of Christ.

Now I used "it" intentionally, because that is how we think about what we own: as an *it*. And when something (or someone) is an *it*, we can do with it what we please – or so we think. But not if it belongs to the LORD! "The Earth is the Lord's and the fullness thereof," proclaims Psalm 24:1 (KJV). "The land is mine and you are but aliens and my tenants," declares Leviticus 25:23.

In Jesus' parable of the tenants, found in all three synoptic Gospels, the tenants say, "This is the **heir**, let us kill him and the inheritance will be ours" (Mt 21:33-46; Mk 12:1-12; Lk 20:9-16, emphasis added).

But do we not slip into a version of this? "Let us kill the Lordship of Christ in our hearts, and all things will be ours. We will turn everything into an it to do with it what we please!"

The Heir was of course killed. But our Colossians passage helps us understand that even Christ's death does not obviate who He is – in fact, it reinforces it. His death and his subsequent resurrection, making him "first-born among the dead," actually reveals who He really is – the Creator, the One through which and for which all of existence was made. Killing the Heir utterly failed. It only reveals Jesus to be Christ the Creator, the One to whom all of creation belongs – including you and me.

But somewhere along the way many Christians lost sight of the fact that Christ is the Heir. Is it because we have focused too exclusively on the fact that Christ is our Savior who provides for the forgiveness of our sins? This brings us back to one of the problems the Colossians were apparently having, that Christ provides for the forgiveness of sins but that other principalities and powers need to be appeased. It is one of the reasons why Paul reminds them what they have confessed about Christ: He is the Creator Who Reigns over all and to whom all of Creation belongs. He is indeed our Savior – but He is much more.

Who has Creation been made for? It has been made **for Him**. It belongs to Him.

THE GIVER AND SUSTAINER OF LIFE

So what type of Heir is Christ? If all of creation belongs to Him, what type of relationship does He have with us and the rest of creation? Is He merely an owner?

It's important for us to understand that *we* are not the owners – we don't own anything, even ourselves. Human laws may say who "owns" what, but it is Christ who is the Heir for whom all things were created.

So how does all of creation belong to Him? And how does He treat what He has created?

According to the human laws he lived under, Jesus apparently did not own much of anything. "Foxes have holes and birds of the air have nests, but the Son of Man has no place to lay his head" (Mt 8:20; Lk 9:58). As Creator, Christ ensured that His other creatures had homes, but when, as Jesus of Nazareth, Christ pitched his tent among us (Jn 1:14) he apparently didn't even have a tent!

And yet, "In Him was life" (Jn 1:4). Christ is the giver of life. Everything receives life from Him. Just as the Risen LORD breathed the Holy Spirit upon the disciples (Jn 20:22), as the Creator he breathed the breath of life into all living things (Gen 2:7; Ps 104: 29-30; Eccl 3:19).

He gives life, but then He does not simply abandon those to whom He gives life. It's not a "good luck with that" type of situation. He not only gives life to all living things, He sustains all life. Hebrews 1:3 says, Christ the Creator is "sustaining all things by His powerful word." The Greek word translated by the NIV as "sustaining" is *phero*. Basic definitions are "to carry some burden"; "to bear with one's self"; "to bear up, i.e., uphold (keep from falling)"; "to bear, bring forth, produce."[396]

Christ created all things *ex nihilo* or out of nothing (Jn 1:3; Col 1:16; Rom 4:17; cf Gen 1:1). Hebrews 1:3 helps us understand that He keeps everything from falling back into nothingness. He keeps living beings from falling into death by sustaining the life in them that He has given to them. Christ actively sustains all life. With His own self He upholds every living thing. As the Creator, as the Giver and Sustainer of all life, every one of Christ's creatures looks to Him "to give them their food at the proper time" (Ps 104:27).

In addition, Colossians 1:17 says that in Christ "all things hold together." Today, scientists and cosmologists speculate about what keeps the Universe from either imploding (falling back into nothingness) or flying apart (reverting to chaos, Gen 1:2, *tovuvabohu*). Hebrews 1:3 proclaims that Christ the Creator prevents the first concern, and our mysterious author(s) of Colossians 1:17, inspired by the Holy Spirit, teach us that it is Christ the Creator who holds all things together, preventing everything from flying apart.

So Christ is the giver, the sustainer and upholder of all life, the One who holds all things together. Yet Jesus owned practically nothing while at the same time being the Heir of everything. Even though Jesus was the Son of God, he "did not consider equality with God something to be grasped, but made himself nothing, taking the very nature of a servant" (Phil 2:6-7).

He does not grasp; he gives. He does not mark off and exclude; he opens up and brings together. He does not hoard; he shares. He does not dominate; he serves.

So creation – which includes us – belongs to Christ not so much as a sense of ownership, but as a sense of relationship. Simply put, all of creation belongs to Christ as a child belongs to her mother.

In sustaining life, in holding life together, Christ has an intimate communion with all things. He is the giver of life; He is the Sustainer of life. He keeps everything from slipping back into nothingness or from flying apart as He holds all things together.

THE TRUTH HAS SET US FREE...TO SEE THE TRUTH

In looking at our New Testament texts we've now seen how nearly 2,000 years ago, a ragtag group of nobodies in a backwater of the Roman Empire propounded a theory about how everything

- came into existence,
- is sustained, and
- is both kept from imploding and flying apart.

It's an astounding claim. More than that, it's actually a ridiculous claim. I can well imagine a response like this from those who were not Christians:

This crazy theory has to do with their leader, a carpenter from a hick town in Galilee, who was convicted by the leaders of his own people and crucified as a criminal by the Romans. This Jewish cult, high on what they term the Holy Spirit, claimed that their executed leader was in fact the Messiah. A crucified messiah? What kind of oxymoron is that? The Messiah is supposed to defeat the Romans, not be crucified by them! Devolving into farce (into a sad delusion, really) these nobodies, in light of the death of their leader ("How do we come to grips with the fact that our messiah was crucified?") actually claim that he was not merely the Messiah, but was in fact

… God … God the Creator, God the Sustainer, the One who holds all things together. They are really nuts.

Is it no wonder that right at the beginning, when proclaiming Jesus to be the Word made flesh, the Gospel of John mercifully reveals to us that *He was in the world, and though the world was made through him,* **the world did not recognize him**. *He came to that which was his own, but his own* **did not receive him** (Jn 1:10-11, emphasis added).

This is why faith is such a gift! To believe such wonderfully outlandish things that make us fools in the eyes of the world. God's ways are clearly not our ways. Our thinking can become so flat, so rigid, so limited. We don't recognize God when He's right in front of our face unless by His grace and mercy He gives us eyes to see – and then He gives us "the right to become children of God" (Jn 1:12). As the Truth (Jn 14:6), Christ mercifully allows us to see Him for Who He really is. The Truth sets us free to see the truth about Him – that He is the Creator, the Heir, the Giver and Sustainer of all life.

IT'S ALL GOOD

So before "the beginning" the Word was with God and the Word was God. In the beginning the Word created the world, and as Genesis 1 tells us, the Word saw that all of His creation was good. How could it not be? He created it.

As the true light (Jn 1:9), Christ the Word created physical light, and saw "that the light was good" (Gen 1:4).

Then Christ created the sky, and land, and seas, and He "saw that it was good" (v. 10).

Then Christ had the land produce all of the vegetation of the Earth, every tree, every seed-bearing plant, and He "saw that it was good" (v. 12).

Then Christ created the sun, moon, and stars, including over 100 billion galaxies, with 1,000 trillion stars, and "saw that it was good" (v. 18).

Then Christ, the Giver of life to all living things, created all of the life in the oceans, including "the great creatures of the sea" which Christ "formed to frolic there" (Ps 104:26). He created all of the birds of the air and he let them "fly above the earth across the expanse of the sky" (Gen

1:20). And Christ "saw that it was good" (v. 21). It was so good that Christ "blessed them and said, 'Be fruitful and increase in number and fill the water in the seas and let the birds increase on the earth'" (v. 22).

After being blessed the waters were "teaming with creatures beyond number – living things both large and small" (Ps 104:25). Christ blessed the birds with homes among the trees, where "they sing among the branches" (Ps 104:12).

Christ had "the land produce living creatures according to their kinds: livestock, creatures that move along the ground, and wild animals, each according to its kind" (Gen 1:24). And He "saw that it was good" (v. 25).

Then Christ the Word, who Himself would become flesh, created human beings "in his own image" so that we would image or reflect Him in the stewardship of the rest of His creation and our love for one another. As He did for the other creatures, Christ blessed us, and He also allowed us to fill the earth.

And Christ "saw all that he had made, and it was very good" (v. 31).

Thus, Genesis 1 tells us that **seven times** Christ the Word saw that what He created was good.
 (1) Christ saw that the light was good…
   (2) … that the sky, land, and seas were good …
     (3) … that all the plants and trees were good …
       (4) … that the Sun, moon, and stars were good …
         (5) … that all the life in the oceans and the birds of the air
              were good…
           (6) … that all of the wild and domestic animals were
                good…
                And when He saw everything that He had made – all of
                the stars, all of the planets, all life on Earth, including
                us – Christ saw …
           (7) … that the whole of His creation was "very good."

It's all good.

GLOBAL WARMING AND CHRIST THE CREATOR

But as Genesis 3 informs us, it's not all good anymore. Human beings have created many disturbances in Christ's good creation since we abused our freedom and fell into sin.

As we saw in chapter 1, Christ, the Creator blessed us with a finely tuned atmosphere, including the greenhouse effect that keeps the planet 50°-60°F warmer on average than it would be without it, making it habitable. But we have been enhancing this naturally occurring greenhouse effect to the point where we have given the planet a fever, so to speak. Initially, we may not have understood that we were doing so – but now we do.

And as we discussed in chapters 3-7, global warming threatens Christ the Creator's plants and trees, the life in His oceans and the birds of the air, His wild and the domestic animals, and many of the world's most vulnerable people – all of whom belong to Him.

Can we begin to try to imagine global warming through the eyes of Christ the Creator? Of what global warming has done and will do to the good creation that He loves?

WALKING WITH CHRIST THE CREATOR

So what does it mean to walk with Christ the Creator?

To be honest it seems a rather outrageous concept. It feels both scary and comforting. It strikes me as both arrogant and humbling. "The Creator of the Universe walks with me? I guess it's all about me after all! But, gee, do I really want the Creator of the Universe right beside me? Yikes!"

It would be ridiculous, arrogant, and scary if we were the ones making the claim.

But it is the Risen LORD Himself who tells us, "and lo, I am with you always, even to the end of the age" (Mt 28:20, NKJV). He said this to comfort us. And it should.

The Risen LORD walks beside us to comfort the afflicted and afflict the comfortable.

Christ the Creator walks beside me saying, "But Jim, what about Galib? I'm walking with Galib, too – though he doesn't know it – and we're walking through waist-high water. And Jim, what about all of the Galibs

who don't yet know me who will face the floodwaters of global warming? I will be walking beside them, too."

And Christ the Heir of everything says to me (as He said to Peter in John 21), "Jim, son of Myrl Guy and Billie Jean, do you love me?"

And I say, "Yes, Lord, you know that I love you."

And the Sustainer of all things says to me, "Feed my lambs."

A second time Christ says to me, "Jim, do you love me?"

And I say to him, "Yes, Lord, you know that I love you."

And the One who holds all things together says to me, "Tend my sheep."

And the Creator of the Universe who walks beside me says, "Jim, do you love me?"

I am grieved because He asks me a third time, and I know that I have failed to truly love Him so many times, and I say, "Lord you know everything; you know that I love you, that I try to love you."

And Christ the Creator says to me again, "Feed my sheep" (Jn 21:15-17).

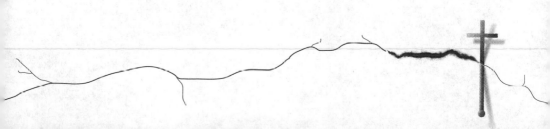

9

WALKING WITH CHRIST –
THE TRUE IMAGO DEI

"I think God is angry at us, but I don't know why." So said Makhabasha Ntaote, a 70-year-old grandmother from Lesotho, South Africa, after experiencing several years of crazy weather – the type of weather global warming brings. The rains have been coming too early, then too late. Over 90% of sub-Saharan Africa is dependent upon rain-fed agriculture, and Makhabasha Ntaote's area is no different. No rain, or rain at the wrong time or in the wrong amounts, means no food. Her fields have produced no crops for three years. Many in Makhabasha Ntaote's village have sold their plows and oxen to buy food. They have eaten the seeds for their crops. The children are too weak to walk to school. And unfortunately, the crazy weather is not the only catastrophe they must deal with. Her area has the world's fourth highest AIDS rate. The disease has decimated the adult population. At least one-third of them are infected. There are over 70,000 AIDS orphans for area families to take care of. Makhabasha Ntaote tries to make money by selling firewood, but her arthritic knees make it difficult. In spite of all these other trials Makhabasha is enduring, it is the crazy, global warming-like weather causing her to believe God is angry with them.[397]

Do you ever wonder why God created us? Why He created finite creatures with freedom? What a combustible mix. Didn't God know it wouldn't work – that we would fall into sin?

It is the greatest drama, the greatest love story, the greatest tragedy, the greatest story of ultimate triumph ever told. Our attempts to capture the drama of our existence in movies and the theater are pale ciphers in comparison to our story in the mind and heart of God. And each of us is part of that story – God's story with us. But we know almost nothing of God's side of the story, because we are not God.

And yet we were made in the image of God, *imago dei*. We are finite creatures made of flesh and bone who will die some day and return to the earth from which God formed us. But we are also creatures endowed with freedom who are made to be *imago dei*.

Why did God do it? This is something for us to ponder deeply to help us orient our lives, but it's a question we will never have a full answer to. That God did it, that He did create us, is a cornerstone of Christian faith.

WE ARE ALL VENUS de MILO

What does it mean to be made in the image of God? Some have argued that it refers to our freedom, our capacity to choose, so it is redundant to say that we are "endowed with freedom" and we are "made in the image of God." But this only comes close to the mark, in my view, if freedom is understood from a Christian perspective. The freest person is the one who completely does the will of God. This is the opposite of a popular understanding of freedom as the freedom to do whatever we want. That's a childish understanding of freedom, frankly. To go against God's will doesn't make us truly free; it makes us slaves of unrighteousness, captured by our own sinfulness.

Our diet is a helpful analogy. We can choose to only eat cake and ice cream instead of a healthy diet, but such choices end up making us unhealthy and therefore less free.

The song, "Big Rock Candy Mountain," apparently a recruitment song sung to entice boys into the life of a hobo, captures the truth of this illusory understanding of freedom quite well. It's a life of no responsibilities where you don't have to work for anything; indeed, the "jerk who invented work" is hung! To go along with the "cigarette trees" and the "lemonade streams," it's a place where you "never change your socks; and the little streams of alcohol come a-trickling down the rocks."

But it's the last stanza of the song that pulls the curtain back to reveal the truth. A boy (the lyrics say "punk") who becomes a hobo to search for Big Rock Candy Mountain turns to the hobo and complains that during his hobo travels he hasn't found any candy. Indeed, it has been a hellish existence that has included sexual exploitation.[398]

A truly free existence comes from making the right choices. We are truly free when we are who God intended us to be. The Apostle Paul puts it this way:

> *When you were slaves to sin, you were free from the control of righteousness. What benefit did you reap at that time from the things you are now ashamed of? Those things result in death! But now that you have been set free from sin and have become slaves to God, the benefit you reap leads to holiness, and the result is eternal life. For the wages of sin is death, but the gift of God is eternal life in [or through] Christ Jesus our Lord* (Rom 6:20-23).

Thus, a Christian understanding of freedom – choosing rightly – is aligned with our being made in the image of God; but being the image of God is not aligned with a mere capacity to choose. God Himself is not free to choose to do evil and remain Himself. But God is free to be Himself. This is what it means to be the image of God: free to be our true selves; free to be who God made us to be – "slaves of God" as Paul puts it starkly in contrast to being slaves of sin. When we are slaves of God, then we are truly free.

And who did God make us to be? Each of us was made by God in our own distinctive way to be ourselves. But as human creatures, who were we made to be?

The third chapter of Genesis reveals that humanity according to God's original design didn't last very long. Adam and Eve sinned; they disobeyed God, thereby diminishing their true selves and limiting their freedom to be themselves in the process. And we've been doing the same thing ever since.

You don't have to have a Ph.D. in theology to know we aren't reflecting the image of God. We were made to do it, but something is wrong. Pick up the local paper and you can read article after article that demonstrates how human beings don't reflect God in our behavior. A quick perusal of human history, or just the twentieth century, or heck, even what we've experienced of the twenty-first century, provides confirmation that we don't image or reflect God. Can we, with our freedom, choose to do God's will every moment of our lives? Sadly, the evidence from our behavior says we cannot. As Paul puts it, "All have sinned and fall short of the glory of God" (Rom 3:23).

To fall short of the glory of God is to fall short of being our true selves, of who God made us to be. We are like damaged or defaced works of art who most of the time pretend we are not. When we gaze upon the statue Venus de Milo, with its incredible beauty, we ignore the fact that she no longer has arms. We look past how weathered she is and the various ways she has faded from her original glory. In the same way, while we retain the image of God, our sinfulness has limited our ability to be the free creatures, the true images of Himself God intended us to be. It's as if we were the world's greatest pianist, or the world's greatest heart surgeon, and we suddenly lost our arms. Suddenly we were no longer able to exercise our greatest gifts.

FREE, BEAUTIFUL, AND GLORIOUS

While human beings have fallen away from true freedom, thereby falling short of the glory of God, one man has in fact lived the freest and most glorious life ever. He was completely free, because he completely did the will of his Father. Even in chains he was freer than anyone else in human history. And nailed to a cross he set us free.

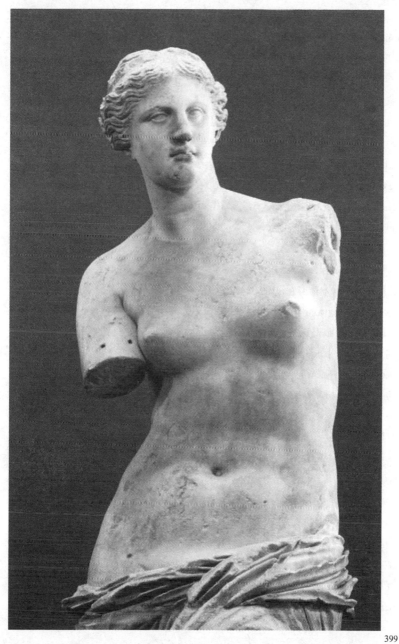

399

Let that one soak in a bit – nailed to a cross, the ultimate symbol of political subjugation, he set us free. As Paul put it, "God made him who had no sin to be sin for us, so that in him we might become the righteousness of God" (2 Cor 5:21).

But before he became sin for us, before he became burdened with all of the sins of human history, before he became the most spiritually subjugated person who ever existed as he bore the unrighteousness of the world on the cross, he was the freest person who ever lived.

Spiritually he was the most beautiful person to walk this earth. He was the most gifted glorifier of God, with the most beautiful spiritual voice ever heard by the heart of God.

To be a true image of God is to be spiritually beautiful, to be gloriously luminescent in the eyes of God. You can see Jesus' spiritual beauty in his compassion and healing.

One of the most moving stories in the Bible is Jesus' healing of the woman with uncontrollable feminine bleeding.

> And a woman was there who had been subject to bleeding for twelve years. She had suffered a great deal ... and had spent all she had; yet instead of getting better she grew worse. When she heard about Jesus, she came up behind him in the crowd and touched his cloak, because she thought, "If I just touch his clothes, I will be healed." Immediately her bleeding stopped and she felt in her body that she was freed from her suffering. At once Jesus realized that power had gone out from him. He turned around in the crowd and asked, "Who touched my clothes?" "You see the people crowding against you," his disciples answered, "and yet you can ask, 'Who touched me?' But Jesus kept looking around to see who had done it. Then the woman, knowing what had happened to her, came and fell at his feet and, trembling with fear, told him the whole truth. He said to her, "Daughter, your faith has healed you. Go in peace and be freed from your suffering."(Mk 5:25-34)

She was considered unclean by Old Testament law (Lev 15:25-27) because of her bleeding. Her disease bankrupted her and she was an outcast for 12 years. Can you imagine the loneliness? She's so desperate that she is willing to make Jesus "unclean" by touching him. And when Jesus searches for the one who was healed, she falls at his feet trembling with fear. She exposes her shame to everyone and admits to making him unclean. Perhaps she thought, "What will he do to me? Will he explode in wrath because I touched him and made him unclean?"

Jesus' response is quite the opposite. He shares her shame and through this healing experience they are now family. He calls her "daughter." He says, "Your faith has healed you." The Greek word for healing here implies both physical and spiritual healing. She has been made clean and whole by the Great Physician of body and soul.

It is not by a "free" act of will that Jesus heals her. The text clearly indicates that it is the woman who initiates the healing. He is simply himself, whom God made him to be – a healer.

But it is Jesus' response to the healing that further reveals who he is and the freedom he exercises by simply being himself. His compassion is not limited by what others might think or the risk of being labeled "unclean." He doesn't allow the views of others make him less than who he is and have him respond in unnecessary judgment. Jesus understands that the situation has been transformed by the woman's act of faith. God has provided for the woman a new way to become clean: faith in Jesus' power to heal her. The simple fact is she is no longer unclean.

Is he not beautiful! Is he not spiritually unblemished! This story reveals Jesus' inner glorious luminescence as he truly images or reflects his merciful Father. In freedom he is simply himself, a healer who then shows mercy. He does not allow the limited views of others to spiritually diminish him, to have him fail to gloriously and beautifully reflect or image God.

Not only can we see Jesus' spiritual beauty in the Gospels, we can hear the spiritual beauty of his voice when he says,

> *Blessed are you who are poor,*
>> *for yours is the kingdom of God* (Lk 6:20).

Or this,

> *It is not the healthy who need a doctor, but the sick. But go and learn what this means: "I desire mercy, not sacrifice." For I have not come to call the righteous, but sinners* (Mt 9:12-14).

We hear the spiritual freedom in his words when he says,

> *You have heard that it was said, "Love your neighbor and hate your enemy." But I tell you: Love your enemies and pray for those who persecute you, that you may be sons of your Father in heaven. He causes his sun to rise on the evil and the good, and sends rain on the righteous and the unrighteous* (Mt 5:43-45).

Where does the freedom come from to say something as audacious as "Love your enemies and pray for those who persecute you"? To teach that we should actually pray for those who persecute us?

For Jesus and his audience this is not some abstract teaching, nor is it a metaphor. Jesus lived in an occupied country. He and his fellow Jews were literally persecuted. It is no wonder that a common teaching of the time was that we could hate our enemies.

But Jesus has the freedom to say pray for the one who systematically and unjustly mistreats you. Pray for the one who executed your friend. Pray for the one who beat your father, who humiliated your mother, who accosted your sister. Jesus was free to understand that the Father loves even those who do evil.

To be a true child of God is to truly reflect or image the Father. A true test of this is whether we can love those who are even our enemies, like God the Father does. In reality, we may find it hard to love even our loved ones, but Jesus says that's not enough to be a true image or reflection of the Father. Jesus points out that even unbelievers love those who love them.

Jesus makes clear that this teaching is part of what it means to image or reflect God when he closes with verse 48, "Be perfect, therefore, as your heavenly Father is perfect."

The word translated "perfect" here does not mean that to be like God you must score 100 on all math exams and win the spelling bee. It means you must be "complete," or have come to full maturity. You are complete in your righteousness, in other words, spiritually beautiful.

Thus, **in freedom we are to be just like God in our choices**, in our behavior. In our particular circumstances we are simply **to image or reflect God** by being the true persons He created us to be every moment of our lives. To do otherwise is to become less free, less spiritually beautiful as we fail to reflect or image God, diminishing our ability to glorify Him. There has been only one person free enough to be himself every moment of his life, free enough to love his enemies, even those who killed him, even those who caused his death, free enough to do the will of God. **Jesus Christ was freedom incarnate**.

As Colossians 1:15 says, "He is the image of the invisible God, the firstborn [or preeminent] over all creation." Jesus Christ is the one, true, glorious, beautiful, freely obedient reflection or image of God. He so completely reflected the Father that he can say in the Gospel of John, "Anyone who has seen me has seen the Father" (14:9). Christ was and is preeminent over all creation because he faithfully reflected the Father.

A SERVANT'S HEART

How did Jesus express his preeminence? Before the start of his ministry Satan tempted Jesus with all the kingdoms of the world, but Jesus rejected them (Mt 4:8-10).

Instead he fed the hungry. He healed whoever came to him: the destitute, the rich, unclean outcasts, foreigners, even the servant of a Roman occupier. He worked with the disciples even when exasperated with them (e.g., Mk 9:19).

So what does Jesus' behavior mean? Amazingly, on the very night of his betrayal and abandonment, the disciples actually have a discussion about which one of *them* is the greatest. This throws in sharp relief how Jesus exercised his "firstborn" status.

Even though as a true image of God Jesus is in fact preeminent, and even though the model of preeminence that the world offers is one where those in authority "lord it over" those below them, he tells his followers that they are not to be like that.

Now some of you already know where we're headed: the Christian idea of servant leadership.

To be honest, it can be hard not to become cynical today about the concept of "servant leadership" because it is rightly talked about so much in Christian circles – and yet servant leaders who consistently get close to the mark are rare. Some leaders are genuinely trying. Others are not. Some try some of the time and fall significantly short at other times. This inconsistency (and sometimes downright hypocrisy) is what we experience of servant leadership from our fellow Christian brothers and sisters: frail, finite creatures with freedom just like us.

So when we hear talk about servant leadership, we can be tempted (and I mean that literally) to tune out. "Here they go, talking about servant leadership again. Yeah, yeah, yeah. Blah, blah, blah. Nobody really does it."

But we can't become jaded or cynical about Jesus' servant leadership example and teachings in the Gospels! We can't give into temptation and give up.

Jesus doesn't give up on the disciples, even as they argue cluelessly about who among them is the greatest on the very night he will be betrayed. As his disciples or followers, we can't give up on our leaders or ourselves. The disciples failed. We fail. Our leaders fail. But Jesus doesn't give up on us.

So we must try to truly hear Jesus' words. In the Gospel of Luke, right after Jesus instituted the Lord's Supper, he told his followers that when

they exercise power they are not to "lord it over" those within their power and influence. "Instead, the greatest among you should be like the youngest, and the one who rules like the one who serves. For who is greater, the one who is at the table or the one who serves? Is it not the one who is at the table? But **I am among you as one who serves**" (22:25-27, emphasis added).

Those were not empty words. And Jesus does not let us get away with projecting a false Christian servant-leader model. Jesus doesn't let us fool ourselves with lame rationalizations by in effect saying in our own minds, "I can be a servant leader in my heart and at the same time continue to lord it over those within my power and influence." His teaching and example are very concrete. To be the youngest is to be last in power and influence. Those who served at table were not those who did such service temporarily. No, they had a full-time, 24/7 servant status. They had masters whose power and influence they were under.

To be crystal clear, the Gospel of John reports that Jesus even washed the disciples' feet. Afterward he said to them, "'Do you understand what I have done for you? ... You call me 'Teacher' and 'Lord,' and rightly so, for that is what I am ... I have set you an example that you should do as I have done for you" (Jn 13:12-15).

In Philippians Paul gives us perhaps the greatest interpretation of Jesus' example of service and how it is to inspire us to live out the same attitude. We should consider others not only equal to ourselves, but **better** than (or superior in authority to) ourselves (2:3) – just like a servant in the first century would. He goes on to say:

> *Your attitude should be the same as that of Christ Jesus:*
> *Who, being in very nature God,*
> > *did not consider equality with God something to be grasped,*
> *but made himself nothing,*
> > *taking the very nature of a servant,*
> > *being made in human likeness.*
> *And being found in appearance as a man,*

> *he humbled himself*
> *and became obedient to death—*
> *even death on a cross!* (Phil 2:5-8)

Jesus as a human being is not only the true image of God; he is also the preexistent Son of God, Creator of all things. He not only empties Himself of His Divine Sonship, "pitching his tent among us" or becoming finite flesh and blood just like us, he becomes servant to the servants (us), even unto death.

As the true image of God, Jesus is the "firstborn," or preeminent. He exercises his preeminence by doing the opposite of the common "lord it over" understanding of preeminence. He serves the other slaves. He even dies for them. In the eyes of the world, this is absolutely crazy. But at the precise moment Jesus took up the cross he was the freest, truest image of God human history has ever beheld. He truly imaged the deepest part of the heart of God. And in becoming sin for us, in freely becoming the most spiritually bound person who ever existed, he set us free to begin our journey of becoming a true image of God with the heart of a servant.

EVER-INCREASING GLORY: GRACE-FILLED, SPIRIT-TRANSFORMED, IMAGING CHRIST

While it is empirically true that all still fall short of the glory of God, and that "there is no one righteous" (Rom 3:10), the actions of God have transformed our situation. Christ's death dealt with the consequences of our sin in our relation to God. "For the wages of sin is death, but the gift of God is eternal life in Christ Jesus our Lord" (Rom 6:23). This gift of grace is available to us as a result of Christ's atoning death. "God made him who had no sin to be sin for us, so that in him we might become the righteousness of God" (2 Cor 5:21).

Our relationship with God has been transformed not so that we can go on sinning, not so we can continue to limit our freedom, not so that we

can blithely fail to image or reflect God in our circumstances and continue to mar our spiritual beauty. No! "We died to sin; how can we live in it any longer?" (Rom 6:2).

We who are in Christ, who have accepted him as Savior and LORD, can't continue to let sin limit our freedom. That is like saying "No" to Christ on the cross.

"Jim, I'm dying on the cross now to set you free from sin."

"Ok, thanks. But let me do a few more things the old-fashioned way just to save a bit of time. Just hang on a bit longer and I'll start living the new life you've made possible for me in a minute."

The "no" cannot be from us to Christ on the cross. No! "For freedom Christ has set us free" (Gal 5:1, NRSV).

Our sinful nature has warped the image of God within us, preventing us from freely reflecting God in our concrete circumstances, preventing us from freely doing God's will and being righteous or in right relation to God. We need to be free to be whom God made us to be. But we cannot be truly free without God's help.

And that help comes from all three Persons of the Trinity. God the Father sent God the Son into the far country of our sinful selves, not so that we would abuse and crucify His son, but so that we would accept him as Savior and LORD. And with Jesus' atoning crucifixion, and with Christ's resurrection and ascension, the way is set for the coming of the One he promised, the Paraclete, the Comforter, the Holy Spirit (Jn 16:7).

Christ became sin for us, so that "in him we might become the righteousness of God" (2 Cor 5:21). For us **to become** the righteousness of God is for us to be who God made us to be – beings who consistently image or reflect God. "Be perfect [complete in righteousness, spiritually beautiful], therefore, as your heavenly Father is perfect" (Mt 5:48).

But how? In our day-to-day reality it is clear that we are not perfect images of God. We don't reflect how He would act in the circumstances we find ourselves in. So how do we begin to live out the freedom He has created in us?

We do so by allowing the power of His grace and love to work inside us, strengthening our will, so that through our spiritual participation in Christ, we become righteous inasmuch as "Not my will, but Thine be done" becomes a reality in our lives. To that degree are we true images of God as we image Christ, the true image who completely did the will of his Father. And we are not left alone by God to fend for ourselves in understanding what it means to image Christ in our circumstances.

In the third and fourth chapters of 2 Corinthians, the Apostle Paul brings our themes together: freedom, glory, and how we can become true images. Paul explains that while the Law of God as delivered to Moses was glorious and good because it came from God, it only brought the awareness of our sinfulness. It did not provide the power to break free from our bondage to sin. We are still warped images, damaged works of God's art. The Law helps us see that spiritually we are like Venus de Milo, still visibly beautiful, but weathered and without arms. The Law is like the best piano concerto ever written, and God has gifted us to be a piano virtuoso spiritually, but we have lost our arms and can only read the music and hear it in our heads – we cannot play it.

With the coming of Jesus Christ, the true *imago dei*, we don't just have a book (what Moses helped provide), we also have a person. We don't just have the written Word; we also have the Word made flesh, a living, full-bodied example of what it means to image God, to be the greatest spiritual piano virtuoso who ever lived, whose performances were recorded for us in the Gospels.

Paul reminds us how incredibly fortunate we Christians are that God has helped us to understand and accept this. "For God, who said, 'Let light shine out of darkness' [Gen 1:3], made his light shine in our hearts to give us the light of the knowledge of the glory of God in the face of Christ" (2 Cor 4:6).

Nonbelievers are not so fortunate: "The god of this age [Satan] has blinded the minds of unbelievers, so that they cannot see the light of the gospel of the glory of Christ, who is the image of God" (2 Cor 4:4). What

can't they see? What spiritual gift of sight has God given us? What has the spiritual light God has shone into our hearts allowed us to comprehend? They are blinded from seeing the glorious news about the gloriously triumphant Christ, "who is the image of God."

Remarkably, this last phrase, that Jesus Christ is the true image of God, almost feels like a throwaway line. But it is one of the concepts that undergird the entire passage: that humanity was made to image or reflect God's glory and that Christ truly did so.

When Moses came back from being in God's presence, his face literally shone with a reflection of God's glory, so much so that the Israelites could not look steadily at Moses' face. But once Moses left the presence of the LORD it faded over time. Paul suggests that Moses put a veil over his face to keep the Israelites from seeing the ever-increasing *dimming* of the glory of the LORD as reflected in his face (2 Cor 3:13).

Jesus' glory does not fade, because he perfectly reflects the glory of his Father. Jesus was the true image of God, a true reflection of His will, and precisely because of this, he can be our sinless Savior who gloriously triumphs over sin and Satan on the cross.

It is because of Christ's glorious triumph on the cross that we don't just have Jesus' example of how to be a true image of God; we also have the power of his healing grace to begin to make it possible.

And yet there is even more help from our merciful, giving God! He has not just given us understanding, not just helped us with acceptance, and not just provided us with the power of His grace. But, grace upon grace, we also have the guidance of Christ's Spirit, the Holy Spirit as we participate in the even more glorious "ministry of the Spirit" (2 Cor 3:8).

As Paul says,

> *Now the Lord is the Spirit, and where the Spirit of the Lord is, there is freedom. And we, who with unveiled faces all reflect the Lord's glory, are being transformed into his likeness with ever-increasing glory, which comes from the Lord, who is the Spirit* (3:17-18).

Christ's Spirit, the Holy Spirit, is transforming us into His image. Because the LORD Himself is glorious, because He perfectly or completely reflects the glory of the Father, the more we image or reflect Christ the more glorious we become. Because where the Spirit of the LORD is, there is freedom, the more we reflect Him the freer we become. It is a process of "ever-increasing glory." And to the degree we do so, to the degree that we image Christ, we become the glorious, spiritually beautiful righteousness of God.

So how do we become more glorious, more free, more spiritually beautiful – how do we become the righteousness of God? We do so through the work of God the Father, God the Son, and God the Holy Spirit. We do so by keeping our unveiled faces turned towards the true Image, Christ. Unlike Moses, whose reflection of God's glory faded once he left the presence of God, the Risen LORD never leaves us. And finally, we do so by allowing the LORD's Spirit to transform us into Christ's likeness. Only in this way will our glory not fade, our freedom not diminish, our righteousness not falter.

A PARTICULAR TASK FOR THE IMAGE

Before we discuss a particular task that Scripture articulates for the image of God, that of caring for the rest of creation, it is important for us to recap what has been said thus far to have a proper Christian understanding of this task.

As we have seen, imaging God is what we as human beings were made to do. It is our calling as creatures. It encompasses all of our behavior, all of our relationships. It's a 24/7 occupation. But our sinfulness has prevented us from being true images. Not so Jesus Christ. Christ is the true image. For those of us who profess him to be our LORD and claim to be his followers and disciples, our lives are to be about imaging or reflecting Christ, his glory, his spiritual beauty, his freedom, his servant heart.

As Jesus had the spiritual freedom to look past the narrow interpretations of the Law and offer words of comfort to the "unclean" woman who had touched him, so too are we to have mercy, compassion, and forgiveness mark our days.

As he had the spiritual freedom to live out and proclaim, "Blessed are you who are poor … I have not come to call the righteous, but sinners … Love your enemies and pray for those who persecute you," so too are we.

As Christ the Creator emptied Himself of His divine power and attributes to "pitch his tent" among us, as Jesus taught and practiced sacrificial service, service to the other slaves, even to the point of death on a cross, so too are we to live lives of service as servants of the Suffering Servant.

Thus to image our Risen LORD Jesus Christ, the true image of God, is what we as Christians are called to do in our concrete circumstances. Another name for it is discipleship: following the Risen LORD and thereby doing the will of the Father. To image Christ is to image the Father is to act in His stead on earth. It is to fulfill the Lord's Prayer: "Thy will be done, on earth as it is in heaven."

And now on to our discussion of a particular task for the image.

If we open our Bibles and start at the beginning to look for the first instance where *imago dei* is discussed we wouldn't have far to go. It's right there in the first chapter of Genesis, and with it is an explicit description of a particular task that is part of imaging the Creator:

> *Then God said, "Let us make man in our image, in our likeness, and let them* **rule** *over the fish of the sea and the birds of the air, over the livestock, over all the earth, and over all the creatures that move along the ground." So God created man in his own image, in the image of God he created him; male and female he created them. God blessed them and said to them, "Be fruitful and increase in number; fill the earth and* **subdue** *it.* **Rule** *over the fish of the sea*

and the birds of the air and over every living creature that moves on the ground" (1:26-28, emphasis added).

The Hebrew word translated "rule" by the New International Version of the Bible (NIV), and "have dominion" by the King James Version (KJV) is *radah*, which can also mean "dominate," "tread down," even "scrape out."[400]

Kabash is the Hebrew word translated "subdue." We are to fill the earth and *kabash* it. In its various meanings it can be translated as to subjugate, subdue, conquer, even to assault sexually.[401]

Over the years I have struggled with these words, *radah* and *kabash*. I've come to think God put them there to emphasize the tremendous power over the rest of creation He was giving us – power enough even to conquer.

Freedom and power can be terrifying things in the wrong hands. God intended for us to rule as He would rule, to image Him in our care of His creation. He took what might seem to us to be an incredible chance in giving us freedom and power. Would we rule as He would rule, or would we try to make ourselves little kings?

Because these words are in the very first chapter of the Bible, and because of our fallenness, it's easy to read these verses, go no farther, and conclude that what is said is pretty straightforward. It says we are to rule and subdue – we are to use our power to put everything under our power. God has given the rest of creation to us to do with it pretty much whatever we want.

This can lead to a certain type of attitude: "We're in charge. We're superior. We're the most important. The rest of creation exists to meet our needs." Let's call this a *kabash* mentality.

Yet this granting of dominion in Genesis 1 occurred before the Fall in Genesis 3, before we became warped images, distorted reflections.

Now to counter the development of a "*kabash* mentality," many Christians concerned about caring for the rest of God's creation point to the

second Genesis account of how the Earth came into existence. There it states:

> The LORD God took the man and put him in the Garden of Eden to work [also "tend"] it and take care of [also "keep"] it (Gen 2:15).

The Hebrew word translated as "work" in the NIV and "tend" in the Revised Standard Version (RSV) is *abad*, which also can mean to serve, including to serve God, to serve another by one's labor, to make oneself a servant.[402] The word translated "take care of" in the NIV and "keep" in the KJV is *shamar*, which also means to guard, keep watch over, protect.[403] It is the same word used when God instructs Moses on how Aaron, the chief priest, is to bless the people, "The Lord bless you and keep [*shamar*] you" (Num 6:24).[404]

Some have emphasized *radah* and *kabash* to make the case that humans are in charge, that we are superior, we come first and get to exploit the rest of creation for our benefit.

Others suggest that we should have a *shamar* mentality that emphasizes that we are to be about tending and keeping the rest of creation, not exploiting it.

Now I think a lot of folks might look at these two Genesis texts and say, "Well, maybe sometimes we are to *kabash* and sometimes we are to *shamar*, and sometimes maybe a little of both."

Just as an exercise let's pretend that Genesis 2:15 doesn't exist. There's no textual balancing act between *kabash* and *shamar*. All we've got is *radah/kabash*. For Christians I think it is very simple. We don't get to read *kabash* any old way we want. We read it as disciples of Jesus, as those who have professed him as LORD. How did Jesus utilize his power? How did he rule? Would Jesus rule over the rest of creation like what I have called a *kabash* mentality suggests? Would he emphasize his superiority? Was Jesus a *kabash* type of guy?

Jesus of Nazareth was a meat-eating carpenter. As a carpenter he was a practitioner of the technological arts of his day. To help put food on the table he turned trees into useful products for customers in his community. He also had tremendous power over the rest of creation. As the disciples remark after he calms a storm, "even the wind and the waves obey him" (Mt 8:26-27; Mk 4:39-41; Lk 8:24-25).

But during his temptation in the wilderness by Satan, Jesus was promised all the kingdoms of the world. He rejected them. He did not grasp at power to dominate for his own benefit. During his time in the wilderness Mark tells us that Jesus was "with the wild animals" (1:13). He doesn't try to rule them. He's hungry, but he doesn't eat them. He is simply with them.[405] He lets them be themselves. He could have accumulated much wealth and indulged in a great deal of consumption, but he chose not to. At the end of his life the only thing he owned was the clothes on his back. He exercised his freedom not to dominate or overconsume.

But Jesus went further. It wasn't that Jesus simply didn't do bad things with his freedom. It wasn't that he didn't abuse his power or try to hoard or increase his power. Power freely flowed out of Jesus, not in. It flowed out of him to do righteousness. In the case of the woman with feminine bleeding, Jesus is so free and giving with his power to heal that it even flows out of him without his conscious choice. Jesus is about giving, not grasping. A way to visualize this metaphorically is that Jesus' hands are constantly open, not closed – open hands, open heart.

What Jesus, the Word made flesh, does with his freedom and power is simply consistent with what He has done as the pre-existent Creator. As the Creator He also gives freedom and power away, in this instance to his human creatures.

In these ways, both as preexistent Son and as the Word made flesh, Jesus Christ reflects the Father, who gives the Son all things. As Jesus says in John, "everything you have given me comes from you" (Jn 17:7). Mutual giving marks the divine Father-Son relationship. "All I have is yours, and all you have is mine" (Jn 17:10). The Trinity is a mutual giving society.

Let's compare God's actions towards us with our actions towards the rest of creation.

It's impossible for us to comprehend how completely Other and incomprehensively superior the Creator is to us. We are not on a continuum with Him, as if we are at one end of the continuum, with angels, say, in the middle, and God at the other end. The Psalmist rightly asks, "What is man, that thou art mindful of him? And the son of man, that thou visitest him?" (Ps 8:4, KJV).

And yet God has given so much to His creation – and to us most of all. Though He is completely Other to us, God has given us freedom and power to be like Him. God the Son even became one of us and died to save us! How much more he could he give?

But we **are** on a continuum with the rest of creation. We are finite creatures made from the earth (Gen 2:7, 3:19). We are children of Adam, whose name is both a generic name for humanity as well as the personal name of the first man. It is also a word play with the word for earth or soil, *adamah*. We are *adam* from *adamah*. We are earth-men and earth-women.

Analogies completely fail us here, but maybe we can use one to "see through a glass darkly" and grope towards an incomplete but helpful perspective. For us to try to begin to understand what Christ the Creator has done for us, it is as if we were to have the capacity to give freedom and dominion to single-celled organisms like amoebas; then when they got into trouble we first became one of them, and then died to save them – even allowed them with the freedom we had given them to kill us – because we loved them so much. (This is the servant's heart of the one we are to image or reflect!) This analogy may strike some of you as ridiculous – that we would do such things for amoebas. And yet as finite creatures we are on a continuum with them, whereas God is not on a continuum with us.

So God, who is completely Other to us, who is incomprehensibly superior to us, gave us everything and died to save us, and we, finite creatures who have the other finite creatures and the Earth under our power do we reflect Him in our care of them?

Sadly, because we are warped images, the answer is no. We can even use Gen 1:28 to justify a "*kabash* mentality" where we use our freedom

THE AMAZON

The Amazon is the largest tropical rainforest on Earth.[406] It is home to around one million people comprising 400 indigenous groups.[407] Over 1 billion people "depend to varying degrees on [the] forests for their livelihoods." [408] The Amazon supports approximately 25% of the world's land-based species.[409] Unfortunately, the synergistic interactions of global warming and deforestation threaten to transform large swaths of Amazon rainforest into much less biologically diverse ecosystems such as seasonal forest or savannah.

Even without global warming, the Amazon is severely threatened by human activities. Thirteen percent of the Amazon has already been lost to deforestation. If current trends continue, 47% could be gone by 2050.[410] This is being driven to a large extent by the forest being converted into cattle ranches and farmland to grow beef and soy for export to the United States and Europe. The expansion of the production of biofuels could add another factor in the acceleration of deforestation.[411]

But a large-scale transformation is not inevitable if major changes are made quickly both in terms of deforestation and reductions in global warming pollution (including emissions from deforestation itself, which currently accounts for more than 17% of global greenhouse gas emissions[412]). A recent study suggests that "intact Amazonian forests [as opposed to frag-

not to care for what is within our power, but to dominate and exploit it to fulfill our wants and desires to the detriment of the rest of Christ's creation.

As we have seen Jesus "did not consider equality with God something to be grasped, but emptied himself." In like manner for us to "grasp" at the status of *imago dei* in order to puff up our importance is to lose it. We focus our reflective capability upon ourselves, instead of focusing it upon God. We become "the image of me" rather than the image of God. To grasp, to hoard, to reflect ourselves instead of Christ, to not have open hands and open hearts, is to lose, not gain, to become diminished, not mented ones] are more resilient, although not invulnerable, to climatic drying" than had been previously thought.[413]   If large swaths of the Amazon are left intact and not carved up by roads and deforestation, then these areas stand a better chance of surviving the drying of global warming. This study suggests that deforestation could be significantly reduced with the proper regulation and market incentives and that with "the expansion of protected areas and effective legal enforcement of private land use, the projections of loss of 47% of original forest area by 2050 could be reduced to 28% loss." [414]

However, it is the caveat that the forests are not invulnerable to droughts brought about by global warming that must be kept in mind, especially since another recent study projects that such droughts will hammer the Amazon if current global warming pollution trends continue. Instead of extensive droughts occurring once every 20 years as they do now – something, apparently, the forests can handle and to some extent even benefit from – by 2025 they will occur every other year; and by 2065 such droughts would occur nine out of every 10 years. [415]

If droughts increase and lead to major wildfires, significant deforestation continues, and deforestation and global warming combine to transform the Amazon into seasonal forests or savannahs, then a tipping point, a point of no return for the Amazon, would be reached.[416]   The consequences would be devastating. Indigenous cultures could be wiped out. The livelihoods of hundreds of millions could be threatened. A catastrophic disappearance in species could occur.[417]   And given that the Amazon is "estimated to contain about one-tenth of the total carbon stored in land ecosystems," [418] a huge amount of global warming pollution would be released into the atmosphere, feeding the cause of its own destruction.

exalted; it is to become spiritually inferior, to become less free, to reflect ever-decreasing glory.

Even if there were no call in Genesis 2:15 to *abad* and *shamar* the rest of creation, the example of the Son of God as captured perfectly in Philippians 2 – the preexistent Son's kenosis, or emptying in becoming a human (i.e., a servant of the Father), and further sacrificial service in dying for the other slaves – tells us we are called to serve those under our power, whether they be other people or our fellow creatures.

It is when we have power over something or someone that we show our true moral character. The more power we have, the bigger the test.

Having dominion over the rest of creation is a true test of whether we will use our freedom and our power to image Christ or ourselves.

The Earth and God's other creatures came before us. They are like the elder brother in the parable of the prodigal son. They too were blessed. But just like God favored Jacob over Esau, Joseph over his brothers, and David over his brothers as well as Saul the King, so too God has blessed us with freedom and power in comparison to the rest of creation.

Why us? Well, why Jacob? Why David? God chooses whom He chooses. But His choice is a cause for an attitude of humility, not one of superiority. Jacob was a liar and David was a murderer and adulterer, yet God found a way to use such earth-men or "jars of clay" (2 Cor 4:7) to accomplish His will.

There is nothing wrong with utilizing the rest of creation to meet our basic needs just as Jesus did. In this way we are creatures like God's other creatures, whom He also blessed and allowed to eat from the vegetation of His good earth (Gen 1:29-30).

But there is something profoundly wrong with stealing God's blessing from His other creatures. We have been doing this at a terrifying speed, making the natural extinction rate anywhere from 100 to 1,000 times faster. Estimates of God's other creatures we have driven to extinction range about 784 in the last 500 years.[419] But to look into the face of the Suffering Servant on the cross is to understand that even one is too many.

Unfortunately, global warming will accelerate and worsen what we are doing to God's other creatures. Indeed, global warming will be the largest single threat to biodiversity in this century. Up to 30% of God's other creatures could be threatened with extinction by 2050 simply because of global warming[420] – not counting all the other ways we are causing our fellow creatures to go extinct.

Christ's other creatures were made by Him to glorify the Father. It is a terrible sin for us to prematurely extinguish their capacity to glorify Him by wiping them off the face of the earth. When we reflect upon how we are to image Christ in the exercise of our freedom and power, to understand that

we are to care for his other creatures as he would, it is enough to drive us to our knees to beg for forgiveness.

CONCLUSION

The Risen LORD is with us on every step of our journey. Through his healing grace our "spiritual arms" have been restored; through the power of his grace the strength to do the Father's will and truly image God flows into our weakened limbs.

To reflect the Risen LORD's glory with unveiled faces (2 Cor 3:18) all we have to do is turn towards him instead of away from him. But honestly, much of the time we turn our backs to the Risen LORD. We don't truly follow him. Because he loves us so, the One with the nail-scarred hands, open hands, follows us into our outer darkness, where we wail and gnash our teeth. It is shameful – shameful! – that we turn our backs on him, that the light of the world follows after us into our spiritual darkness. We are to follow Him!

Unbelievers may be blinded to the fact that Jesus is their Savior (2 Cor 4:4), but many believers do not completely understand the full implications of Jesus' Lordship, of our servanthood, of imaging Christ. They may be both blind and in the dark, but in the darkness of our failure to image Christ we do not fully see.

But to turn to the Light of the world! To turn to the Risen LORD is to fulfill the deepest desires of our hearts. As Augustine put it, "Thou hast formed us for Thyself, and our hearts are restless till they find rest in Thee."[421]

To turn to him is to have the light of His glory shining upon our unveiled faces! To walk with Him along the path of righteousness God has chosen for us is to image Christ with ever-increasing glory. It is to become luminescent with His glory, to shine like the Son. To image or reflect the Risen LORD is to become spiritually beautiful, to become truly free – free, beautiful, and glorious.

As I think about Makhabasha Ntaote, the 70-year-old African grandmother mentioned at the beginning of this chapter, and as I consider God's other creatures whose very existences are in our power, as I contemplate the journey we will have to take to overcome global warming, it's hard for me to think of anything more inspiring than all the tremendous opportunities we have to become more free, more spiritually beautiful, more glorious, as we image Christ, as our hearts are transformed into servants' hearts, as we reflect the Risen LORD in overcoming this scourge that will hurt so much of His creation.

In my own imagination, Makhabasha Ntaote's heart is like that of the woman with feminine bleeding – waiting, waiting for the Risen LORD to come by so she can just touch the hem of his cloak so that her parched land will be healed.

But actually the Risen LORD is waiting with her, waiting for **us** to come by, waiting for us to turn towards Him and accept the redeeming power of His grace and reflect His love, His freedom, His beauty, His glory.

Overcoming global warming – it's not ultimately a story about windmills and photovoltaic cells. It's a grace-empowered story about ever increasing love, freedom, beauty, and glory; it is about reflecting the Risen LORD with unveiled faces.

10

WALKING WITH CHRIST
OUR SAVIOR –
OUR NEED FOR A SAVIOR

I felt all the other kids were looking at me as I boarded the old converted school bus that would take us to our homes. My hair was wet, but my clothes were dry. I had changed back into my own clothes from the donated ones the church had for just such occasions. Those temporary clothes didn't fit very well and were quite worn, but they were the clothes I was baptized in. When the pastor baptized me he got my name wrong – but no matter. Jesus knew who I was.

The old bus was repainted to say "Jupiter Road Baptist Church" on the outside. It roamed our local neighborhood on Sunday mornings to pick up children and bring them to the church. Sometimes I would ride the bus to make it for Sunday school and then see my parents at the worship service. This was one of those mornings.

It was March 10, 1974; I was 12 years old, and I had just publicly confessed what I had privately accepted in my heart – that Jesus Christ is my personal Savior and LORD.

Several months before, the Associate Pastor of the church, Dr. Odum, had come to our home to explain how I could accept Christ into my heart and be saved from my sins. That night I did so, but I didn't tell anyone – although I knew I needed to. I had been taught the verse, "if you confess with your mouth, 'Jesus is LORD,' and believe in your heart that God raised him from the dead, you will be saved" (Rom 10:9). I knew I was a sinner. I knew I needed a Savior. I knew I needed to publicly confess Christ.

Although I didn't know it at the time, Jupiter Road Baptist Church was an independent, fundamentalist Baptist church. When we moved to Richardson, Texas, about a year earlier we had started attending Jupiter Road because my older brother, Danny, liked the youth group. My mom wasn't crazy about the church. She had grown up with a more formal worship style at the First Baptist Church of McComb, Mississippi, a church affiliated with the Southern Baptist Convention that was filled with many of the prominent citizens of the town.

The pastor of Jupiter Road, James Starkes – or Brother Jamie as he was known – was essentially an evangelist (right down to the white shoes and white suit) who had settled into a local church. His message each Sunday always boiled down to one statement he would yell in his hoarse, cracked voice: "You've got to be saved!" Brother Jamie would woo, coax, plead, cajole. He would hold altar call after altar call and we would sing again and again the last verse of the closing hymn to allow time for sinners, such as myself, to make the fateful decision and walk the aisle. "Don't let this moment slip away. Today is the day for you to be saved."

For months I had kept the decision to accept Christ private. But on this morning I couldn't resist the wooing of Brother Jamie and the promptings of the Holy Spirit any longer. I walked the aisle with several others and told one of the elders at the front that I had accepted Christ.

Now at most Baptist churches in the South during this time, once you walked the aisle, your decision would be recognized by the pastor that morning and the next week or so you would meet with the pastor to discuss being baptized and a date would be set in consultation with the family.

Not so at Jupiter Road.

Brother Jamie was not about to let the fires cool. When you walked the aisle at Jupiter Road you were baptized that very morning. Once I came forward I was taken into the back and given donated clothes to wear. The congregation waited while we were prepared. Brother Jamie baptized me in the name of the Father, and of the Son, and of the Holy Ghost. I was buried with Christ in baptism, and raised to walk in newness of life.

  Global Warming and the Risen LORD

Unfortunately, none of my family had made it to church that morning, so I took the church bus home. When I arrived and told my mom, she was glad for my decision, but upset and disappointed that she and other family members did not witness my baptism. In her family such events were special occasions attended by extended family members.

Not long after, when the youth leaders kept after my older brother to get his hair cut short, he soured on Jupiter Road and the family moved our memberships to a more mainstream Southern Baptist church across town, Richardson Heights.

As I entered the baptismal waters in 1974 I was dressed as if I were Oliver Twist in a school play. My family wasn't there, and Brother Jamie got my name wrong when he baptized me. But my walk with the Risen LORD had begun. I had been raised to walk with Him in newness of life. That's all that ultimately mattered.

JUMO AND GRANHELEN

The fact that my family regularly attended church was due to a large extent to my mother's parents: my grandfather, whom we grandkids called "Jumo," (James or Jimmy Wilkinson, after whom I was named) and my grandmother "GranHelen" (Helen Wilkinson). These two served as the spiritual patriarch and matriarch of our family. They were indispensible in how I came to confess Christ and grow in discipleship.

My grandfather Jimmy was one of the prominent merchants in town, owning his own store in downtown McComb, a small Mississippi city of approximately 12,000 around 80 miles south of Jackson. The store was called the "Dixie Auto-Lec," essentially a combination hardware, appliance, and auto parts store. It sold just about anything using electricity, including refrigerators, washers, TVs, toasters, waffle irons, and clocks, as well as automotive supplies and tools. The transportation needs of children were not forgotten, as my grandfather also sold bicycles, tricycles, and red-painted wagons.

In addition to being a leading merchant, Jimmy Wilkinson was also a deacon at the First Baptist Church. Even though I was quite young when he died at age 59, I have fond memories of a man who was loving, kind, and devout.

My grandmother, Helen Wilkinson, lived nearly 40 years without her Jumo but never remarried. They were childhood sweethearts: Jumo was the love of her life. No one could replace him.

GranHelen's home was filled with Bibles, bible study guides, and framed reproductions of Warner Sallman's famous devotional portrait of Jesus and Heinrich Hofmann's *Christ in the Garden of Gethsemane*. For fun at her house she would play Bible games with us, including a board game of a trip around "the Holy Land" that involved answering questions from the Bible.

It was soon after the death of her husband Jumo that she took a trip to the real Holy Land. It was to be one of the highlights of her life. As GranHelen recounted it, Dr. Taylor, then the pastor of her church, came and visited her one day and told her of a trip he was organizing to the Holy Land. "And you're going to come with us," he announced with authority. "I've already signed you up." And so she went. She saw where Jesus walked, and where her Savior died.

She devoutly read her Bible all her life, even during the latter years when her mind was wracked with dementia. During her last days she used a stroller to get around that had a cloth pocket hanging down from its center bar. Inside was a well-worn, dog-eared, thoroughly underlined copy of the New Testament and Psalms. She read her Bible to her dying day.

GranHelen was keen to focus on the spiritual development of her grandchildren – but not only ours. At First Baptist she taught fifth grade Sunday school for over 40 years.

Years later I had the honor of preaching at her funeral. I recounted a story that had occurred a few months prior when my brother, sister-in-law, and I were moving her from Mississippi to Virginia to allow my cousins to oversee her care. On our journey we stopped to spend the night at a motel, and I shared a room with her.

That night I discovered one thing I didn't know about my grandmother. She talked in her sleep. Rather loudly. At four in the morning.

Global Warming and the Risen LORD

It wasn't mumbling. What she said was clear and distinct, and spiritually profound. For the most part she kept repeating one question, and it was a prayer addressed to God. "O Lord, what can I do now, Lord?" Her question was filled with emotion, and by her inflection it was clear that she wasn't simply bemoaning her present state or using it as an excuse. She was asking how, in her current condition, could she serve. She was sharing her suffering with the LORD, to be sure. But she wasn't asking that it be taken away. She was asking for guidance as to how she could still serve the LORD. She was in the Garden alone with her LORD.

She was saying, in effect, "Lord, I'm over 90 years old. My mind and body are worn out and broken. I'm no longer in charge of my life. But I still want to serve you. What can I do now, Lord?"

Her question was its own answer. Unbeknownst to her she was witnessing once again to her grandson (and now to you) of her profound faith and devotion to God. These words spoken in her sleep attest to their purity. It was a prayer straight from her heart to the LORD.

The Danish Christian philosopher Soren Kierkegaard said that purity of heart is to will one thing. Kierkegaard of course never met Helen Wilkinson, but if he had he would have learned a thing or two about purity of heart. In the end, her life reached its full glory in her unadorned, pure devotion to the LORD. She wished for one thing – to serve her LORD.

MOSES COMES TO TOWN

Of course my grandparents were not perfect by any means, not perfect in the sense of Jesus' command to "Be perfect [complete in righteousness, spiritually beautiful], therefore, as your heavenly Father is perfect" (Mt 5:48). This hit home for me in a Walden bookstore about 25 years ago when I opened to the Table of Contents of Taylor Branch's book, *Parting the Waters: America During the King Years*. What did I find there? The 13th chapter entitled "Moses in McComb." It was quite a shock to find the hometown of my parents and grandparents – a sleepy little town that I thought hardly anyone had ever heard of – in a Pulitzer prize-winning book.

The rather boring little town I knew as a child in the late 1960s and 1970s was in fact a major flash point nationally in the struggle for civil rights. Numerous nationally known magazines and newspapers, such as *Time* magazine and the *New York Times*, wrote headline stories on it. There were nightly stories on the national newscasts during certain key periods and constant mentions of events in McComb by the country's best-known syndicated newspaper columnist. Briefings by the Attorney General were given to the President (who subsequently met with local activists whose homes had been bombed and who apparently threatened to send in federal troops). Events transpired that would later inspire the Hollywood movie *Mississippi Burning*.

In 1961, a few weeks before I was born, the struggle for civil rights was about to flair forth in my family's hometown. The activities in McComb over the next several years would play an important role right at the apex of the struggle for passage of the most important pieces of civil rights legislation in our country's history: the Civil Rights Act of 1964 and the Voting Rights Act of 1965.

In mid-July of 1961, civil rights activist Bob Moses came to McComb to register African-Americans to vote at the invitation of local NAACP leader C. C. Bryant. Moses came to what a front-page *New York Times* headline called a "Hard-Core Segregationist City" and what *Time* magazine labeled "the toughest anti-civil rights community in the toughest anti-civil rights area in the toughest anti-civil rights state. … consistently the most intransigent of the intransigent."[422]

On October 4, the struggle burst into public view in downtown McComb.[423] The city's residents, white or black, had never seen anything like what was about to happen: 120 African-American high school students marching in protest through downtown to City Hall where they attempted to hold a prayer vigil. This peaceful protest march shocked white McComb and led to mob violence by some of the local whites – who were not about to let a large group of young black students get "uppity" in their town and pray in front of City Hall.

The immediate precipitating event was the refusal of the black high school to readmit a student, Brenda Travis, who had been jailed and suspended for trying to desegregate several lunch counters in town.

The simple truth was that the African-American students were fed up with how they were treated. Those involved in the local NAACP youth chapter had heard famed civil rights leader Medgar Edgars speak in McComb. The local head of the NAACP, C. C. Bryant, had mentored them. They saw the differences between the white schools and black schools in terms of the textbooks and the school buildings. They saw and experienced consistent police brutality against African-Americans, and constantly lived with the inequalities of segregation. And they were tired of the injustice of it all.[424]

By the time the 120 African-American students reached City Hall over 200 whites had spontaneously surrounded them. They pulled one of the white civil rights activists, Bob Zellner, from the protesters and beat him. Law enforcement officials ignored this, but arrested over 100 African-American students.

As I read the accounts of this protest and the white reaction, many questions came into my mind, including those having to do with my namesake. What was my grandfather, Jimmy Wilkinson, doing when the African-American students marched through the streets? Did they go by his store? Did white customers rush from his store to help form the mob? Did my namesake join them, even as a bystander? Did he see the white mob beat Bob Zellner while the police and FBI agents stood by? Did he watch the activists and students get arrested? Or did he simply hear about it afterwards? What were his thoughts about the outright lawlessness of the white mob, and the complicity of the local law enforcement officials and the FBI agents? What did he say to others about these dramatic occurrences? When he and my grandmother talked about it what did they say?

RED AND MALVA RUN OUT OF THE BOMBING CAPITAL OF THE WORLD

The struggle in McComb would peak three years later in the summer and fall of 1964, when McComb would be called "the bombing capital of the world."[425] This was not hyperbole, as upwards of 20 bombings took place during this time, with homes and churches destroyed. Oliver Emmerich, the moderate and courageous editor of the local paper the

Enterprise-Journal, later explained, "Almost everybody was hysterically afraid."[426] Emmerich himself was a target, with shots fired into his office window and a burning cross planted in his yard.

Any local white person who was even friendly to the civil rights activists was literally run out of town. A prime example is the case of the Heffner family, made famous by syndicated columnist Drew Pearson as well as a book by Pulitzer prize-winning Hodding Carter called *So the Heffner's Left McComb*.[427]

Albert "Red" Heffner and his wife Malva were solid, even exemplary, citizens who had lived in Mississippi all of their lives and had lived in McComb since 1954. In 1957 Red received the Community Service Award from the Chamber of Commerce. In 1963 one of their daughters was crowned Miss Mississippi – about as close to being a princess as one could get.

In a span of weeks they would be gone.

As Red Heffner explained to columnist Drew Pearson, "It all happened because I had bought six dozen hot tamales and had invited our pastor and his wife to come for dinner."

At the last minute they couldn't come, so the Heffner's invited Don McCord of the National Council of Churches (NCC), whom they had met during the summer at the home of their pastor. "He asked if he could bring a friend, and of course we said yes." The friend turned out to be white civil rights activist Dennis Sweeney.[428]

That very night local thugs surrounded the Heffner's house with their vehicles, blocking the cars and following McCord and Sweeney back to their residences. Over the next several weeks the Heffner's received up to 400 harassing phone calls, some of them death threats. Their tires were slashed. White supremacists would surround their house at night and the police would not respond to requests for help. Even though Red Heffner had done $2 million worth of insurance business the year before, his business dried up to nothing. His landlord for his office canceled his lease. Their dog was poisoned and died.[429]

Global Warming and the Risen LORD

What had Red Heffner and his family done to deserve having their lives destroyed?

Besides inviting Don McCord and Dennis Sweeney to dinner when his pastor couldn't make it, their other "crime" was that Red had earlier tried to help facilitate communication between the NCC's Don McCord and local churches. He helped arrange for McCord to speak to Sheriff Warren about the violence and lack of protection. Heffner told columnist Drew Pearson, "I had been a champion of absolutely no cause. I did not believe in integration."[430] And yet this former winner of the McComb's Chamber of Commerce Community Service Award, the father of Miss Mississippi, was run out of my family's hometown.

During this period extremists held sway and were backed by well-placed members of the establishment. The city council member who oversaw the police chief was a member of Mississippi's white Citizens Council,[431] a group that Heffner himself aptly described as "the KKK in cufflinks."[432] Law enforcement officials were infiltrated by the KKK to the point that on September 20 one of the bombers went home after bombing the house of local activist Alyene Quin,[433] put on his deputy's uniform, and showed back up at the Quin house to help his colleagues investigate. The Sheriff subsequently accused Mrs. Quin of bombing her own house.[434]

WAVING A WHITE STATEMENT

By the fall of 1964 moderate white civic leaders in McComb knew something needed to be done to save their community. They were not willing, "to burn down the city to save it" from what hard-core segregationists feared most: even a hint of integration. The solution crystallized in the form of a statement by over 600 prominent white citizens of McComb that a subsequent *New York Times* editorial described as "the most encouraging news out of Mississippi since the civil rights struggle reached fever heat" given that McComb had been "the center of bombings and other violence."[435]

The federal officials themselves were prodded into action when a civil rights organization took local restaurant operator and activist Alyene Quin and two other bombing victims to Washington a few days after

Alyene's house had been bombed. They spoke a press conference that led to a private meeting on Thursday, September 24, with President Johnson. Mrs. Quin asked the President to send federal troops to McComb and the President assured her that action would be taken.[436]

Whether he had inside information or not, the next day, on September 25, the *Enterprise-Journal*'s editor, Oliver Emmerich, wrote his first editorial cautiously calling for change, warning that if the federal government imposed martial law there would be serious economic consequences. This was the first time any white leader had spoken out.[437]

A week after Alyene Quin's house was bombed, on September 27, officials from the Justice Department met Emmerich and other white moderates in McComb. The Justice Department officials recommended a statement by leading citizens, subsequently telling their superiors that many powerful white citizens wanted to use whatever means necessary to protect the status quo, but that white businessmen were fearful of federal intervention. In other words, things probably wouldn't change without more outside pressure (of the type Alyene Quin asked the President for), especially from the federal government.

Just two days later, on September 29, the local district attorney, Joseph Pigott, received a call from someone he knew in the Justice Department warning him that the federal government was about to send in troops and declare martial law in McComb. Five minutes later Pigott received a phone call from the Governor of Mississippi, Paul Johnson, informing him he would meet with Pigott and other local officials in the district attorney's office that afternoon. The Governor had received a similar call from the Justice Department. He told the local officials that he planned to beat President Johnson to the punch and send in troops himself.[438]

As Emmerich had warned in his editorial on September 25, such action could have dealt a crippling blow to McComb's already distressed economy. The bombings and violence had created a situation where white housewives were afraid to come downtown and shop. Local African-Americans were taking their business elsewhere. National businesses were avoiding McComb. Martial law would have been economically devastating.

Local officials asked the Governor for 48 hours to produce some signs of restoring law and order. Twenty-four hours later the first bombing suspects were arrested. In a little over three weeks the bombers were in court pleading guilty.

But this was apparently only a show created to avoid martial law, not real justice. The maximum sentence was death, but the judge gave the bombers suspended sentences, saying "outside agitators" provoked them. They didn't even spend one day in jail.[439]

On the same day the bombers were given probation, the City Council member in charge of the McComb police, who had ties to the KKK, ordered the Chief of Police to arrest the civil rights activists for cooking meals where they lived.[440] The charge was operating a food establishment without a health permit. They were put in jail on the very day that those convicted of attempted murder were set free.

Previously such injustice and outright lawlessness under the guise of law might have been tolerated by the citizens of McComb and gone unnoticed by powerful interests from outside the city. But at this time McComb was under a microscope, with national media, federal officials, and outside business interests all waiting to see whether McComb would clean up its act.

With the violence and intimidation continuing, the economic situation perilous, and the threat of federal intervention still hanging over their heads, the situation finally grew intolerable for white moderate leaders. They followed up on the suggestion by an official with the Justice Department and issued a statement calling for law and order to be reestablished.

Thus a group of whites from McComb calling themselves "Citizens for Progress" overcame their fears and prejudices and draft what they called a "Statement of Principles." Near the beginning they state:

> *We believe the time has come for responsible people to speak out for what is right and against what is wrong. For too long we have let the extremists on both sides bring our community close to chaos.*

Aside from calling the civil rights activists extremists, this was an accurate and forthright appraisal of their situation. It was about right and wrong. They had waited too long and if responsible people don't stand up for what is right and allow extremists to rule the day, then the consequence can be chaos.

The Statement of Principles has been aptly described by historian John Dittmer as "an impressive document" that represented "a remarkable change in the public position of white people of social standing and political and economic influence."[441] It began as follows:

The great majority of our citizens believe in law and order and are against violence of any kind. In spite of this, acts of terrorism have been committed numerous times against citizens both Negro and white.

It is shocking in this post-9/11 world to see the word "terrorism" accurately applied to what was going on in my family's hometown: nearly 20 bombings and countless other acts of violence and intimidation.

The Statement proclaimed that there was "only one responsible stance we can take: and that is for equal treatment under the law for all citizens regardless of race, creed, position or wealth." Everyone must obey "the laws of the land regardless of our personal feelings." Some laws (including federal laws such as the Civil Rights Act of 1964, passed that summer) "may be contrary to our traditions, customs or beliefs, but as God-fearing men and women, and as citizens of the United States, we see no other honorable course to follow."[442]

In order for "peace, tranquility, and progress" to be "restored" to McComb (note that the African-American citizens of McComb were fighting to achieve these things, not restore them), the statement said that law enforcement officials should "make only lawful arrests" and no law enforcement officials should be a part of the KKK or similar groups.

The Statement of Principles was in fact an admission that law enforcement had become corrupted, that terror rather than the rule of law prevailed in McComb. The trial of the bombers – brought about only because of the threat of federal troops – their being let out on probation, and

Global Warming and the Risen LORD

the jailing of the civil rights workers for cooking in their home, demonstrated that the judicial system itself was corrupt.

BUT I STILL HAVEN'T FOUND WHAT I'M LOOKING FOR[443]

Since my family is from McComb, I'm searching for something more in my review of McComb's civil rights struggle. Did any white Christians from McComb stand up and do the right thing? More specifically, did my grandparents, Jumo and GranHelen, do the right thing?

We are all products of our time and place, and there can be a collective moral denial that groups and communities participate in. But in 1964 in McComb there was no place to run and hide morally when it came to the struggle for civil rights. There was no way to claim ignorance when upwards of 20 bombings occur in your community; when your small town is regularly featured on the national newscasts; when the nation's best-known columnist, whose column is published in your local paper, visits your town and features the civil rights struggles occurring in your town at least 14 times during the fall of 1964; when even your own local newspaper's editorials during September and October are highlighting the issue and calling for change. If you own a store in downtown McComb like my grandfather did, there is no way to ignore the economic impacts of people being afraid to shop downtown.

It was clear to those who drafted the statement that they were dealing with issues of right and wrong. Equally apparent was the fact that the actions of those violently opposing change were evil, "acts of terrorism." And finally, the statement made clear that there must be "equal treatment under the law" for everyone, regardless of one's personal views or feelings.

If it was all too clear to the drafters and signatories, surely it was clear to my grandfather, no? Surely as a prominent merchant, a deacon at First Baptist, a respected member of the community, he would have been asked to sign, right?

As I began to scan the list of signatories I started seeing names that looked familiar.

Is that the father of my mother's best childhood friend? Is so-and-so the father-in-law of my aunt, in whose home I often played with my cousin? Is that the son of the lawyer we consulted as we looked into guardianship for GranHelen? Are any of these folks with the last names of Brewer (GranHelen's maiden name), Wilkinson, or Ball related to me?

But mainly I was searching … hoping … to find my grandfather's name. If Jumo's name was there then maybe the moral questions raised when I first learned nearly 20 years ago that Moses had come to McComb could be assuaged.

And there, at the top of the middle column of the next to last page was a strange entry. No other name looked this way. The last name was right, "Wilkinson." But the first name was tauntingly strange: "J..i..i..y." My grandfather went by "Jimmy." Could this be him?

Did my grandfather sign the Statement of Principles on behalf of himself and my grandmother? After a long investigation the answer is … I don't know.

Such ambiguity, such lack of clarity, such seeing through a glass darkly, is a plaintive, restive, metaphor for the moral standing of southern white Christians like my grandfather and grandmother and their spiritual offspring. But not knowing if I had found my namesake's name made me face a simple truth: absolution would not have come from finding a name on a list.

What I do know is that I never remember my grandparents or my parents making anything close to racist statements. We were taught to treat all people with respect, to treat others equally regardless of race or religion – indeed, as Christians, to love everyone.

But righteousness is not the absence of bad acts, of sinful, hurtful acts. It is the presence of good acts, the presence of right relations with everyone and with God.

The Heffner's put one foot on this road in terms of race relations and they were run out of town. The Statement of Principles was a good statement for its day. But while it may have signified a change of heart by some of those who signed, it was external political and economic pressure that brought it about. True righteousness doesn't require external pressure. It flows out of a righteous heart.

It is easy to judge my grandparents and the other white residents of McComb during this time.

But what would you have done?

Let me ask myself that question about the cause to which I have dedicated my adult life: how far would I go to stand up for the poor in overcoming global warming?

Would I be willing to change a basic tenet of my worldview? (For them, it was segregation and the racial views underlying it.) I've done some of that over time, as I've tried to reshape how I understand the world and live in it to conform to a more biblical approach to global warming.

But would I go so far as to have my life destroyed in a matter of weeks like the Heffner's? To be beaten in broad daylight in the street? To be shot at? To have a cross burned in front of my house? To have my home dynamited and my family possibly murdered? These things happened in McComb – and could have happened to my grandparents. I have faced some strong criticism and ridicule for my global warming activities – I've even been called a heretic by people who meant it – but nothing like this.

Would I face such trials to help overcome global warming? The honest answer is, I don't know. I would hope I would stand up and do what's right. But maybe I wouldn't. That's not an excuse. It's an admission.

I had not known the full extent of what happened in my family's hometown in the 1960s before I started writing this book. I was pleased to discover the story of the Heffner's – that some local white Christians tried to begin to do the right thing. I was also pleased to discover the 1964 Statement of Principles. Neither of these are examples of the fullness of righteousness, but at least they were steps in the right direction.

At the same time I have been saddened to learn in more detail the sheer extent of the systematic corruption of justice. I am shocked by the outright terrorism, brutality, and lawlessness that took place in the town where I knew so much love. In our post-9/11 world it was quite jarring for me to read that white southerners were aptly describing the acts of other white southerners as terrorism. One thing became clear; it would have not been easy to have been in the shoes of my grandparents when faced with such threats, but also to overcome so much of my own culture.

I may have been unaware of the extent of the violence and lawlessness in McComb, but I had known about the moral failure of most southern white Christians from the time of slavery through the civil rights era. Over 200 years ago, before evangelical Christians were the dominant religious force in the South, many of them were against slavery. But as they became a part of the establishment, as they began to see the economic livelihood of their region connected to slavery, the views of white evangelical Christians began to shift until the vast majority held a pro-slavery stance. Many had a well thought-out biblical and theological position defending their views. The Southern Baptist Convention, the denomination of my grandparents and the one I grew up in, was created in 1845 by those who broke away from their denomination because it passed a resolution saying slaveholding missionaries would not be appointed.

In 1995 on the eve of the 150thcelebration of the Southern Baptist Convention (SBC), the conservative leadership recognized the moral failure that was the impetus for the SBC's founding and the racism that shadowed its history:

> *Our relationship to African-Americans has been hindered from the beginning by the role that slavery played in the formation of the Southern Baptist Convention ... Many of our Southern Baptist forbears defended the right to own slaves ... In later years **Southern Baptists failed, in many cases, to support, and in some cases opposed, legitimate initiatives to secure the civil rights of African-Americans** ... we apologize to all African-Americans for condoning and/or perpetuating **individual and systemic racism** in our lifetime; and we genuinely repent of racism of which we have been guilty, whether consciously (Psalm 19:13) or unconsciously (Leviticus 4:27).*[444]

For me there is a deep, deep shame associated with this "individual and systemic racism" of my spiritual ancestors.

Just like civil rights in the 1960s, global warming is one of the great moral challenges of our time. Will we as Christians rise to the challenge? Will our witness encourage our spiritual offspring in the faith? Or will they have to issue a statement to the poor asking for forgiveness for what we failed to do?

In the end, the failure of my grandparents and other southern white Christians to do all that righteousness required in the treatment of African-Americans simply proves once again the truth of the Apostle Paul's declaration that "There is no one righteous, not even one ... for all have sinned and fall short of the glory of God" (Rom 3:10, 23). Jumo and GranHelen, my beloved grandparents, the saints of our family and the familial spiritual foundation of my life, certainly wouldn't argue with this. They knew full well that they were sinners in need of a Savior.

I NEED A SAVIOR

If there is one thing I know in life for certain, if there is one thing I know in my gut, if there is one thing I know in that small little room in my mind where I tell myself the truth about myself, it is that I am a sinner in need of a Savior to save me from my sins. Not somebody else's sins. My sins. All of my education, all of my accomplishments, well, they make me an educated, accomplished sinner. Have I done some good in my life? Then I'm a good sinner.

Does this make me feel bad about myself to admit that I am an educated, accomplished, good sinner? No, it makes me feel real about myself. It makes me feel just a small touch honest with myself. Not too honest, because I don't have enough spiritual courage and strength for too much honesty about myself, but honest enough to know I need a Savior, that I desperately need a Savior every moment of my life, for there is not a single moment when I am not a sinner.

And I'm not just talking about "original sin" in the sense of being born a sinner. I'm not talking about an idea where I could say, "Sure, I'm a sinner every moment. But that's that original sin stuff that I didn't have anything to do with. Otherwise, I'm a pretty good guy."

I personally know that as a consequence of original sin I have an inclination towards sinning and believe that every human being does. But I also believe that I continually sin, that I could choose not to, but that I don't. My bent towards sinning overwhelms my feeble will as I sinfully sleepwalk my way through my day.

And these are conscious, hurtful thoughts and actions. I'm not even talking about my failure to do righteousness. As I've said, righteousness is the presence of good acts, not simply the absence of bad ones. I am inconsistent in righteousness, but consistent in my sinfulness. And even most if not all of my righteous acts are mixed with varying degrees of sin.

So I personally know deep in my shallowness the truth of Paul's anguished cry:

> *For I have the desire to do what is good, but I cannot carry it out.*
> *For what I do is not the good I want to do; no, the evil I do not want*
> *to do—this I keep on doing ... What a wretched man I am! Who will*
> *rescue me from this body of death?* (Rom 7:18-24).

And, frankly, I'm in worse shape than Paul, because much of the time I don't even desire to do what is good. I'm just too spiritually lazy, too set in my sinfulness to make the effort to even desire what is good. There is a part of me – the goodness God created within me that survived the Fall – that yearns to be righteous (for otherwise I wouldn't even know of my sinfulness). But there is a big part of me that basically says, "To hell with it."

Boy, do I need a Savior!

This is certainly not popular to talk about. "Geez, stop being so negative! Don't talk like this! It can lead to low self-esteem, which can lead to all types of bad things."

Well, it is *not* talking about it that can lead to the baddest thing – not understanding that we all need a Savior! It is not talking about it that can have us remain in the exact place that makes low self-esteem look like a picnic – unforgiven, lost.

While we are all sinners, as Christians we must never treat people **as** sinners. While we need to help people understand that "there is no one righteous" (Rom 3:10) and that "all have sinned and fall short of the glory of God" (Rom 3:23), we must never lead them to feel that they are worthless. To do so is essentially to treat Christ's death on the cross as worthless, to deny God's inestimable love.

Paul states the truth of it in this way: "But God demonstrates his own love for us in this: While we were still sinners, Christ died for us" (Rom 5:8). And so our worth is not ultimately based upon the judgments of other sinners, but upon the love of God.

Therefore, we must treat everyone as who they are: someone Jesus died to save – more worthy of our tiny gift of love than we could ever imagine! We must love everyone. We must love, love, love with God's love. We must build people up and constantly encourage them precisely because they are sinners.

But we cannot live in sinful denial and find our way to the path of righteousness God has laid out for us. We cannot lie to people about our sinful condition. That's not the "nice" or the loving thing to do; it's the cowardly, cruel thing to do.

So we must help people acknowledge that they are sinners so that they can also acknowledge their need for a Savior. But we must do so mercifully, always motivated by love, always in recognition of our own sinfulness and moral frailty, always in a way that doesn't lead to defensiveness or denial, but rather to recognition of the truth.

I know I'm a sinner in need of a Savior. But I also know that if my grandparents were sinners in need of a Savior, then everybody is a sinner in need of a Savior.

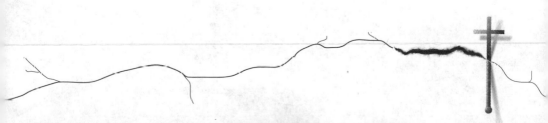

11

WALKING WITH CHRIST OUR SAVIOR –
SET FREE TO LOVE HIM BACK

CHRIST THE CREATOR IS OUR SAVIOR

I don't know how to say this strongly enough – it is impossible for us to comprehend the perfection, the joy, the love that exists between God the Father, God the Son, and God the Holy Spirit. Perfect contentment. Perfect peace. Lacking nothing. Sharing everything. As we compare this with fleeting moments in our own lives, we catch a faint, childish glimmer of a whisper of a rumor of what this might be like. Nothing like what it truly is, but enough to imagine what we cannot know; enough to stumble our way towards the truth of this.

And because something rather than nothing exists, because Creation, the Universe, exists, because we exist, we know that this sharing, giving, joyous love has spilled out from God the Father, Son, and Holy Spirit into the nothingness that preceded everything. On behalf of God the Father and God the Holy Spirit, God the Son, "calls things that are not as though they were" (Rom 4:17).

Just as Christ the Creator of all things continuously keeps the Universe itself from slipping back into nothingness, "sustaining all things by his powerful word" (Heb 1:3), so too are we kept from slipping back into our spiritual nothingness when we accept Christ into our hearts as our Savior.

As John 1 puts it,

In him was life, and that life was the light of men. The light shines in the darkness, but the darkness has not understood it ... The true light that gives light to every man was coming into the world. He was in the world, and though the world was made through him, the world did not recognize him. He came to that which was his own, but his own did not receive him. Yet to all who received him, to those who believed in his name, he gave the right to become children of God—children born not of natural descent, nor of human decision or a husband's will, but born of God (Jn 1:4-5, 9-13).*

Christ the Creator calls to each of us: "Come out of nothing into life, come out of your spiritual darkness into the light. I am the Creator. Let me be your Savior, too." The world did not recognize who He really was (and still doesn't). Even "his own" did not receive Him and accept Him for who He really was.

But to those of us who do receive Him, Christ our Creator and Savior gives us "the right to become children of God."

And yet there are profound spiritual forces at work to keep us in our own spiritual darkness, to keep us from either answering or fully answering Christ's call to come into His light, to make those of us who do answer into blind spiritual infants in Christ. These malevolent forces want us to stay in our own spiritual nothingness, to be like them in their misery. It is the positive desire for communion that has been turned into its opposite. They were made for spiritual communion, and while that desire is still there, it works its way towards the exact opposite: total spiritual isolation, each in our own spiritual nothingness. It is the exact opposite of the spiritual communion that the Father, Son, and Holy Spirit share within the Trinity and desire to share with us. Christ the Son can be our Savior precisely because He is our Creator and the Creator of all things, because as such He has the power to overcome these malevolent spiritual forces.

But while these malevolent forces of nothingness shepherd us into our own darkness, we stumble there on our own accord. Even as these forces do everything they can do to deceive us and entrap us within ourselves, we never lose the freedom to cry out, for Someone to save us and help us become whom He created us to be. Each of us freely gives up true freedom to become a selfish "image of me" (i.e., our own self-centered selves) rather than what we were made to be, images of God, reflections of the communion of the Trinity.

Christ is that Someone to Whom we can cry out. He can be our Savior precisely because He created us and thus can overcome the darkness within our own hearts. He does not fear our fear. He does not recoil from the stench of our hate. He does not turn away from our spiritual ugliness, our centering upon ourselves, which leads to our own emptiness. Rather, he joins us in our nothingness. Even though we are spiritual lepers – who should be shouting to any who come close, "Unclean, unclean! You cannot have communion with me! If you touch me you will become what I am!" – he embraces us. Even though spiritually we are not, he calls to us as if we were. By the power of His own Presence, He creates within us a true spiritual presence, our own unique reflection of His glory, His beauty, His freedom. The Apostle Paul reminds us that nothing can separate us from the love of Christ:

> Who shall separate us from the love of Christ? ... I am convinced that neither death nor life, neither angels nor demons, neither the present nor the future, nor any powers, neither height nor depth, nor anything else in all creation, will be able to separate us from the love of God that is in Christ Jesus our Lord (Rom 8:35, 38-39).

The only thing that can separate us from Christ is ourselves.

TEMPTED AS WE ARE

> She will give birth to a son, and you are to give him the name Jesus, because he will save his people from their sins (Mt 1:21).

But when we first find out about Christ as Savior in the New Testament, right in the first chapter of Matthew's Gospel, he seems anything but powerful, his ability to be the savior of anything, anything but sure. His very survival hangs by a thread from one moment to the next, dependent upon the protection of a man who is not his real father, a man who could have easily thought him to be the bastard child of an unfaithful wife-to-be. Indeed, Joseph could have had Mary stoned to death for unfaithfulness (Dt 22:20-22), and her womb would have become the baby Jesus' grave. In Matthew's Gospel from nearly the moment of conception the Son of God faces the threat of death. Our savior's protection depends upon Joseph literally believing his dreams in understanding the pregnancy of his fiancée – whom he has not yet known intimately – to be the work of God. (Yeah, right. Uh, huh. An "A" for originality – and audacity.)

At first blush it certainly doesn't inspire confidence. But wait! It gets worse.

This unwed pregnant peasant girl lives in a small, obscure part of the Roman Empire presided over by a paranoid ruler, Herod, who had three of his own sons executed for fear that they were plotting against him. His legitimacy to rule ever haunts Herod. He was not born a king, nor was he a full-blooded Jew. Indeed, his title, "King of the Jews," had been given to him by the Romans and not by the people he ruled. And so, when the Magi come to him and ask about the birth of "the King of the Jews," the exact title the Romans had given him, Matthew's Gospel is quite understated in saying, "When King Herod heard this he was disturbed, and all Jerusalem with him" (Mt 2:3). After all, if a star has announced the birth of "the King of the Jews," then this must be the real King! And if Herod had three of his own sons killed out of fear they were plotting against him, then he wouldn't hesitate to kill the baby Jesus.

Once again Jesus' protection depends upon Joseph believing his dreams.

So what is Joseph's response? In reflecting upon this we must remember that as a carpenter Joseph's craft required planning, careful preparation, precision. But upon receiving the angel's message in his dream Joseph awakens and in the middle of the night abruptly starts a long, arduous journey to Egypt with Mary and the baby Jesus. What is required of him appears to go against his training, but there is no second-guessing. No waiting to see evidence of soldiers coming with swords drawn. Trust. Obedience. Decisive action when told about a deadly threat. What could have appeared rash or impulsive to outsiders, and out of character to those who knew Joseph was in fact what saved our Savior's life. It is the faithfulness, the proper exercise of freedom of an ordinary man who was no one important in the eyes of the world, which ensures the survival of our Savior and changes the course of human history.

Why would God go about it this way? Why so precarious a journey? Why is Christ the Creator's early life dependent upon the faith of one insignificant peasant in Palestine?

I don't know. But Hebrews 2:17 might give us a clue: "He had to be made like his brothers [i.e., us] in every way, in order that he might become a merciful and faithful high priest in service to God, and that he might make atonement for the sins of the people."

All babies are in fact at the mercy of the decisions of their parents. Through their care – or lack of it – they have the power of life and death over their children. But even with the best of care, babies can die. Death ever threatens life. In Matthew's Gospel this was especially true for the baby Jesus.

Our incredible health care, public sanitation, abundance of food, and political stability, makes it hard for us in the United States to comprehend how precarious was the baby Jesus' existence. Approximately one out of three children in first century Palestine died before reaching puberty.[445] But the odds of the baby Jesus' survival were worse given that the Holy Family were poor refugees pursued by a paranoid homicidal tyrant. In this way, the threat of death in Jesus' early life parallels or exceeds the vast swath of humanity throughout history. And for the poor in this century, Jesus knows what it means to have one's meager chances of survival made worse by the actions of others, as global warming has done and will do.

And so the Word left the perfect Communion of the Trinity to become flesh and pitch his tent among us as an infant pursued by the vicissitudes of his time and place for one reason, explained by the Apostle Paul: "though he was rich, yet for your sakes he became poor, so that you through his poverty might become rich" (2 Cor 8:9).

But significant threats to life expectancy didn't end with puberty. During Jesus' time the chances for survival from birth until the age of thirty were not good, only one out of three.[446] However, if one did, there was a good chance one could live to be 60 or more years old.[447]

And so when he begins his public ministry Jesus has beaten significant odds and survived. But rather than becoming cautious and protective of his next 30-plus years, Jesus intentionally puts himself in greater danger simply by fulfilling his ministry and doing the will of his Father.

Indeed, right before his public ministry began, Jesus allows himself to be led by the Spirit into the wilderness to encounter both physical and spiritual danger (Mt 4:1-11; Mk 1:12-13; Lk 4:1-13). Satan's temptations are captured by the challenge,

"*If* you are the Son of God …"

By simultaneously recognizing and questioning Jesus' true identity Satan tries to feed on one of humanity's greatest weaknesses: insecurity combined with pride. He also tempts Jesus to use his power and his freedom outside the will of the Father, in effect saying "True freedom is when you defy God, not when you obey him. Only then are you exercising true choice." But during the temptations Jesus encountered and resisted Satan, the great nothingness, the great absence, who tempts us to embrace our own emptiness and become "images of me" rather than images of God.

Jesus' battles with Satan and his minions don't end there. His first recorded miracle in Luke involves the casting out of a demon in a synagogue on the Sabbath, who screams, "Let us alone! What have you to do with us, Jesus of Nazareth? Have you come to destroy us? I know who you are, the Holy One of God" (Lk 4:34, NRSV). That same evening, at the home of Simon Peter's mother-in-law, Jesus cast out more demons, who shouted, "You are the Son of God" (Lk 4:41). While Jesus' contemporaries don't have a clue that he is the Son of God, Satan and his demons know exactly who Jesus is and want to expose him before the proper time to thwart Jesus' ministry.

This spiritual warfare between Jesus and the same malevolent forces that want to keep us separated from God and in contradiction of our true selves does not end until his great and terrible battle with them in Gethsemane and on the cross, the place they most don't want him to be (Col 2:14-15; 1 Jn 3:8; Jn 12:31-32). Satan is even able to ensnare Peter in his attempts to keep Jesus from the cross (Mt 16:23; Mk 8:33).

But it is not only spiritual forces that oppose Jesus. After his first recorded sermon that took place in his hometown, Nazareth, the congregants are so enraged at his message that they attempt to kill him by throwing him off a cliff (Lk 4:16-30).

The violent opposition from his hometown was spontaneous. However, that from the scribes and Pharisees (the respected local religious leaders of his day) began mildly in comparison near the beginning of Jesus' ministry; but when he continued to heal on the Sabbath and challenge their teachings and authority, their opposition grew quickly into plots to kill him (Mk 2:6-7, 16, 24; 3:6).

Throughout his ministry, Jesus is continuously tempted to deny his calling and the will of his Father. His hometown church tried to kill him. His family heard that he was crazy and tried to stop him (Mk 3:21). He is continuously criticized, endlessly plotted against, constantly misunderstood, even by his own disciples who also try to stop him. The threat of death is ever present, and he knows the cross awaits him in the end (Mk 8:31; 9:31-

32). Indeed, the Apostle Paul said that Jesus faced death for our sake "all day long" (Rom 8:36, quoting Ps 44:22).

His faithfulness to the Father, and to us, allows him to be our Savior. As Hebrews puts it: "For we do not have a high priest who is unable to sympathize with our weaknesses, but we have one who has been tempted in every way, just as we are—yet was without sin" (4:15).

TAKE THIS CUP

Indeed, Jesus' temptations go far beyond anything we could ever imagine.

According to the Gospels, the only time Jesus ever asked if he could get out of doing his Father's will was in the Garden of Gethsemane. "Take this cup from me" (Mk 14:36; cf Mt 26:39; Lk 22:42). The cup is the cup of God's wrath against our sins (Isa 51:17, 22).

I imagine that in the frail humanity he shares with us, Jesus shrank from the sheer terror that lay before him. Not only did he not want to die a horrible physical death – one of the most gruesome ever devised; again, I can only imagine that he, who knew no sin, also did not want to become sin and face the wrath of God against all of the sins of the world. Jesus, the most wondrous reflection of the Father's glory, did not want to be stripped of that glory. The freest person the world has known or will ever know did not want to become the most enslaved person to ever exist. The most spiritually beautiful person to grace the Earth did not want to be transformed into the most grotesque spiritual horror ever. Most profoundly, he did not want to face the absence of the Father's presence, to no longer be in communion with Him. Jesus shrank from facing something so spiritually terrifying that he himself could not imagine it – God's wrath against the sins of the whole world, as he became everything he was not.

"Take this cup from me." Here was Satan's greatest chance, with Jesus at his most vulnerable. Here was the greatest test of true freedom: would the son deny the Father's will at the moment of truth. Here was the greatest opportunity to reflect the Father's glory, as the salvation of the whole world hung in the balance. Here was the supreme moment of spiritual beauty: would love win?

These brief moments held our eternity and the reconciliation of the universe.

THE CROSS

That very night one of his own betrayed Jesus with a kiss. That very night all of his disciples fled and left him alone to face his persecutors. That very night the disciple who said he would never deny him – Peter, "the rock" – denied him three times when simply challenged by a servant girl and others around her.

It was the highest religious and political leaders of his own country who conspired to hand him over to the occupiers. They pressed repeatedly to have him wrongly executed, even when Pilate – the head occupier, of all people – found him innocent and wanted to release him.

When Pilate brought Jesus, whom he called "the King of the Jews," before the people, his own people called for his crucifixion.

> *Wanting to release Jesus, Pilate appealed to them again. But they kept shouting, "Crucify him! Crucify him!" For the third time he spoke to them: "Why? What crime has this man committed? I have found in him no grounds for the death penalty. Therefore I will have him punished and then release him." But with loud shouts they insistently demanded that he be crucified, and their shouts prevailed. So Pilate decided to grant their demand. He released the man who had been thrown into prison for insurrection and murder, the one they asked for, and surrendered Jesus to their will* (Lk 23:20-25).

Betrayed, abandoned, denied, unjustly convicted, Jesus is scourged, mocked by the Roman soldiers, who press a crown of thorns upon his head, and led outside the city gate to the place called "the Skull," or Golgotha, where he will be crucified between two criminals.

To be outside the city gate is to be cut off from the protection of one's community, to literally be cast out. This is also where blasphemers were to be executed (Lev 24:14), a charged leveled by the High Priest (Mk 14:64; Mt 26:65-66). As Hebrews 13:12 says, "Jesus also suffered outside the city gate to make the people holy through his own blood."

Death chased after him all of his life, and when he was nailed to a cross, death finally caught up to him.

An honorable burial at the time would have included mourners and burial in a family tomb. Both were denied Jesus, resulting in a burial of shame, dishonor even in death.[448]

The physical suffering of Jesus, especially the scourging and the crucifixion, are horrific in and of themselves.[449] The psychological torment and social death he experienced were cruel in the extreme.

But it is the spiritual suffering that is the most profound, which appropriately is only hinted at in the New Testament, being beyond our experience and comprehension. The Gospels report that after three hours on the cross, and right before he dies, Jesus cries out:

> *"Eloi, Eloi, lama sabachthani?"– which means, "My God, my God, why have you forsaken me?"* (Mk 15:34; Mt 27:46).

He who knew no sin became sin to save us, and experienced what sin does to us all – separates us from God. The Triune communion was broken.

Jesus died socially, physically, and spiritually – abandoned even by God the Father, as all the sins of the world poured into his heart. "Then Jesus gave a loud cry and breathed his last" (Mk 15:37, NRSV).

FOR OUR SINS

I truly believe that because God loves each of us so much, Jesus would have died just to save me or just to save you. His love for you is that deep. As Psalm 86:13 says, "For great is your steadfast love towards me; you have delivered my soul from the depths of Sheol" (NRSV). In Galatians

2:20 Paul speaks very personally, saying, "I have been crucified with Christ and I no longer live, but Christ lives in me. The life I live in the body, I live by faith in the Son of God, who loved me and **gave himself for me**" (emphasis added). And in Jesus' parable of the lost sheep, the shepherd leaves the 99 and seeks to find the one, saying that the Father "is not willing that any of these little ones should be lost" (Mt 18:10-14).

But if Jesus would have died just to save me, then that means my sins alone drove the nails through his hands. I am responsible for my Savior's death, death on a cross. What am I to do! What can I do! Woe is me! My Savior died because of me! How can I live?

Now while I believe it is theoretically true that Christ would have died for me alone, the Bible teaches us that Jesus died to save all sinners, even that Jesus' death had cosmic ramifications.

Right at the beginning of John's Gospel, when he first sees Jesus, John the Baptist says, "Look, the Lamb of God, who takes away the sin of the world!" (Jn 1:29). It could not be any clearer than this: Jesus is God's sacrificial sin offering to take away the sins of the world.

Why? Again, John's Gospel makes it abundantly clear: "Because God so loved the world, he gave his only begotten Son" (3:16 KJV). So the Father gives the Son because they love the world.

Because of this love, Jesus has the heart of a servant. As the Father gives the Son, so the son gives himself: "For even the Son of Man did not come to be served, but to serve, and to give his life as a ransom for many" (Mk 10:45; also Mt 20:28). As the Good Shepherd, he gives his life for the sheep, which is a reason the Father loves him (Jn 10:11, 14-18).

On the night he was betrayed, Jesus created a way for the Church to understand and be reminded of his death and what it meant, known as the Lord's Supper. Its significance cannot be overstated, as it is the only ritual that Jesus himself instituted. In Matthew's version of what has come to be called the "words of institution," Jesus states, "This is my blood of the covenant, which is poured out for many for the forgiveness of sins" (26:28).

Here in Matthew's Gospel Jesus clearly and simply tells us why his blood is to be poured out: "for the forgiveness of sins."

For the Apostle Paul this belief was at the very heart of the gospel (good news) he preached. It is not something he himself came up with, but was part of the message that he received – meaning it is part of the oldest message of the early church. As he told the Corinthians, "For what I received I passed on to you as of first importance: that Christ died for our sins according to the Scriptures, that he was buried, that he was raised on the third day according to the Scriptures," (1 Cor 15:3-4). For Paul, Jesus' death "for our sins" and his subsequent resurrection are at the core of the Christian faith; they are "of first importance."

RECONCILED

What is the result of Jesus dying for our sins? As Jesus himself said at the Last Supper, it is for the forgiveness of our sins. His death, his blood "purifies us from all sin" (1 Jn 1:7). The book of Hebrews says "we have been made holy through the sacrifice of the body of Jesus Christ once for all ... he has made perfect forever those who are being made holy." (10:10, 14).

The result of this forgiveness, purification, and being made holy is humanity's reconciliation with God. Our sins – the ways we have harmed our relationship with God by acting contrary to His will for our lives – have ensnared and imprisoned us in a situation of alienation from God. We are powerless to do anything about it. We are slaves of sin (Rom 6). We are God's enemies, powerless to become His friends. Clueless, we are lost in our own sinfulness with no hope of coming up with a solution. We were made to be in communion with God, to image or reflect the Triune communion of Father, Son, and Holy Spirit. But we can't.

The good news, the glorious news, is that "when we were still powerless ... while we were still sinners, Christ died for us ... when we were God's enemies, we were reconciled to him through the death of his Son" (Rom 5:6, 8, 10). As Paul famously says in 2 Corinthians 5:19, "God was reconciling the world to himself in Christ, not counting people's sins against them (TNIV)." And as he also says in Colossians 1:22, "But now he has reconciled you by Christ's physical body through death to present you holy in his sight, without blemish and free from accusation."

All the hurt, all the harm, that we have done and that has been done to us – that therefore has been done to those whom God loves, because He loves the whole world, including you and me – has been overcome by the sacrificial death of the one who cried out, "Eloi, Eloi, lama sabachthani? … My God, my God, why have you forsaken me?" (Mk 15:34; Mt 27:46). The perfect communion between Father and Son was broken to restore communion with us, so that we could once again image or reflect Their communion.

If anyone were to do serious harm to those you love, they do harm to you, too. Because God the Father so loves the world, all the hurt and harm is ultimately done to Him – and was searingly done to Jesus on the cross. We made ourselves into God's enemies. But Jesus, taking our sins upon himself, overcame the alienation we caused and led us by the hand to be reconciled to the Father, presenting us to Him "without blemish and free from accusation" (Col 1:22).

BUT WHY THE CROSS?

The Apostle Paul says in Romans that the Father "presented him as a sacrifice of atonement" (3:25) and that Jesus "was delivered over to death for our sins and was raised to life for our justification" (4:25). Again, the Father "did not spare his own Son, but **gave him up** for us all" (8:32, emphasis added). And in the first sermon preached after Jesus' death, Peter tells his fellow Jews at Pentecost that "This man was handed over to you by God's set purpose and foreknowledge; and you, with the help of wicked men, put him to death by nailing him to the cross" (Acts 2:23).

But why did the Father have to give His son up to all that he experienced on the cross? Why did Jesus' Heavenly Father fail to protect him like his earthly father Joseph did?

I have been struggling with this question for at least 25 years, and I'll no doubt be struggling with it until the day I die. I don't like it that Jesus suffered and died on the cross! That he loved me so much he died on the cross makes me not want him to die on the cross!

Is there not something a bit scary in the idea that the Father would give up what he loves most? There is a seeming intellectual and emotional contradiction that eats into my heart. Couldn't an all-powerful, loving, forgiving God have found another way? Why does forgiveness require a crucifixion? I'm called to forgive, after all. Isn't the Father big enough to let bygones be bygones? Let's just move on, for Christ's sake. Why the cross? Why did blood need to be shed?

I don't know. There is mystery here that is beyond human comprehension. But let us try to find some clarity where we can.

We did it. God allowed us to do it. The Father gave His son into our hands. If freedom is really freedom, if sin is really sin, then we are responsible, not God.

The cross is the price of our freedom.

If God gave us real freedom and this is not just some game where we have "pretend freedom," then we can alienate ourselves from God. Free beings can choose to do things that require reconciliation. But what if one party can't bring that reconciliation about? "God in Christ was reconciling the world." The Triune God took the initiative.

Surfacing all such questions is important for our spiritual growth, otherwise they can become barriers. But if we are not careful it can also become a convenient way to take the focus off ourselves and make God the problem, make God the Father the bad guy.

It's the oldest trick in the book of human sinfulness – blame someone else. Make excuses. Don't accept responsibility. "Am I my brother's keeper?" Cain rhetorically asked God (Gen 4:9). In other words, "Don't ask me. That's Your job. It's your fault." Blame God.

"Save yourself," mocked Jesus' detractors. "*If* you are the Son of God …" Satan said.

In my heart I say to the Father, "If you are the loving, all-powerful Father, then save your son." It's essentially the same temptation Satan offered Jesus. But then I am reminded of Jesus' response to Peter, when Peter wanted him to forego the cross. "Get behind me, Satan."

I childishly but understandably ask the Father, "Get me off the hook. Don't let my sin be so costly." It is ultimately unbearable for me to think that Christ died just to forgive me – and it should be. Because while I believe he loves me (or you) enough to have died just for me, he didn't. "For God so loved the world." He loves me, yes, but not just me. He loves the world.

In both the Old and New Testaments, King David is described as a man after God's own heart (1 Sam 13:14; Acts 13:22). When his own son, Absalom, died, King David cried, "O my son Absalom! My son, my son Absalom! If only I had died instead of you—O Absalom, my son, my son!" (2 Sam 18:33).

Is God the Father's love not infinitely greater than David's? How much did it cost the Father to give His Son into our hands, into the hands of sinners like us? How much did it rend His heart?

It's literally unbearable – unbearable! – for us to think about. But for the sake of our own souls we must catch a glimpse of it, to begin, to start, to try to fathom the Father's love for us, for the whole world.

It's ultimately about love, real love. The love between the Father and the Son and the Holy Spirit is a love so joyous and giving that They naturally wanted to share it. But only created beings with true freedom can reflect such love as true images of God created in relationship as male and female. Because of our disobedience, our sin, we have freely failed to image or reflect God and the love between Father, Son, and Holy Spirit.

But Jesus is the true image. "Anyone who has seen me has seen the Father" (Jn 14:9). "For God so loved the world" does not just reflect the Father's love. It's the son's love, too. On the cross he freely reflects the

Father's love; he freely gives himself. As Jesus says in John's Gospel, "I lay down my life—only to take it up again. No one takes it from me, but I lay it down of my own accord. I have authority to lay it down and authority to take it up again. This command I received from my Father" (10:17-18).

Yes, in his fragile, finite humanity, Jesus said, "Take this cup away from me." But remaining true to his true humanity, Jesus immediately says, "Thy will be done" (Mk 14:36; cf Mt 26:39; Lk 22:42). And in this one moment Jesus' imaging of the Father's love was ultimately tested and found to be supremely true. The Son and the Father's wills remained one: "Our will be done."

The Father loves the Son. The Father and Son love the world. Because of this love the Father gives the Son; the Son gives Himself.

But, again, why, specifically, the cross? Hebrews 9:22 says, "without the shedding of blood there is no forgiveness of sins" (NRSV). OK, but why does blood need to be shed for God to forgive?

I don't know. God knows. But I do know that God loved us so much He broke His own heart to save us. And for that I owe Him everything. For that I am a fool for Christ (1 Cor 1:18-31).

We may not like the cross. Indeed, we should not. It should never have happened. But because it did we can never forget. We must look back even as we look away.   At the foot of the cross is where we should lay our hearts. In whatever childish and incomplete way we can image Christ and do the Father's will, we should have Jesus' words in our minds: "Do this in remembrance of me" (Luke 22:19). Our lives must be the ultimate sacrament. The doing of His will is, indeed, an outward sign or manifestation of an inward and invisible grace.

CHRIST THE RECONCILER OF ALL THINGS

Because of God's love I am a fool for Christ, at least I try to be. But sometimes I'm just a fool – better make that most of the time. And when I'm a fool I'm a spiritually blind fool, to boot.

As I alluded to in the Preface, much of my life I was spiritually blind to God's relationship to the rest of His creation. Now I shouldn't wallow in a perverse, egocentric solipsistic echo chamber thinking I was the only one – somehow special even in my spiritual stupidity. Ah … no. I've had plenty of company.

It took a gentle challenge from a fellow Christian who urged me to explore what the Bible had to say about this that helped to begin to open my eyes. And it was when I came to Colossians 1:15-20, especially verses 19-20, that I began to see.

We have already begun to explore this early Christian confession about Christ in Chapter 8. Here I would like to concentrate on these last verses. Because these two verses – which are actually one sentence split into two verses – are at the very center of what I have been taught all my life to believe about Christ.

First, verse 19 says, "For God was pleased to have all his fullness dwell in him." Here in the first half of this sentence (repeated by Paul in 2:9) is one of the foundational verses that confirm the Church's orthodox understanding of who Jesus Christ really was and is: the Son of God, Second Person of the Trinity.

This sentence continues with verse 20: "and through him to reconcile to himself all things, whether things on earth or things in heaven, by making peace through his blood, shed on the cross."

This second half of this astounding sentence is something I have heard numerous times in sermons and hymns growing up. Jesus' blood, shed on the cross, covers my sins and reconciles me to the Father.

But how have many of us missed one particular part of the message? It's right in front of our eyes! This truth that has been missed is enshrined in one of the earliest confessions or summaries of who Jesus Christ is by the early Church, pre-dating even the Apostle Paul's teachings. It is found in one of the biblical cornerstones of an orthodox Christian understanding of who Jesus Christ is. To be frank, while all of Scripture is God-breathed, while all of it is God's word, we read some texts more than others. And this Colossians text certainly isn't one that the Church has neglected over the years.

It isn't one I had neglected either. Growing up I heard countless sermons on one aspect of its message; that Christ's blood, shed on the cross, reconciles me to the Father. In addition, back in the mid-80s when I became involved in my church's peacemaker group, verse 20's discussion of reconciliation and peace was foundational. As God in Christ took the initiative to make peace with us while we were still his enemies, so too were we to do the same with our enemies.

And yet I applied this reconciliation only to humanity, as did the vast swath of the Church before me, as most still do today.

Maybe precisely because the sentence that makes up verses 19-20 says so much, precisely because it is so deep, precisely because it tells us that Jesus Christ is in fact God, precisely because it proclaims that the blood of Jesus brings about reconciliation – maybe all of this dazzling richness is part of the reason we have missed one of its central messages.

Maybe, to be frank, we have missed this message precisely because we have been focused on our own salvation and our rank in the cosmic scheme of things. (It's all about **us**, right?)

And maybe, just like for most of the contemporaries of the insignificant nobodies who wrote this text, its claims are just too audacious to be believed.

Well, it is time to follow the lead of our spiritual forbearers and be audacious; it is time to let the Holy Spirit give us "eyes to see." It's too important not to.

So what am I saying we have missed? What is hidden in plain sight in two well-known verses? What did God reveal to the wonderful unknown nobodies who wrote this incredulous early Christian confession that Paul quotes precisely to remind the Colossians what they know and have confessed? What do we need to allow the Holy Spirit to give us eyes to see?

   Global Warming and the Risen LORD

It is this: Jesus' blood does not just reconcile me, not just you and me, not just humanity, but in fact Jesus' blood, shed on the cross, reconciles "all things."

To think that a crucified peasant from a hick town in a puny, insignificant part of the Roman Empire could be both the Creator (v. 16) and the Reconciler of all things? This is the glorious foolishness that Colossians 1:15-20 is calling us to confess.

And the scope of Christ's reconciliation is possibly the most intellectually "certifiable" (i.e., crazy) aspect of all. Because nothing exceeds the scope of "all things" (*ta panta* in the Greek), as the phrase "whether on earth or in heaven," the biblical phrase for all creation makes clear. "All things" means exactly what it says, all things.

Verse 20 sees more than the individual Christian standing at the foot of the cross in need of salvation, more than confessing believers huddled outside the gates of Jerusalem at Golgotha. It is audacious in its claims, universal in its scope, bursting the bounds of time and space, limitless in its effects.

I have realized that my vision of the reconciliation brought about through Christ's blood was too small. Christ's reconciliation has no horizon. Verse 20 and the Holy Spirit prompt those of us gathered at the cross, at the feet of our crucified Savior, to look around, to look behind us, to look at all that has gathered there, farther than the eye can see. The whole creation is gathered at the foot of the cross.

When I finally was given eyes to see this profound, glorious truth, it made perfect sense for those who believe. Of course the Creator of all things (v. 16) would also be the Reconciler of all things. Of course the One who declared all of creation "good" at the beginning would not allow it to wallow forever under its curse brought about by our sin. But somehow for years I was too blind to see it. I haven't been the only one.

In my discussions with one senior colleague he kept on stumbling over the idea that other creatures could not intellectually understand the literal preached word nor freely and consciously accept Christ as Savior and

LORD. Thus, Christ's reconciliation could not be actualized in them. As if the Creator would hold their created limitations against them, disqualifying them from receiving the benefits of His grace!

But of course the rest of creation does affirm His Lordship, that He is the Creator. "Who is this? Even the wind and the waves obey him!" exclaim the disciples (Mk 4:41; cf Mt 8:27; Lk 8:25).

And in its own way the rest of creation acknowledged his passion, his suffering on the cross to bring about the reconciliation of all things.

During the crucifixion and at the time of his death, the Gospels picture the rest of creation in symbiotic suffering with its dying LORD. "From the sixth hour until the ninth hour darkness came over all the land" (Mt 27:45; cf Mk 15:33; Lk 23:44). At the moment of his death, "The earth shook and the rocks split" (Mt 27:51b).[450]

Such signs from the rest of creation were in fact the first proclamation of the gospel, for they lead a Gentile unbeliever, one of the Roman centurions, to proclaim who Christ really is. "When the centurion and those with him who were guarding Jesus saw the earthquake and all that had happened, they were terrified, and exclaimed, 'Surely he was the son of God'" (Mt 27:54).

The Risen LORD Himself declares that the good news about His reconciliation is to be shared with the rest of creation. He commands His disciples to "Go into all the world and preach the good news to all creation" (Mk 16:15).

To make sure the Colossians get the connections, the Apostle Paul immediately follows up Colossians 1:20 by proclaiming:

> *Once you were alienated from God and were enemies in your minds*
> *because of your evil behavior. But now he has reconciled you by*
> *Christ's physical body through death to present you holy in his*
> *sight, without blemish and free from accusation – if you continue in*
> *your faith, established and firm, not moved from the hope held out*
> *in the gospel. This is the gospel that you heard and that has been*

proclaimed to every creature under heaven, and of which I, Paul, have become a servant (1:21-23, emphasis added).

As with verses 19-20, these are packed with meaning because they are a summary of the gospel: our own sinful behavior has made us God's enemies; but God has reconciled us by means of Christ's death to present us holy, sinless, and "free from accusation." Yet Christ's reconciliation is not just for us. Significantly, also packed in here is the message that this gospel or good news, of which Paul himself is a servant, has been "proclaimed to every creature" as well.

Because of our sin, the rest of creation was living under a curse (Gen 3:17). While it had not the freedom to fall, it suffered because of our Fall. Paul says that the rest of creation is "groaning as in the pain of childbirth" (Rom 8:22) as it awaits the complete realization of Christ's reconciliation, of Christ's Shalom, of the Kingdom of God, where "the creation itself will be liberated from its bondage to decay and brought into the glorious freedom of the children of God" (Rom 8:21).

Isn't it wonderful that God's glorious vision of freedom is so much bigger than ours? God's reconciling love is even broader and deeper than our little finite minds have comprehended! It's exciting and comforting all at once. Our God is indeed an awesome God.

HEALED FOR RIGHTEOUSNESS, SET FREE TO BEGIN TO LOVE HIM BACK

Whatever challenges we face, we are empowered by Christ's grace. It is important not only for our spiritual growth for us to reflect deeply on what our LORD had done for us, but also for the betterment of the world.

We must understand profoundly in our hearts the cost of his suffering, the depth of his love, to guide and inspire us as we struggle to become more like him in our everyday lives. This includes making the changes necessary to overcome global warming. We must continually remind ourselves that the Risen LORD with the nail-scarred hands walks besides us. By his wounds we have been healed (1 Pet 2:24), healed to be transformed into true images of God as we reflect His glory, as we become spiritually beautiful in reflecting the will of the Father on Earth, as we fulfill the Lord's prayer, as we momentarily and fleetingly bring in His Kingdom when we do His will. The blood of our Savior touches everything we do. There is indeed wonder working power in the precious blood of the Lamb – to forgive, to heal, to transform, to empower for righteousness. No problem we face, no problem the world faces, cannot be transformed in the here and now by the wonder-working power of his blood.

As we reflect on making changes, we may fall into the trap of thinking that small efforts don't matter. Our Risen LORD is big enough to care about the seemingly trite, mundane challenges of our everyday lives. He taught us that those who are faithful in little can then be faithful in much (Lk 16:10; Mt 25:21) because we become more like him. Even these seemingly small, insignificant moments are filled with potential glory, filled with the opportunity for spiritual transformation, filled with the chance to love those we will never know. Nothing our LORD wants us to do is small. One look at the cross should cure us of that.

Not only will becoming more like Him make us happier and more content (because we are becoming what we were made to be), not only will it make us freer and more glorious and more spiritually beautiful, in doing so we will begin to love him back; this is the true story of our lives, the true journey of our hearts.

Thus, our seemingly small decisions as individual Christians have profound spiritual implications. Each one is a way to love our Savior back. They also matter individually in overcoming global warming. As we shall see, this threat will be overcome by billions of individual decisions around the world, such as someone choosing efficient light bulbs, or an engineer using his talent to invent solutions. It may be an investment banker financing a "green" project or a government leader using her power to create new laws. The necessary changes will come down to individuals making decisions.

Christians should be the ones leading this collective wave. Every time we make a change to overcome global warming, we are becoming more like our Risen LORD, the spiritual leader of all efforts to achieve the will of the Father. Purchasing LED lights or combining trips in the car to save energy is a spiritual act of love towards our Savior. To choose to purchase clean, green electricity is a Kingdom act. It is all seen and known and empowered by our Risen LORD, whose nail-scarred hands ever remind us of the price he paid. No righteous act is too small to the one who bore our sins on the cross. By his physical and emotional and spiritual wounds we have been healed; we have been set free from the crippling infirmity of our own sinfulness and powerlessness to begin to love Him back. Our Risen LORD pulls us up to walk again on the path of righteousness for His name's sake.

CONCLUSION

On March 10, 1974, the newspapers were full of stories about the first OPEC oil embargo and Watergate. On this date another small and obscure event occurred, one that even those who witnessed it may fail to remember. Brother Jamie at Jupiter Road Baptist Church in Garland, Texas baptized a 12-year-old boy dressed in hand-me-downs. That is the day and the place when and where I publicly confessed for the first time that Jesus Christ is my Savior and LORD.

Absent that confession I wouldn't be writing this book, I wouldn't have become a leader of Christians overcoming global warming.

On that day I certainly had no idea where that public confession would take me. I only knew I was a sinner who needed a Savior. I still do. I am thankful Jesus taught us, "unless you change and become like little children, you will never enter the kingdom of heaven" (Mt 18:3; cf Mk 10:15; Lk 18:17). That confession is not something I needed to grow out of, it is something I still need to grow into. We who have been baptized are "buried with him through baptism into death in order that, just as Christ was raised from the dead through the glory of the Father, we too may live a new life" (Rom 6:4).

Whether or not I would have become a Christian absent the spiritual foundation provided by my parents and grandparents is unknown. But what I do know is that they did provide that foundation. They were my spiritual guardians. They themselves were nurtured by the Southern Baptist Church, as was I.

In light of this it has been difficult for me to reconcile the fact that those who helped to bring me to Christ also were caught up in the injustice of racism in the South, whether as participants or bystanders.

At a collective level, the admission of failure by the Southern Baptist Convention in their 1995 Resolution on Racial Reconciliation is helpful for those of us in the present so we can learn from the mistakes of these spiritual forbearers. To honor their confession means to strive to make the right choices when it comes to the great moral issues of our day.

As for my grandparents, as I have recounted, there is no conclusive evidence that my grandfather signed the *Statement of Principles* by McComb's white establishment. Even if he did, my grandparents could have done more to begin to travel down the path of righteousness when it came to civil rights (as could have the vast majority of other white Christians at that time, to be fair).

By me owning up to their failures to be proactive in righteousness, as well as my own failures in this regard, I pay tribute to the spiritual seeds they planted in me that led to my confession that Jesus Christ is my Savior and LORD. We both learned from countless sermons what Romans 3:10 proclaimed: "There is no one righteous, not even one."

The juxtaposition of civil rights and global warming has found its way into these pages because of my own spiritual journey. But it is helpful for us to briefly compare these two important problems. Our history with one can offer us guidance and encouragement with the other – as long as we understand the differences and similarities, both of which are instructive.

Racism (prejudice plus power) is a problem that manifests itself at particular times and places like McComb in the 1960s. As a social contagion, racism can be contained and treated and defeated. Global warming is a global problem whose consequences will be felt all around the world for centuries. While its impacts cannot be limited to one area, it too can be overcome.

Whereas in many instances the perpetrators of racism know their victims or see some evidence of the consequences of their acts, with global warming it is not nearly so immediate. Moreover, while racism is in many instances institutionalized, individuals can commit racist acts against other individuals. It can be very personal, face-to-face encounters. As such, there can be a visceral recognition in our own gut that we are doing something wrong.

In contrast, no one person can cause global warming let alone target its impacts to harm a particular individual. It is and will continue to be the result of billions of individual acts that collectively have caused a problem heretofore unforeseen by most. And we will never know the vast majority of those who have been and will be harmed by our collective activities. To us they are silent, unknown, numbers without faces, families without names. Nevertheless, they are people who love their families and friends and hope for a better life. They are people known and loved by our Risen LORD just like we are. Global warming's consequences will be quite personal and devastating to them, but to us they are abstract and hard to imagine – and therefore easy to forget.

Where global warming may in fact be an easier challenge to face than civil rights is in the cultural and social intensity of resistance. My grandparents, for instance, would have had their lives ruined like the Heffner's if they would have tried to make changes on their own. And yet only a few months after the Heffner's were run out of town, change did come when over 600 of McComb's leading white citizens signed the *Statement of Principles*. Even in a seemingly intractable situation literally infused with terrorism, glimmers of righteousness can begin to shine forth if people work together to do what is right.

Both civil rights and global warming are great moral issues of our times. Both are controversial and enmeshed in economics and politics, and in the way we live our lives. They require personal changes in attitudes and behaviors as well as economic and political changes by society. Both, therefore, have been denied and resisted by powerful forces – but also by Christians.

With slavery, white, Bible-believing Christians in the South let false economic conclusions harden into a "way of life" that continued after the Civil War. This form of institutional racism denied African-Americans their civil rights. Perpetrating this injustice helped to prevent them from being all that Christ created them to be: free, beautiful, and glorious. By enslaving and then oppressing others they themselves became enslaved in systemic unrighteousness.

In 1964 it became clear to the 600 white citizens of McComb who signed the *Statement of Principles* that institutional racism was no longer acceptable. Today it is clear that global warming is a profound threat. The United States has been a major contributor. As mentioned in chapter 2, we didn't know. But now we do. We cannot deny the facts any longer.

Christians in the United States, of every race, can't allow the same to thing happen to us with the challenge of global warming that happened to white, Bible-believing Christians in the South in regards to slavery and civil rights. We cannot allow barriers, be they economic or cultural, to prevent us from participating in Christ's reconciling activities.

Let me ask you a question to get right down to it. Would you knowingly commit a racist act today? Even if you messed up and did, would you keep doing it? I think the answer is clearly no. Now, can you knowingly not do your part to help overcome global warming?

This is where believing that Jesus Christ is my personal Savior and LORD is crucial to helping me be able to change my behavior to address whatever major problem I confront. I am overwhelmed both with despair and gratitude by the thought that Jesus would have died just to save me. By all rights I should have died for him, not he for me.

And yet knowing he died for me allows me to begin to explore and to admit the depths of my own unrighteousness, because his love is light

amidst my darkness, strength amidst my weakness, beauty amidst my ugliness. It is upon this foundation that I can see the true, flawed, horrible, beautiful, fragile, tender, savage nature of myself and the world. As terrible as threats like global warming might be I have a Savior who died for me. As I face the truth about myself with the courage Christ's love provides, so too can I face the truth of global warming or any other threat the world or I might be facing. Again to quote the Apostle Paul (because we need to hear this over and over),

> *What, then, shall we say in response to this? If God is for us, who can be against us? ... Who shall separate us from the love of Christ? Shall trouble or hardship or persecution or famine or nakedness or danger or sword? ... I am convinced that neither death nor life, neither angels nor demons, neither the present nor the future, nor any powers, neither height nor depth, nor anything else in all creation, will be able to separate us from the love of God that is in Christ Jesus our Lord.* (Rom 8:31-39).

Christ the Creator and Reconciler is calling you and me into the light, calling us out of our selfish nothingness, creating and sustaining within us freedom, beauty, and glory, calling things within us that are not as though they were (Rom 4:17).

When we were his enemies, when we were helpless and powerless to bring about even a partial reconciliation, Christ took the initiative. He came to us. He even became one of us. And though we rejected him and even crucified him, he reconciled all things by our very act of ultimate betrayal and viciousness. He transformed the ugliest and most brutal act ever perpetrated into the most glorious, the freest act ever known.

Christians can face any truth about ourselves or the world because God the Father loved us so much He sent His Son; because Jesus Christ loved us so much he came to earth and died to save us. We can now work to overcome whatever problems come our way – no matter how big they seem at first. Because Christ is our Savior, we are spiritually equipped to not deny the truth, not to deny major threats, but to face them. The Truth has set us free to face the truth with gloriously luminous "unveiled faces" – **and** unveiled hearts and minds.

But we are not just to face them. With global warming, to image Christ, the perfect reflection of the Triune God, is to take the initiative. To lead. The Risen LORD is in fact already leading the way. To follow is to lead. We are compelled in the true freedom of love to help to transform this challenge (2 Cor 5:14-15), regardless of what others do or don't do, regardless of what others say or don't say, regardless of what others think or don't think, regardless of whether anyone thinks we will succeed.

Global warming has now become part of the love story between ourselves and our Savior and Our Father Who Art in Heaven, whose Kingdom comes to the extent we do His will on Earth as it is in Heaven. The Father and Son have set us free to love Him back, and in response we have laid our hearts at the foot of the cross, where the whole of creation is gathered and reconciled. Global warming works against Christ's reconciliation, against Christ's blood. To say "NO!" to global warming is one of the ways to say "YES!" to our Savior today. It is part of what it means today to follow or image our Risen LORD with unveiled faces, hearts, and minds and become more beautiful, more glorious, more righteous – to become our true selves. Forgiven by Jesus' blood, transformed within by the Creator and Reconciler who calls into being spiritual capabilities where none existed, empowered by Christ's grace, walking with our Risen LORD, *We Shall Overcome* … starting today.

12

WALKING WITH CHRIST OUR LORD –

RESISTANCE & OPPOSITION

It is not the healthy who need a doctor, but the sick. I have not come to call the righteous, but sinners.

Matthew 2:17; cf Luke 5:31-32

Overcoming global warming is at its root a spiritual challenge. How we will meet it as Christians is found in our relation to God, in our loving response to all He has done for us. We will overcome global warming by living the glorious lives He intends for us, by imaging Christ through free obedience to His will. For Christians, playing our part in overcoming global warming comes down to Jesus' Lordship in our lives. It's that simple.

But no one can image Christ or actualize Jesus' Lordship without the empowerment of Christ's grace and the presence of His Spirit in our hearts. That's why the previous two chapters on Christ being our Savior came before the chapters on His Lordship. Because Jesus Christ is our Savior, He can become our LORD. Then His Lordship can be actualized in our lives.

Many Christians describe Jesus Christ as their "Lord and Savior," putting the title "Lord" first. For years I have been very intentional in saying "Savior and LORD," because I know I can't begin to actualize His Lordship without the spiritual empowerment provided by his saving grace and the guidance of His Spirit. Our spiritual reality is this: Savior first, then LORD. As the Apostle Paul says, "I can do all things through Christ who strengthens me" (Phil 4:13, NKJV).

As regards the person of Jesus himself, there was never a time when he was not both. As the angels proclaimed to the shepherds, "For unto you is born this day in the city of David a Saviour, which is Christ the Lord" (Lk 2:11, KJV).

The Bible is quite clear; salvation is contingent upon confessing Jesus Christ as LORD. If you want him to be your Savior, then confess Him as LORD.

The archetypal texts in this regard are from Acts 16 and Romans 10. In Acts 16, the Roman jailer asks Paul and Silas, "'Sirs, what must I do to be saved? They replied, 'Believe in the Lord Jesus, and you will be saved— you and your household'" (Acts 16:30b-31).

"What must I do to be saved?" In Romans 10:9, Paul gives an even more explicit answer: **"If you confess with your mouth, 'Jesus is Lord,' and believe in your heart that God raised him from the dead, you will be saved"** (emphasis added).

Paul goes on to quote Joel 2:32, "Everyone who calls on the name of the Lord will be saved." For Joel, "the Lord" of course meant God. And Joel's understanding is representative of the Old Testament generally. As my former teacher Alister McGrath puts it, "for the Old Testament there is only one who could save, the Lord God of Israel … It is the Lord, and the Lord alone, who will save Israel from their sins."[451]

Thus, for Paul, anyone who confesses "Jesus is Lord" is confessing that Jesus is God. Simply put: Jesus saves us from our sins; Jesus is the LORD; Jesus is God. Another way to view it is that because Jesus is God, then Jesus is LORD, and because Jesus is the LORD God, He can save us from our sins.

He is simultaneously Savior and LORD. As a sinner I need for Him to be my Savior first so that empowered by His saving grace I can let Him be my LORD, too.

LORDSHIP IN GOSPELS

God has been profoundly, profoundly merciful in designing the Gospels so they simultaneously present the reality of Jesus' Lordship from the beginning, as if letting us in on a secret, and yet showing through the experience of the disciples how His Lordship can be misunderstood, even by his followers.

The Gospels also demonstrate, by showing the opposition to Jesus, that the reality of His Lordship can be denied and resisted by the religious authority figures of his society (e.g., the Chief Priests and the Pharisees) who should have "eyes to see." Just so we don't miss the fact that the latter's examples apply to us as well, let me draw the parallel explicitly. As an ordained minister I am a religious authority figure. Each of you is called to express spiritual leadership in some way. Their blindness is a warning to us. We are not immune to such spiritual failure.

God has been merciful in the way the Gospels present the reality of Jesus' Lordship because He knows that our misunderstandings, denials, and resistance continue even after we are let in on the secret, so to speak, that Jesus is in fact the LORD, even after we have professed Him to be so. God knows that all Christians fail to allow Jesus to really, completely, truly be the LORD of our lives. We continuously need the mercy provided by our pilgrimage with the Gospels simply to have us stammer slowly forward in our journey with the Risen LORD. This is especially true for challenges like global warming that are well beyond our everyday experience and thus can go "unseen" or easily forgotten or flat out denied in our walk with the LORD.

And so immersing ourselves in the Gospels and the rest of the New Testament is essential for the necessary, never-ending task of recognizing moment by moment what is probably the most ancient of Christian confessions.[452] It's one Paul says no one can truly utter except if inspired by the Holy Spirit (1 Cor 12:3). It is the simple but profound confession that once made ever after defines our lives: **"Jesus is LORD."** As one prominent New Testament scholar put it: "This brief formula expresses the whole faith of the early church."[453]

But this confession cannot be a true one if it is merely an intellectual assent. We cannot simply recognize His Lordship; we must begin to realize it. One cannot truly say, "Jesus is LORD" if there is not a corresponding attempt to actualize His Lordship. Otherwise our confession is only empty words, which do not count with the LORD (Lk 19:46). And Jesus cannot simply be LORD of only part of our lives. That defies the very confession itself. He must be the LORD of the whole of our existence.

Our lives, rather than our lips, must be the ultimate confessor of Jesus' Lordship.

And so in our journey with the Risen LORD, especially with "unseen" problems like global warming that require daily obedience and yet don't capture our conscious attention very frequently, our constant traveling companion must be the New Testament, particularly the Gospels. For it is here that we meet the LORD Jesus Christ. If, as Christians, we are to have any hope of playing our part in overcoming global warming, it is here that we must center ourselves; and we must do so firm in the knowledge, conviction, and experience that he is our Savior, that his grace empowers us to live out His Lordship. We must discover and confess Him ever anew.

All spiritual challenges require both that Jesus is in actuality our Savior **and** our LORD, but this is especially true for such spiritual challenges like global warming.

Do you want to know what the spiritual reality of global warming is for Christians? It is this. The Risen LORD has proclaimed Lordship over our contributions to the problems and the solutions to global warming even when we don't recognize His Lordship in these areas of our lives. Like the disciples in the Gospels, we don't completely understand nor accept His Lordship, and to that extent we fail to live it out. But just as Jesus doesn't give up on the disciples, neither will He give up on us.

And so, to inform and inspire our own pilgrimage with the Risen LORD, let us once again experience in this chapter and the next the pilgrimage to confession of his Lordship as found in the Gospels.

LORDSHIP DISPLAYED IN LIFE AND MINISTRY

Early on, the Gospels not only let us in on the secret of Jesus' true identity as the Son of God and therefore the LORD of our lives, but they also display His authority from the beginning. I've chosen to briefly highlight four dimensions of His life and ministry to illustrate this: (1) His teaching; (2) His Lordship over the Sabbath; (3) His casting out of evil spirits; and (4) His forgiveness of sins.

First, His teaching naturally exudes **authority** right from the start. As described in the first chapter of Mark, "The people were amazed at his teaching, because he taught them as one who had authority, not as the teachers of the law" (Mk 1:22; cf Lk 4:32).

At the end of the Sermon on the Mount, almost the exact same language is used (Mt 7:28-29) to state what should be obvious, given what he is saying and how he says it. It was the custom of the time to quote a revered Rabbi to justify one's moral position. But six times in the Sermon on the Mount, Jesus explicitly challenges past teaching with his own authority: "You have heard it said ... But I say to you." This was in fact profoundly, incredibly bold to say "but I say to you." Yet Christ can do so because of who He is.

Second, his boldness in interpreting the Law, in understanding God's will, is captured quite well in his proclaiming Himself **Lord of the Sabbath** as he challenges the authority of the scribes and Pharisees. In doing so, He earns their enmity and for the first time they begin to plot how they can kill Him.

The account in Matthew 12 captures this well (Mt 12:1-14; cf Mk 2:23-3:6 and Lk 6:1-11). The rigid interpretations of the Pharisees had blinded them to legitimate human need. Jesus quotes Hosea 6:6, "I desire mercy, not sacrifice," to help them see that if they truly understood the heart of God they would not be criticizing Jesus' acts of mercy on the Sabbath. Instead, they move even further in the wrong direction by trying to trap Jesus in such an act of mercy. A man with withered hand is present in the synagogue, and so the Pharisees actually ask Jesus, "Is it lawful to heal on the Sabbath?" Jesus replies, "If any of you has a sheep and it falls into a pit on the Sabbath, will you not take hold of it and lift it out? How much more valuable is a man than a sheep! Therefore it is lawful to do good on the Sabbath" (vv. 10-12). Jesus healed the man, and "the Pharisees went out and plotted how they might kill Jesus" (v. 14).

The third example is possibly the supreme demonstration of the exercise of His spiritual **power** and authority as the Son of God and LORD. It is his **casting out of evil spirits** and his ability to give this authority to His followers to do so. (On the latter see Mk 3:14-15; Mt 10:1; Lk 9:1.)

Mark has his first example of this in a dramatic scene in the first chapter of his Gospel. Jesus was in Capernaum teaching in the synagogue on the Sabbath, and immediately after the people exclaimed he taught with authority "a man in their synagogue who was possessed by an evil spirit cried out, 'What do you want with us, Jesus of Nazareth? Have you come to destroy us? I know who you are – the Holy One of God!'" (Mk 1:23-24). Jesus commands the evil spirit to leave the man and the people are amazed and exclaim: "'What is this? A new teaching—and with authority! He even gives orders to evil spirits and they obey him'" (v. 27; cf Lk 4:36).

The people recognize that Jesus has a new type of authority both in His teaching and in His power over demons, yet while the evil spirit knows Jesus is "the Holy One of God" the people don't yet fully comprehend who He is.

The final dimension of His authority, His ability to **forgive sins,** is perhaps the supreme example of his spiritual authority over our lives. Indeed, it confirms He is God. As is a common pattern, it is Jesus' opposition – this time human rather than demonic – that recognize (but reject) the implications of his authority even as the disciples remain clueless.

At the beginning of the second chapter of Mark (also told in Lk 5:17-26 and Mt 9:1-8; see also Lk 7:36-50) is the story of the paralytic. His friends cannot get him into the house where Jesus is due to the crowd, so he is lowered through the roof and set before Jesus to be healed. Because of their faith, Jesus says to the paralytic, "Son, your sins are forgiven" (2:5). Immediately the teachers of the law who were present began thinking to themselves, "Why does this fellow talk like that? He's blaspheming! Who can forgive sins but God alone?" (2:7).

Quite right – only the LORD God has authority to forgive sins.

From the beginning of his ministry, Jesus' extraordinary authority is displayed but only partially recognized by the people and His followers. However, it is fully recognized by the opposition.

AUTHORITY CHALLENGED FROM WITHOUT

Precisely because His authority is recognized, it is also immediately and consistently challenged. The main challengers are those who themselves feel threatened by Jesus' authority: religious leaders. It's pretty straightforward as to why. Jesus is on their turf (e.g., the synagogues for the scribes and Pharisees, the Temple for the Chief Priests) talking about their stuff and doing things that go against what they teach. To them he's just some guy from Nazareth, one in a long line of itinerant preachers and wonder workers. Who is he to challenge their authority?

Lest we think we have nothing in common with those opposing Jesus – think again. Precisely because Jesus is LORD He is on our "turf" (i.e., our daily lives) talking about our "stuff" (i.e., our stuff) and asking us to do things that go against our sinful nature. These stories we read in the Gospels are warnings and examples of what we could become even though we know who He really is. Our own resistance can become so strong that we put ourselves outside of Jesus' Lordship and outside the circle of those living out His Lordship.

During His ministry, the opposition often asked Jesus for a miracle or sign from heaven. Both Mark's and John's Gospels report one such occasion where Jesus had just fed thousands of people (4,000 men in Mark and 5,000 men in John, not including women and children). And in John's account, Jesus had also just walked on the water (Mk 8:11-12; Jn 6:30).

Such signs are gifts of compassion that naturally flow out of Jesus within a context of need, as do all of his miracles. They are gifts of compassion not only in the acts themselves (e.g., feeding people) but also for us who in our weakness lack faith and yet yearn for Him to be our Savior and LORD. They are to help us have eyes to see.

Now if anything would be miraculous signs, it would be feeding thousands and walking on the water. But apparently Jesus' opponents were not satisfied.

To be fair, they may not have actually seen these signs. Perhaps they only heard about them from others. But isn't that the point for us the reader? We too did not actually see these miraculous signs being performed. We are being told "secondhand" through the witness of the Gospels. But if we have come to believe that the Gospels are true, if we trust them, then such demands for a sign after we know that thousands have been fed and Jesus has walked on the water appear ridiculous. "He just walked on the water for God's sake!"

John's account captures the situation well: "What miraculous sign then will you give that we may see it and believe you? What will you do?" (Jn 6:30). But as Jesus says several verses later: "**You have seen me and still you do not believe**" (6:36, emphasis added).

Even when some of his opponents believed He had performed miraculous signs like casting out demons, they give what in reality is a blasphemous interpretation: "He is possessed by Beelzebub (i.e., Satan)! By the prince of demons he is driving out demons" (Mk 3:22; cf Mt 12:24 and Lk 11:15).

So for these folks, Jesus could never do enough signs. The apex of their unbelief is when they see him on the cross. "'He saved others,' they said, 'but he can't save himself! Let this Christ, this King of Israel, **come down now from the cross, that we may see and believe**'" (Mk 15:31-32, emphasis added; cf Mt 27:41-43 and Lk 23:35). As Jesus pointed out in the Gospel of John, they see Him and yet do not believe. Here they actually affirm that Jesus saves others, but essentially they want a bigger magic trick to be performed, something akin to what Satan tempted Jesus with before His ministry began when he told Jesus to throw Himself off the Temple – a big, gaudy public display of their understanding of supernatural power.

The thing is, they still would not believe because they had not surrendered to the LORD. They still want to be in charge. They still think they are right. They want Him to become their type of messiah. They cannot accept a suffering Savior, a servant Savior, as their LORD. They are not open to the reality of Who He is, a reality that was never more real, never more clear than on the cross. If Jesus were to have come down from the cross no doubt they would have again concluded he was in league with Satan. (Ironically, this time they would have been right.)

Again, we must be honest with ourselves and recognize that we too do not completely understand nor accept the fullness of His Lordship. If Jesus' opposition is blind to who He really is, so that Jesus can say to them, **"'You have seen me and still you do not believe'"** (Jn 6:36, emphasis added), then could not the Risen LORD say to us," You have confessed me and still you do not do as I say."

AUTHORITY RESISTED FROM WITHIN

Similar to Jesus' opposition, His disciples also don't understand his type of messiahship. What's the difference between the disciples and the religious leaders? The disciples have put themselves under His Lordship, however their understanding is incomplete and fragmentary. Thus, they are willing to be taught and to have their eyes opened. Yet it is only after Jesus' resurrection that they begin to truly see.

DON'T UNDERSTAND HIS TEACHINGS

During Jesus' ministry, there are many times the disciples don't understand his teachings. After Jesus proclaims the parable of the soils, the disciples come up privately and ask for an explanation. Jesus exclaims: "Don't you understand this parable? How then will you understand any parable?" (Mk 4:13).

When Jesus' disciples question him about His teaching that what was most important were the qualities within a person as opposed to external observances like religious rules about food, He replied, "Are you so dull?" (Mk 7:18).

The disciples fail to understand how to enter into His Kingdom (Col 1:13). When they try to keep people from bringing their children to Him to be blessed, He became "indignant" and said, "Let the little children come to me, and do not hinder them, for the kingdom of God belongs to such as these. I tell you the truth, anyone who will not receive the kingdom of God like a little child will never enter it" (Mk 10:14-15). It is such childlike trust and obedience in Jesus' Lordship that are the keys to Kingdom living.

HAVE DOUBTS AND LITTLE FAITH

The disciples fail to grasp Jesus' true identity, and because of this have fear and doubt. When he calms a storm, they are "terrified." Jesus says to them, "'Why are you so afraid? Do you still have no faith?' They…asked each other, 'Who is this? Even the wind and the waves obey him!'" (Mk 4:40-41; cf Lk 8:25).

When Jesus metaphorically warns the disciples to "'Watch out for the yeast of the Pharisees." they think He's referring to the fact that they didn't bring literal bread. Jesus reminds them that He has fed thousands and says, "Why are you talking about having no bread? Do you still not see or understand? Are your hearts hardened? Do you have eyes but fail to see, and ears but fail to hear?" (Mk 8:17-18; cf Mk 6:51-52). They don't yet understand who He is.

In the Gospel of John Jesus plainly states who He is, and yet Philip still does not understand:

> "I am the way and the truth and the life. No one comes to the Father except through me. If you really knew me, you would know my Father as well. From now on, you do know him and have seen him." Philip said, "Lord, show us the Father and that will be enough for us." Jesus answered: **"Don't you know me, Philip, even after I have been among you such a long time?** Anyone who has seen me has seen the Father. How can you say, 'Show us the Father'? Don't you believe that I am in the Father, and that the Father is in me? The words I say to you are not just my own. Rather, it is the Father, living in me, who is doing his work" (Jn 14:6-10, emphasis added).

After the crucifixion and before they had encountered the Risen LORD, the disciples doubted the reports of others about His resurrection because "their words seemed to them like nonsense" (Lk 24:11). Mark reports, "Later Jesus appeared to the Eleven as they were eating; he rebuked them for their lack of faith and their stubborn refusal to believe those who had seen him after he had risen" (Mk 16:14). In Luke's account the LORD is gentler. When he appears among them He begins by saying "'Peace be with you.'" The disciples "were startled and frightened, thinking they saw a ghost." Jesus says to them, "'Why are you troubled, and why do doubts rise in your minds? Look at my hands and my feet. It is I myself! Touch me and see; a ghost does not have flesh and bones, as you see I have.'" To offer even more reassurance, the Risen LORD eats a piece of fish in front of them (Lk 24:36-42).

Astoundingly, at the end of his Gospel, Matthew recounts that even at His final post-resurrection appearance with the disciples, "some doubted."

FAIL TO UNDERSTAND MESSIAHSHIP AND MISSION

But their most consistent fears are associated with their biggest failure of understanding, indeed the biggest failure of their discipleship – the failure to grasp the true nature of His messiahship and mission. As Jesus comes nearer to Jerusalem, nearer to danger, the disciples become fearful. Why is he walking into danger with nothing but his good intentions? Why are we not rallying the troops? Is he simply going to call down fire from heaven?

It appears that the disciples had a fairly standard understanding for the time of what type of messiah Jesus was going to be: a political messiah who would overthrow the Romans, through force or supernatural intervention or both, and establish his own political kingdom.

A crucifixion of Jesus by the very Romans the messiah was to overthrow is beyond their comprehension. Equally incomprehensible would be the idea that one person – messiah or not – would be raised from the dead before the rest of us.

Repeatedly Jesus tries to break through to them (Mt 16:21-26; 17:22; 20:18-19; Mk 8:31-33; 9:31-32; 10:32-34; Lk 9:18-27; 9:44-45; 18:31-34). He tells them exactly what is going to happen:

(1)   He will be betrayed;

(2)   He will be arrested by the Chief Priests and condemned to death by the Sanhedrin, who will hand him over to the Romans;

(3)   the Romans will mock Him, spit on Him, flog Him, and crucify Him;

(4)   three days later He will be raised from the dead.

The first time Jesus tells His disciples about His impending passion and resurrection is right after Peter, inspired by God the Father, proclaims Jesus to be the messiah. It produces the sharpest exchange recorded in the Gospels between Jesus and one of His apostles. The reason? Just as one of them says for the first time that Jesus is the messiah He shatters their dreams of messianic glory.

Jesus has asked them who others are saying He is, and they provide various answers: John the Baptist come back to life, Elijah, Jeremiah. Then He asks them directly,

But who do you say that I am? (Mk 8:29).

This question is one of the climactic moments of the Gospels. It is in fact a question the Risen LORD poses to every single person on Earth.

The disciples have left their families, left their former way of life, to follow Jesus. "Could he be the One?" they have asked themselves. They have already risked and invested so much. And now Jesus lays it on the line. He asks them to speak aloud their fondest dream.

It is Peter, of course, who speaks up: "You are the Christ [i.e., messiah], the Son of the living God" (Mt 16:16).

Jesus does not merely confirm that Peter is right. He says that God the Father, who Jesus now says is actually His real Father, has revealed it to Peter. "Blessed are you, Simon son of Jonah, for this was not revealed to you by man, but by my Father in heaven" (Mt 16:17).

For that brief moment the disciples could have been thinking, "Our fondest dream is actually God's dream too! Isn't God wonderful! His dream matches ours! What a great God we have! He has sent His Son to

     Global Warming and the Risen LORD

defeat the Romans, and I'm going to be with him in our glorious, victorious battles and become an important ruler in his kingdom. They will tell tales of our adventures for years to come" (cf Mk 10:35-40 and Mt 20:20-23).

But it is this precise moment when their fondest dreams appear to even have divine sanction that Jesus first reveals to them the true nature of His messiahship and mission. Eventually they will have victory and glory beyond their wildest dreams. But the Father's will is not going to be achieved through military might; rather, it will be achieved through the opposite. The ultimate battle will be fought, but not with weapons of war. It will be fought in the heart and soul of Jesus as He accepts the Father's will in Gethsemane and defeats sin and Satan on the cross.

Jesus' prediction and description of His death is far too much for Peter to comprehend. He has just been told that, inspired by God, he has revealed Jesus to be the messiah, and Jesus has blessed him, and renamed him "Peter" (rock). Jesus has even said He will give Peter "the keys to the kingdom of heaven" (Mt 16:19). This is the most exalted moment in Peter's life up until that point, and the Son of God Himself has revealed his purpose in life to him.

But then all of a sudden Jesus lays out a nightmare scenario for Himself. He will be betrayed, unjustly tried, mocked, spit upon, beaten, flogged, and crucified by the very Romans he is supposed to defeat.

And so we can understand why Peter, a passionate, heartfelt man of action, does something quite astounding. He actually rebukes Jesus (Mt 16:22; Mk 8:32), the One he has just confessed through a revelation from the Father to be the Son of God. Peter rebukes the Son of God.

Just when Jesus finally reveals His true mission, something that sounds crazy, His right-hand man rebukes Him and rejects His mission. And so we can begin to understand Jesus' strong reaction, his harshest rebuke of an apostle recorded in the Gospels. He says to Peter, "Get behind me, Satan! You are a stumbling block to me; you do not have in mind the things of God, but the things of men" (Mt 16:23; cf Mk 8:33).

Peter goes from the highest heights to the lowest lows. He got Jesus' titles right, but he had no earthly idea what those titles actually meant. He got the mission profoundly wrong.

I identify with Peter. There is no way I would have been able to grasp Jesus' mission. Heck, I still struggle with it today. To be honest, I hope I would have had the courage to say what I was thinking to the LORD like Peter did so I could be shown the error of my ways. But I don't see myself understanding what Jesus was saying. After all, why would anyone who loves Him want Him to go through His passion, something hopefully we wouldn't wish on our worst enemy? And so in my own pride and spiritual ignorance I can readily see myself saying to Jesus like Peter did, "Never, Lord! … This shall never happen to you!" (Mt 16:22).

The question I would be asking myself if I were in Peter's shoes would be "Do I love him enough to die for him?" This would be my gut-check question, which I would assume would be the ultimate question, in other words, God's question. Later during the last supper all four Gospels have Peter stating emphatically that he would die for Jesus (Mt 26:35; Mk 14:31; Lk 22:33; Jn 13:37).

But the real question is "Do I love him enough to let him die for me?" For the disciples this then follows, "Do I love him enough to let my messianic vainglory dreams of power and ambition be nailed to his cross? Do I love him enough to let him be humiliated by our enemies and die a cruel death as an abject failure?"

Of course the disciples never grasp the true nature of His mission until they meet the Risen LORD. But that is not because Jesus didn't try to help them understand. Right after Jesus rebuked Peter he told the disciples and the crowd, "If anyone would come after me, he must deny himself and take up his cross and follow me. For whoever wants to save his life will lose it, but whoever loses his life for me and for the gospel will save it" (Mk 8:34-35; cf. Mt 16:24-25; Lk 9:23-24).

But, again, such denial of self doesn't mean dying while protecting Jesus on a battlefield. That is essentially what Peter tried to do in the garden of Gethsemane when he drew a sword and attacked. The Gospel of John tells us that once again Jesus responded strongly to Peter. It states, "Jesus commanded Peter, 'Put your sword away! Shall I not drink the cup the Father has given me?'" (Jn 18:11; cf Mt 26:50-54 and Lk 22:49-51). Peter still doesn't understand, and once again his actions even work against Jesus' true mission. Peter was ready to die for Jesus, but not to let Him die. For the disciples, denying oneself means literally following the One who will in fact deny Himself and take up His cross, not trying to prevent Him from doing so.

The messiah's true mission is one of sacrificial service.

In Jesus' day I'm sure that when people thought about the messiah they thought of him as "the greatest." He would be the greatest warrior, the greatest king, the greatest man alive. And so it is both ironic and fitting that in three Gospel accounts where Jesus predicts His death, what follows is an argument amongst the disciples about who is the greatest. These scenes once again demonstrate the complete cluelessness of the disciples – and probably their need to be in denial – about who He is and His mission. For our benefit, they are wonderfully contrasted with Jesus' teachings about what true greatness actually is. The Messiah really is the greatest, but being the greatest is the opposite of what they think it is.

In Mark's account, Jesus succinctly states, "If anyone wants to be first, he must be the very last, and the servant of all" (Mk 9:35).

In Luke's first instance of the disciples arguing about who is the greatest, Jesus took a little child and stood him beside Himself and said, "He who is least among you all—he is the greatest" (Lk 9:48).

To lose one's life is to save it. To be first one must be last. The least is the greatest.

It is the third instance of the pairing of Jesus' prediction of His death and the disciples' argument over who is the greatest that is the most dramatic, centered as it is in Luke's recounting of the last supper. It happens immediately after the institution of the Eucharist, the only ritual instituted by Jesus Himself. In other words, it comes in one of the most crucial settings in the Gospels to highlight its importance.

In response to the disciples' arguments, Jesus succinctly captures the seemingly contradictory nature of true greatness, of true Lordship:

> *The kings of the Gentiles lord it over them; and those who exercise*
> *authority over them call themselves Benefactors. But you are not to*

*be like that. Instead, the greatest among you should be like the youngest, and the one who rules like the one who serves. For who is greater, the one who is at the table or the one who serves? Is it not the one who is at the table? But **I am among you as one who serves**.* (Lk 22:25-27, emphasis added).

What is true Lordship? What is true greatness? Jesus turns the common understandings on their head. True Lordship and true greatness are found in sacrificial service.

Caravaggio, *Taking of Christ*, public domain

DENIAL AND DESERTION

Given how they completely misunderstand Jesus' messiahship and mission, it is understandable that when Jesus is arrested and led away, the disciples desert him and flee.

Jesus predicts their failure, of course. But He also gives them hope, a hope they don't yet grasp. "This very night you will all fall away on

Global Warming and the Risen LORD

account of me, for it is written: 'I will strike the shepherd, and the sheep of the flock will be scattered.' But after I have risen, I will go ahead of you into Galilee" (Mt 26:31-32; cf Mk 14:27).

Like soldiers on a field of battle who have lost their leader, the courage of the disciples fails when Jesus prevents them from fighting with violence and He is subsequently led away. They are left completely desolate and disoriented.

It is in Peter's story, once again, that God wants to capture our imaginations. All four Gospels recount Jesus' prediction of Peter's denials and the denials themselves. Forty-six verses are devoted to these vivid scenes. The only accounts to rival this story are the last supper, Jesus in the Garden of Gethsemane, the crucifixion and the resurrection – all where Jesus is the central figure. For the Gospels to focus this much attention on someone other than Jesus in the passion narratives is astounding and drives home the story's importance. Few biblical scenes match them for poignancy and drama. As such, this story is one of the most unforgettable in the Bible. And for all of this we should be truly grateful.

Just as when God was able to use Peter's boldness to reveal who Jesus actually was (even as Peter did not really understand his own confession), and then that same boldness had Peter rebuking the LORD and receiving a stinging rebuke in return, so too does Peter go from adamant protestations of ultimate loyalty, even unto death, to denials when challenged by a servant girl, resulting in heart-wrenching failure and despair.

And so our fellow disciple Peter is our true companion in personal failure and redemption. That the Gospels reveal the complete failure of the leader of the followers is an act of profound mercy to us, his fellow followers of Jesus.

After Jesus has told all the disciples that they will desert Him, Peter declares himself to be exceptional, indeed superior, in his courageous devotion: "Even if all fall away on account of you, I never will" (Mt 26:33; cf Mk 14:29).

In an act of supreme compassion, even as He is about to sweat drops like blood in the Garden, Jesus tells him, "This very night, before the rooster crows, you will disown me three times" (Mt 26:34; cf Mk 14:30).

He then adds "But I have prayed for you, Simon, that your faith may not fail. And when you have turned back, strengthen your brothers" (Lk 22:32).

Peter, full of love and full of himself, replies with a bold assertion recorded variously in all four Gospels: "**Even if I have to die with you, I will never disown you**" (Mt 26:35, emphasis added; cf Mk 14:31, Lk 22:33 and Jn 13:37). And yet only a few hours later he will do exactly that – not once, not twice, but three times.

Here is Matthew's version of Peter's three denials of his Lord:

Now Peter was sitting out in the courtyard, and a servant girl came to him. "You also were with Jesus of Galilee," she said. But he denied it before them all. "I don't know what you're talking about," he said. Then he went out to the gateway, where another girl saw him and said to the people there, "This fellow was with Jesus of Nazareth." He denied it again, with an oath: "I don't know the man!" After a little while, those standing there went up to Peter and said, "Surely you are one of them, for your accent gives you away." Then he began to call down curses on himself and he swore to them, "I don't know the man!" Immediately a rooster crowed. Then Peter remembered the word Jesus had spoken: "Before the rooster crows, you will disown me three times." And he went outside and wept bitterly (Mt 26:69-75).

And here are the three denials one after the other:

"He **denied it before them all**. 'I don't know what you're talking about.'"

"He denied it again, **with an oath**: 'I don't know the man.'"

"Then he **began to call down curses on himself and he swore to them**, 'I don't know the man!' Immediately a rooster crowed."

As can be seen, each denial becomes more intense: first, a straight denial; then, backed with an oath; finally, a denial accompanied by curses upon himself. We don't know exactly what those curses were, but it could have been easily something to the effect of "May God strike me dead if I am not telling the truth. I don't know the man!"

It is at this moment that the rooster crows. It is at this moment that Luke tells us, "The Lord turned and looked straight at Peter" (Lk 22:61). It is at this moment, amidst the disorientation of Jesus' allowing himself inexplicably to be captured and the fight-or-flight adrenalin rush flooding Peter's brain concentrating him single-mindedly on his own survival, that the reality of his greatest failure pierces his heart.

When everything he believed about himself and whom he thought Jesus was supposed to be came crashing down, proud and courageous Peter, who was truly ready to die in a violent struggle to protect his Lord, but who could not fathom allowing his Lord to die a humiliating death on a cross for him, "went outside and wept bitterly" (Mt 26:75 and Lk 22:62; cf Mk 14:72).

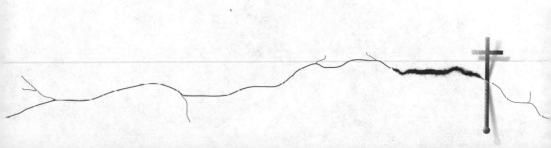

13

WALKING WITH
CHRIST OUR LORD –
STOP DOUBTING AND BELIEVE

Thomas said to him, "My Lord and my God!" John 20:28

Always remember, I was the one who doubted. I was the one who did not see who He really was until I saw His wounds. I had not understood what the prophet Isaiah had said, "By his wounds we are healed." When he healed needy sinners like me, I did not see. When he fed five thousand with a few loaves, I did not understand. When he taught, "Blessed are the poor, for theirs is the Kingdom of God," I thought they were inspired words, but the full reality of who He was eluded me. We had called him "Lord" as a sign of respect. But it was only after seeing the Risen LORD that I was able to profess: "My LORD and my God." And He said to me, "Have you believed because you have seen me? Blessed are those who have not seen and yet have come to believe." I did not see until I saw His wounds.

an Imaginative rendering of a message from
Thomas the Apostle to all future believers

In our pilgrimage to confessing, "Jesus is LORD," the stories of the disciples' failures within the Gospels are merciful mirrors into our own minds and hearts and souls. We fail to grasp His teachings; we doubt that the LORD will save us from the storms of life; we still don't completely understand or accept the fullness of His Lordship and mission; we even deny and desert Him as He tackles the myriad threats that impact "the least of these" with whom He has identified Himself, including, as we have seen, the threat of global warming.

But as we journey with the apostles through the Gospels, our God – the One who inconceivably resurrects a crucified messiah – is about to raise up within us a confession that will set us upon the right path. It is a moment of rare clarity when human perception meets spiritual reality.

And when this light dawns, when we allow the truth of who He is to establish a foothold in our hearts, we can allow Him to transform our lives. That truth was able to dawn in the hearts and minds of the apostles when they encountered the Risen LORD. Our pilgrimage to confession leads to this encounter.

For us to see the Truth, for us to then do the truth, we too must encounter the Risen LORD.

It is the story of the Apostle Thomas in the New Testament that most dramatically captures the utterly transformative spiritual breakthrough necessary for us to become whom God created us to be.

Again, God is quite merciful to have the story of Thomas in His word, whom we have come to know as "doubting Thomas." Thomas would have been quite at home in the twentieth century and the first part of the twenty-first as we are constrained by a culture shaped by thinking that says, "Prove it to me with scientific findings that can be repeated and observed and checked by others."

Thomas essentially asks for this and the Risen LORD patiently and compassionately provides it.

As recounted in the Gospel of John, Thomas is not with the other apostles when Jesus first appears to them, and this sets the stage for his dramatic encounter with the Risen LORD.

For us to journey with the disciples, we must try to imagine their situation after Jesus' crucifixion. Their dreams of accompanying Jesus into messianic glory were crucified with Him on the cross. After swearing they would never abandon Him, they do just that. They are in hiding, afraid for their lives. Jesus had told them repeatedly that He would be raised on the third day, but this is so far out of their experience and their understanding of how the resurrection would work that they cannot grasp it. It is literally incomprehensible to them.

They are leaderless and without hope. They have come face to face with some ugly truths about themselves – their cowardice, faithlessness, and disloyalty. They have been reduced by fear to the most basic of instincts: survival.

And so when the first reports from Mary Magdalene and the other women come in on the morning of Jesus' resurrection, the male disciples don't believe them. It's as if they view such fanciful reports as an old wives tale. As Luke puts it, "But they did not believe the women, because their words seemed to them like nonsense" (Lk 24:11; cf Mk 16:11).

In effect, the disciples were saying to Mary Magdalene, someone from whom Jesus had cast out seven demons (Mk 16:9), "Get real!"

His crucifixion shattered their messianic dreams of glory. His appearance as the Risen LORD will overcome their inability to perceive the Truth. They are about to experience a mind-altering spiritual transformation that will forever change them and launch them into lives of bold action and sacrificial service for their LORD. Without such transformation leading to service and action, Christianity would not exist and you would not be reading this book.

It is Sunday evening, the evening of the resurrection. That morning, Mary first proclaimed to them the Good News that God had raised His Son. They are together in a locked room, hiding from the Jewish authorities (Jn 20:19). Suddenly Jesus appears among them. Luke reports that "They were startled and frightened, thinking they saw a ghost" (Lk 24:37).

Jesus then says to them, "Why are you troubled, and why do doubts rise in your minds? Look at my hands and my feet. It is I myself! Touch me and see; a ghost does not have flesh and bones, as you see I have" (Lk 24:38-39).

Even as He appears before them, they still have doubts.

To comfort them, John reports that the first words the Risen LORD says are "Peace be with you!" Jesus then showed them His hands and side so that they could see that it was He. But Thomas was not with them. When his fellow apostles joyfully report they had seen the Risen LORD, Thomas rejects their accounts just as they had dismissed Mary's.

"Unless I see the nail marks in his hands and put my finger where the nails were, and put my hand into his side, I will not believe it" (Jn 20:25).

The following Sunday, the Risen LORD once again appears to the apostles in a locked room. Jesus comes to Thomas and says, "Put your finger here; see my hands. Reach out your hand and put it into my side. **Stop doubting and believe**." (Jn 20:27, emphasis added).

It is shocking to think someone would actually put his finger into the wound of our Risen LORD, as Caravaggio depicts in his painting, *The Incredulity of St Thomas*. John's account never claims Thomas does so, although the Risen LORD mercifully invites him to. But it matches the gritty doubt and brash moxie of Thomas to demand that he do so before he would believe. It certainly matches our spiritually crass age in the early twenty-first century, where many would expect to see some type of amateur video on YouTube of this holy and profound moment depicting something like Caravaggio's crass painting.

I cannot imagine that Thomas, once the Risen LORD was in his presence showing him His wounds, would put his finger into the wounds of His hands or put his hand into the wound in His side, as Thomas said he would need to do to believe.

No, I think Thomas' first encounter with his Risen LORD – seeing his crucified messiah in His gloriously resurrected body was so revelatory, so transformative, that it shattered all the mental, emotional, and spiritual barriers that kept doubting Thomas from believing.

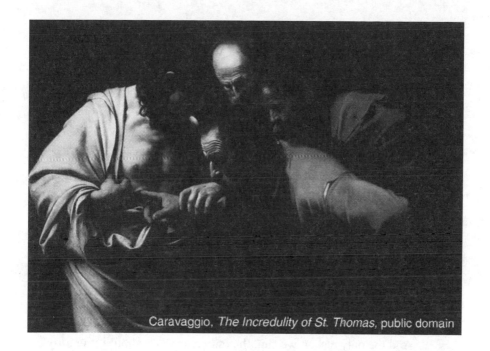

Caravaggio, *The Incredulity of St. Thomas,* public domain

This is the moment that the Gospel of John has been building to. Indeed, it is this type of moment that God has been patiently working towards in all of our lives: to be transformed by the One who Himself was transformed; to be transformed by the One who did the transforming of He who has been raised from the dead.

With the Risen LORD before him, the same man he had followed for years but now resurrected, Thomas utters the confession that for all who confess it is the start of our true lives, the first moment we chose to begin to become our true selves.

Face to face with the Risen LORD, seeing the wounds He bore for us, Thomas confesses,

My Lord and my God! (Jn 20:28)

Here we have finally come to it, the belief that must change our lives forever if we truly believe it: the Risen LORD is God; the Risen LORD is the One to whom we owe total, absolute, uncompromising obedience.

Now we see that to call Him "Lord" is not simply some respectful manner of speech, a mere social convention, the equivalent of "Your Honor," or "sir." It is not a way of bestowing upon him a title marking him a wise teacher whose teachings could later be proven wrong. Nor is this is a mere recitation of a fact, as if the acceptance of its truth did not necessarily have a corresponding claim on our lives.

Rather, this is a life-changing *confession*.

To truly confess this most ancient of Christian confessions, something Christians have done for more than 2,000 years, that "Jesus is **LORD**," is freely to become the servant of His will. His will *freely* becomes ours, and in such moments we become what we were created to be: glorious, spiritually beautiful images or reflections of God.

And the only reason we can, the only reason we should, **the only reason we must make such a confession** about this Jesus of Nazareth – an illegitimately born carpenter's son from a dusty hickville within the Roman Empire; a failed, crucified messiah whose own followers deserted him – **is that He is in fact God**. Not a "son" of God in some adoptive sense but God Himself. He is LORD above all lords because He is God. Anyone or anything else that claims our true obedience is a pretender to the throne – and ultimately a tool of Satan.

And so because doubting Thomas sees the Risen LORD he is able to make the confession that we all must make about Jesus to begin to become our true selves: "My LORD and my God!"

Jesus' response to Thomas' confession is filled with compassion for those of us who would confess Jesus as LORD:

Because you have seen me, you have believed; blessed are those who have not seen and yet have believed (Jn 20:29).

"Blessed are those who have not seen and yet have believed." That's us.

The writer of the Gospel of John immediately helps us understand that his Gospel has been written precisely so we may receive such a blessing, precisely so that we "may believe that Jesus is the Christ, the Son of God, and that by believing you may have life in his name" (v. 31).

John's Gospel mercifully brings us to this point, to a moment when our lives can be forever changed as we understand who Jesus Christ really is, and accept Him as the Son of God - the rightful LORD of our lives.

In the ultimate case of theological foreshadowing, John told us from the beginning who Christ was and is in the very first chapter. Our journey through John's Gospel, including all the miraculous signs, especially the accounts of the apostles seeing the Risen LORD, is intended to help us *confess* what John told us from the beginning – not merely run our eyes over the words, not merely hear the words with interest or even incredulity – but to confess them just as doubting Thomas did so that we can have "**life in His name**."

"Stop doubting and believe."

We can have life in His name, because from all eternity "the Word was with God and the Word was God" (Jn. 1:1). We can have life in His name because "Through Him all things were made" (v. 3). We can have life in His name because "In him was life, and that life was the light of men" (v. 4). Because He is God the Creator, in Him is the life of all of life. He is the Creator, the giver of all life – and something even more. This life that death cannot ultimately conquer is "the light of men."

It is only when the apostles encounter the Risen LORD that the light fully dawns on them about who He really is: "My LORD and my God." He is the one Who has defeated death, because "in Him was life." Only by seeing Him in His victoriously resurrected body does the light finally dawn for the apostles, and thus the life within Him that could not be conquered by death "was the light of men."

The preexistent Son of God, the Creator, the Lifegiver, came into the world as an illegitimate carpenter's son from Nazareth. "He was in the world, and though the world was made through him, the world did not recognize him" (v. 10). As all the Gospels attest, even his own people "did not receive Him" (v. 11). Indeed, the Jewish leaders handed him over to their enemies to crush the dreams of his follows by making him a failed, crucified messiah. But to those who have seen the light of life in the Risen LORD, for those who stop their doubting and believe, for those who confess "my LORD and my God," to us our LORD has given us "the right to become children of God" (v. 12). As such, we have "life in His name" (20:31).

We were made to have true life by freely imaging or reflecting God's glory and spiritual beauty, i.e., by doing His will on earth as it is in heaven. But as we have seen, before Jesus' atoning death and resurrection, even the disciples did not image the One Who was right in front of them, the One Who did fully image or reflect the Father's will on earth as it is in heaven.

Since we are unable to reflect God's will on earth, because we are, as it were, Venus de Milo, still beautiful but without arms, "The Word became flesh and made his dwelling among us," as the prologue to John's Gospel famously states (v. 14a). And the apostolic witnesses of the Risen LORD "have seen his glory, the glory of the One and Only, who came from the Father, full of grace and truth" (v. 14b).

It was seeing the glorious presence of the Risen LORD that gave the apostles eyes to see, that allowed doubting Thomas to confess, "My LORD and my God." Having seen the Risen LORD's glory, even Thomas could stop his doubting and believe.

For those of us who confess along with doubting Thomas "my LORD and my God," we begin our pilgrimage into being reflections of the Risen LORD's glory as we fulfill His Lordship, as we are obedient to His will, as we image Christ in our everyday lives.

Are you tired of living an inauthentic life, a "less-than" life, a fake existence? Do you want true life? Do you want to be the real you, the authentic you, the one God made you to be? Do you want to be truly free? Do you want to be the glorious, spiritually beautiful creature you were intended to be from all eternity?

Such a life does not begin when we die and go to "heaven," as if that is when eternal life begins. Certainly those of us who have accepted Christ as our Savior and LORD will be with the LORD for eternity after we die and are subsequently raised to life in the resurrection where we will dwell with the LORD on His New Earth (Rev 21:1-4).

Rather, this life begins with an authentic confession that He is the Son of God and therefore our true LORD. And we know our confession is true when obedience to His Lordship is displayed in our lives. To the extent we are obedient, we experience here and now, right now, "life in His name." It is then that we become really real, truly authentic, glorious, free, and spiritually beautiful.

And yet true obedience and communion with our LORD is so fragmentary and fragile due to our sin, so fraught with the shadow of futility because of the sinfulness of the world and the opposition of Satan, that we experience the true life like faint whispers of eternity. It is like driving in car on a long journey and briefly catching the signal of a radio station before it fades away, lost in the static of our sinfulness. But the spiritual beauty of the music was like something beyond this world. It is like the scene in the movie *The Shawshank Redemption*, when the hard-bitten criminals at Shawshank prison suddenly hear the aria from Mozart's *Marriage of Figaro*. As Red, the character in the movie played by Morgan Freeman, narrates, "and for the briefest of moments – every last man at Shawshank felt free."[454] True obedience to the LORD is like hearing Him sing an aria like this, and listening to Him melts our sins and our resistance away and we find ourselves joining Him and our voices are beautiful together! At that moment we are truly free; we are truly ourselves.

To do our part in overcoming global warming is one way today to experience "life in His name." It is to join our LORD is singing an aria so beautiful that for the briefest of moments helps others feel free. And not only can they feel free, our actions can help them to become free from the consequences of climate change. As we will see in Part 3, in the twenty-first century, overcoming global warming is a profound opportunity to help billions of people we will never meet.

Global warming is one of the greatest spiritual opportunities of the twenty-first century; it is also simultaneously a great spiritual challenge to our continual *confession* of the Risen LORD's Lordship in our lives given that the causes are so entwined in everything we do.

The Risen LORD clearly does not want the harm that global warming will bring to the "least of these." And yet our actions are contributing to the creation of such harm. The Risen LORD wants us to change and our society to change.

But there are so many convenient excuses for denying that we have to change: "Aren't there a few scientists who still have questions?" We can be driven (consciously or not) by inordinate fears that to do something could mean a material lessening of our lifestyles. It is such grasping as this that is the opposite of what Jesus did. He "made himself nothing, taking the very nature of a servant …he humbled himself and became obedient to death – even death on a cross" (Phil 2:7-8).

In a sinful world such obedience can bring temporary suffering. But as the Apostle Paul says to the Corinthians in his first letter, "Stand firm. Let nothing move you. Always give yourselves fully to the work of the Lord, because you know that your labor in the Lord is not in vain" (1 Cor 15:58). And in his second letter to the Corinthians Paul reminds them that we are a people who do not lose heart. "For our light and momentary troubles are achieving for us an eternal glory that far outweighs them all" (2 Cor 4:16-17).

To obey the Risen LORD in helping to solve global warming may at most bring some "momentary troubles" for Christians in the United States but nothing compared with the suffering to be endured by the poor in developing countries.

And such "momentary troubles" for us are also opportunities to serve the Risen LORD by serving "the least of these" all around the world. In so doing we can experience "life in His name" today as we are being "transformed into his likeness with ever-increasing glory" (2 Cor 3:18) and achieve "eternal glory" when His Kingdom comes in its fullness and we become the spiritually beautiful reflections of His glory that He created us to be.

Today, our continual pilgrimage of confession of the Risen LORD's Lordship includes helping Him to overcome global warming. This is one way today to reflect His glory as we walk with Him and freely become our true selves.

14

WALKING WITH CHRIST OUR LORD –
FULFILLING THE FIVE GREAT LOVES

> **Why do you call me, "Lord, Lord" and do not do what I say?**
> Luke 6:46

For Christians, in every moment of our lives there is something happening that goes unseen by the world. The spiritual reality of the Risen LORD walking with us encounters the gritty, harsh reality of the world we have helped to make. Today, this includes the impacts of global warming. Our understanding of this reality is not meant to have us mentally escape into some spiritual fantasy world (whose theme song might be something like, "Jesus and me, together for eternity.") Its purpose is the exact opposite: to allow us to face our current reality in order to transform it, to turn it in the direction of eternity, of God's reality, which is the real reality.

Jesus' Lordship in our lives is the active presence of the Father's will in creating the existence here on Earth that God desires. Anything that slows us down, that stops us from doing His will – misunderstandings, doubts, fears, denials – are a challenge to Christ's Lordship in our lives.

As you begin your journey into this chapter take hold of the key spiritual fact proclaimed in this book: the Risen LORD is LORD over the problem of global warming. Together with the Father and the Holy Spirit, He is currently engaged in overcoming it. There are principalities and powers, both human and demonic, which are fighting against Him. We ourselves have been resisting His Lordship in this area.

Hear this: now is the time to wake up to this spiritual fact and re-acknowledge His Lordship and get behind His leadership.

But our decision to walk with the Risen LORD in overcoming global warming – and we must make that decision in order to begin – is simply the start of a lifelong journey. To stay with Him on this particular quest will require spiritual resources and spiritual discipline.

God provides both, of course, but we must choose to avail ourselves of them. He provides spiritual resources in the form of His grace, His empowerment, the presence of the Risen LORD and the Holy Spirit, the Bible, and our fellow believers. But we must choose to allow these spiritual resources to form us into the image of Christ as we walk with Him. (This is where spiritual disciplines come in, including constant prayer, the daily devotional study of the Scriptures, and the consistent gathering together with other believers for worship, fellowship, and study.)

Like the herald who came before the King and shouted, "'*Prepare the way for the Lord, make straight paths for him*'" (Mk 1:3; Mt. 3:3, Lk 3:4, cf Jn 1:23), for those of us who have confessed, "Jesus is LORD," and know that He is the Risen LORD, our lives should prepare the way for Him. Yet when it comes to overcoming global warming, it is the Risen LORD who blazes the trail before us – and for those of us who choose to follow, we are simply scrambling to catch up.

The previous chapter reviewed how believers, starting with the disciples, come to confess "My LORD and my God," that the Risen LORD is our LORD. A true confession that "Jesus is LORD" must result in obedience to His will. But how? That is what we will explore in this chapter, first in the spiritually transformed lives of the apostles; then in some of our LORD's

most important teachings. Together these will help us better understand how we can live out our purpose in life: to freely love Him back.

RECOGNITION OF HOW WE CAN

It can be strangely comforting and reassuring to catalogue the failings of the 12 disciples as I did earlier. Even these incredible spiritual leaders blew it over and over, just like we do. But it is also admirable that during Jesus' ministry the disciples stick with Jesus through their doubts and denials and misunderstandings; they leave everything; they face persecution. Yet ultimately, on their own, they fail.

This is the point. On our own, we will fail. How can we turn things around? Once again, the examples of God's dealings with the apostles show us the way. Three things turn frightened disciples hiding from the authorities into apostles defying these same authorities to proclaim the Gospel and do God's will:

(1)   the power of Christ's saving grace through his precious blood;

(2)   the presence of the Risen LORD; and

(3)   the outpouring of the Holy Spirit.

All three are available today to help us walk with the Risen LORD as He overcomes global warming.

We have covered the first, the empowerment of grace, fairly extensively in the preceding chapters. We also explored in chapter 13 what a profound effect the visible presence of the Risen LORD had on the disciples, helping them to comprehend more fully who Jesus Christ was and is and to confess Him as LORD.

But now we must briefly explore the transformation from disciples to full-fledged apostolic leaders – wrought by both the appearances of the Risen LORD and the coming of the Holy Spirit.

No better confirmation of Jesus' resurrection could be offered than encounters with the Risen LORD Himself. As we have seen, the disciples don't believe He had been raised until He appears before them.[455] It was the apostles' belief in His resurrection combined with the guiding and uplifting presence of the Holy Spirit that propelled their apostolic mission forward, even in the face of death.

Consider this. Before His appearances the disciples are like small children whose primal fear of death at the hands of the Jewish leaders has them focused on their own survival, hiding behind locked doors, like children hiding under their beds.

After His appearances and the outpouring of the Holy Spirit, the Temple guards arrest Peter and John. Only a short time prior, fear of arrest had driven them into hiding. But they are not found cowering in a locked room. Rather, they are arrested in the Temple boldly healing in the name of Jesus and preaching about Jesus' resurrection and salvation through Him alone.

When questioned by the Chief Priests about how they were able to heal a crippled man, Peter replies, "It is by the name of Jesus Christ of Nazareth, whom you crucified but whom God raised from the dead" (Acts 4:10).

Remembering that a servant girl got him to deny his Lord three times not so long before, this statement from Peter is simply astonishing. His message is simple, succinct, straightforward, and true. While Peter is not trying to be rude or disrespectful, his proclamation is a profound challenge to the Chief Priests. To bluntly state that the healing power came not only from a man they had executed but a man vindicated by God through his resurrection (a theological concept they did not believe in) exposes their moral, theological and spiritual bankruptcy and their illegitimacy as leaders.

It is so bold, so unexpected, that it temporarily has the Sanhedrin at a loss. How could these "unschooled, ordinary men" have such courage (4:13)? Not knowing what else to do, they command them to cease and desist. Instead of due deference in the face of their power to do them harm, Peter is once again as respectfully "in your face" as he can be, "Judge for yourselves whether it is right in God's sight to obey you rather than God. For we cannot help speaking about what we have seen and heard" (Acts 4:19-20).

It is the last sentence that is so telling for our current purposes: these "unschooled, ordinary men" could not help themselves; they had to tell others the good news of Jesus' resurrection, about what they "have seen and heard." Acts 4:33 concludes, "With great power the apostles continued to testify to the resurrection of the Lord Jesus, and much grace was upon them all."

The warnings of the Chief Priests had absolutely no effect against the "resurrection effect." The apostles continued in the footsteps of Jesus by preaching, teaching, and healing, drawing larger and larger crowds. Once again, the Chief Priests have them arrested and brought before them, where they again order them to stop. Once again, Peter testifies that the LORD is risen and is now exalted at the Father's right hand. "We are witnesses of these things, and so is the Holy Spirit, whom God has given to those who obey him" (Acts 5:32).

Again, the unaffected, unselfconscious fearlessness of Peter's proclamation is a result of him being a "witness of these things," meaning Jesus' resurrection and glorification, as well as the gift of the Holy Spirit. Peter's message was deadly dangerous to the apostles, demonstrated by the fact that the Chief Priests wanted to have them killed but were dissuaded from doing so by Gamaliel, one of the members of the Sanhedrin (Acts 5:33-40).

When Jesus is taken from them, the apostles fail. But after the appearances of the Risen LORD and His presence with them through the power of the Holy Spirit, they cannot help themselves, even if it means death. Peter, the one who was ready to kill with a sword in the Garden but later denies Jesus three times to save his own skin, is now ready to die proclaiming the Risen LORD, to follow in the path of his LORD wherever it leads. In the Garden, Peter's own misguided, vainglorious dreams of messianic conquest gave him a temporary burst of false courage that had him willing to die fighting. But now Peter is ready to die living out his *true* calling: "Thy will be done." His experiences with the Risen LORD and the gift of the Holy Spirit have transformed Peter into a more faithful and therefore more spiritually beautiful and glorious reflection of God's will.

If the transformation wrought in these "unschooled, ordinary men" by seeing the Risen LORD and receiving the Holy Spirit is remarkable, then the transformation of Saul of Tarsus is world shattering. After Jesus, Saul of Tarsus is the second most important figure in the history of Christianity. As the supreme Apostle to the Gentiles, more than any other person he is responsible for the rise of Christianity into today's most populous religion. I am a Christian and you are reading this book because of the radical transformation of Saul of Tarsus.

When we first meet Saul, he is the last person we would think would become the most important apostle. He is introduced to us as one who is present when Stephen is stoned to death (Acts 7:58; 8:1). As Paul later recounts, "I stood there giving my approval and guarding the clothes of those who were killing him" (Acts 22:20).

What would it have been like to watch in approval someone being stoned to death for preaching the Gospel? But Saul is not done. Acts 9:1 finds him "still breathing out murderous threats against the Lord's disciples." This is no hyperbole, given Stephen's stoning. Saul's persecution of Christians could have easily resulted in more killings. His anti-Christian zeal is so strong that he asks the High Priest for the authority to go to Damascus and arrest Christians to bring them back to Jerusalem for trial.

Because Christians love and respect Paul, it is easy to not think too much about the actual human suffering Saul perpetrated against Christians. His persecution had one basic goal: to kill the early Church. Before his own conversion, to Saul the early Church was like an arm or a leg with gangrene. It must be cut off and destroyed so that it does not infect the entire body. Saul was actively trying to destroy the early Church by jailing and even killing its members.

It is this man – someone who never met Jesus before his resurrection, who was not one of the 12 – who becomes the Apostle to the Gentiles, who helped spread Christianity farther and wider than any other apostle.

How does one become the other? How does Saul the destroyer become Paul the Apostle? In 1 Corinthians 15 Paul states "I am the least of the apostles and do not even deserve to be called an apostle, because I persecuted the church of God. But by the grace of God I am what I am," (vv. 9-10). All of his apostolic achievements are due to "the grace of God that was with me" (v. 10).

If anyone understands from personal experience what grace or "unmerited favor" is, it is Paul. He was the last person who deserved to become an apostle. But out of His extravagant grace the LORD chose Saul to become Paul. As the LORD states in a vision to Ananias, "This man is my chosen instrument to carry my name before the Gentiles and their kings and before the people of Israel. I will show him how much he must suffer for my name" (Acts 9:15-16).

Saul becomes the last of the apostles as a result of seeing the Risen LORD. As he famously puts it after listing appearances of the Risen LORD to the other apostles, "Last of all, as to one untimely born, he appeared also to me" (1 Cor 15:8, NRSV). By "untimely born" Saul meant his becoming born again had to be induced by the LORD, as a doctor would induce labor to keep a child from dying in the womb.

This appearance of the Risen LORD to Saul occurs as he is on his way to Damascus to persecute Christians. Luke records their encounter thus:

"Saul, Saul, why do you persecute me?"
"Who are you, Lord?" Saul asked.
"I am Jesus, whom you are persecuting" (Acts 9:4-5; cf 22:7-8 and 26:14-15).

The Risen LORD has such a deep spiritual communion with His believers that what Saul does to them Saul does to Him.

As a result of the appearance of the Risen LORD, Saul is rendered blind and is led by the hand into Damascus. There, after three days of fasting, Ananias comes to him, lays his hands on Saul and says "Brother Saul, the Lord – Jesus, who appeared to you on the road as you were coming here – has sent me so that you may see again and be filled with the Holy Spirit" (Acts 9:17). Saul has been graciously chosen; he has encountered the Risen LORD, and he has now been filled with the Holy Spirit. Once Saul recovered from his fast he immediately "began to preach in the synagogues that Jesus is the Son of God" (Acts 9:20). The world would never be the same. The supreme persecutor becomes the supreme defender – and to some, the supreme offender who as a result is severely persecuted himself. As he recounts in 2 Corinthians 11:24-29:

> *Five times I received from the Jews the forty lashes minus one.*
> *Three times I was beaten with rods, once I was stoned, three times I*
> *was shipwrecked, I spent a night and a day in the open sea, I have*
> *been constantly on the move. I have been in danger from rivers, in*
> *danger from bandits, in danger from my own countrymen, in danger*
> *from Gentiles; in danger in the city, in danger in the country, in*
> *danger at sea; and in danger from false brothers. I have labored*
> *and toiled and have often gone without sleep; I have known hunger*
> *and thirst and have often gone without food; I have been cold and*
> *naked. Besides everything else, I face daily the pressure of my*
> *concern for all the churches. Who is weak, and I do not feel weak?*
> *Who is led into sin, and I do not inwardly burn?*

This account doesn't even cover all of Paul's experiences of persecution and suffering for the sake of the Gospel.

So how does Saul become Paul? How is the persecutor of the LORD transformed into the LORD's faithful servant who himself suffers persecution?

As we have seen, it is the same three transformative spiritual catalysts that turned fearful disciples into fearless apostles: (1) the power of Christ's saving grace; (2) the presence of the Risen LORD; and (3) the outpouring of the Holy Spirit. We too can allow these three spiritual catalysts to transform us into dynamic and steadfast agents of the LORD's will, including overcoming global warming.

Rest in this: it is not by our power, but by the Risen LORD walking beside us and working through us that we will do our part in overcoming global warming and other great challenges. The LORD is risen. The Lord is the LORD. The Risen LORD is present to us in the power of the Holy Spirit. As the Apostle Paul says, "Where the Spirit of the Lord is, there is freedom. And we, who with unveiled faces all reflect the Lord's glory, are being transformed into his likeness with ever-increasing glory, which comes from the Lord, who is the Spirit" (2 Cor 3:17-18).

When the damage wrought by our dangerous enhancement of the greenhouse effect encounters the transformative power within us of the resurrection effect, then will we Christians begin to follow our Risen LORD in overcoming global warming.

But just as the disciples did not grasp the hope Jesus offered before his crucifixion – the hope of His resurrection – so too is this a hope we haven't yet fully grasped. Like the disciples, **it is time for us to stop our doubting and believe**. All things are indeed possible through the LORD who gives us the strength (Phil 4:13). With the Risen LORD walking beside us leading the way, Christians can lead the world in overcoming global warming. The words of Jesus our LORD beckon us forward.

FOLLOWING THE LORD'S EXAMPLE AND TEACHINGS

Our pilgrimage to confession of Christ's Lordship has led us to this point: we must discover what obedience looks like so we can put it into practice as it relates to overcoming global warming.

The best way to understand what following or imaging Christ looks like is to see for ourselves in the New Testament how Jesus images the Father's love and to hear for ourselves in His words what obedience means.

We have already explored Jesus' example in earlier chapters. Though He was the Creator of the Universe, "the Word became flesh and dwelt among us" (Jn 1:14, NKJV). He even "made himself nothing, taking the very nature of a servant" and "humbled himself and became obedient to death – even death on a cross" (Phil 2:7-8). Jesus himself says, "I am among you as one who serves" (Lk 22:27). That service was supremely demonstrated upon the cross, where he who knew no sin became sin for us, "so that in him we might become the righteousness of God" (2 Cor. 5:21).

It is this becoming the righteousness of God, or doing God's will, or imaging Christ, that is the calling of those of us who have confessed with doubting Thomas, "my LORD and my God."

As we have seen, we have the spiritual resources to activate His Lordship within us. Now we must gain further guidance from his teachings concerning how to put His Lordship into practice. And this guidance comes from Jesus' teachings about five great loves:

1) to love one another as Christ loves us;
2) to love our neighbors;
3) to love our enemies;
4) to love "the least of these"; and
5) to love God back.

If we are honest, most of us are only part-time practitioners of these five great loves, and living them out can appear daunting at times given our level of spiritual maturity. So let us keep these words of encouragement from Jesus himself before us:

Come to me, all you who are weary and burdened, and I will give you rest. Take my yoke upon you and learn from me, for I am gentle and humble in heart, and you will find rest for your souls. For my yoke is easy and my burden is light (Mt 11:28-30).

At times His teaching may not seem easy or light, due to the sin within ourselves and the world. But it leads to righteousness, which leads us to become more spiritually beautiful, glorious, and free as we become our authentic selves, the ones whom He created us to be. In so doing will we find rest for our souls.

1. *LOVE EACH OTHER AS I HAVE LOVED YOU* – Jn 15:12

While it seems almost unnecessary for Jesus to *command* Christians to love one another – shouldn't this be obvious and natural? – anyone who has been seriously involved in a local Christian community knows that this commandment is very much needed! Just like our biological families, sometimes it can be hardest to love our immediate Christian family. (Indeed, Paul's letters contain a good number of admonitions to members of the first churches to get along with one another.)

This command from Jesus comes in his discourse with his disciples the night he was betrayed. It is the apex of his moral teaching in the Gospel of John. But before Jesus gives this command, he centers it in our love for Him. Even more than that, Jesus makes us an incredible promise.

> *If anyone loves me, he will obey my teaching. My Father will love him, and we will come to him and make our home with him* (Jn 14:23).

Just as confession of Christ's Lordship must result in a life in keeping with that Lordship, those who love him – and one cannot truly confess Him as LORD without loving Him – will obey His teaching. If one does so, Jesus promises that "My Father will love him, and we will come to him and make our home with him."

For us to love Him, to strive to obey His teaching, is to have the Father and the Son become our family. Even though the fullness of this union will occur in the future (Rev 21:3-4), their spiritual presence – and that of the Holy Spirit as promised by Jesus further on in this same discourse (Jn 16:7, 13) – is promised here and now to those who strive to be obedient.

But lest we think we have to love God to receive His love, Jesus continues, "As the Father has loved me, so have I loved you" (Jn 15:9). Jesus' love for them is also the Father's love. As he said earlier on, "it is the Father, living in me, who is doing his work ... I am in the Father and the Father is in me" (Jn 14:10b-11). Jesus is the true son, the true *imago dei*, the true reflection of his Father's will, which flows from the Father's love. Our love of God is simply returning to Him what he has given to us – His love – because "God is love" (1 Jn 4:8, 16). The Father is the fount of all love, including our love back to Him.

Jesus has shown us what it means to be a true image or reflection of the Father's will, which flows from the Father's love. "As the Father has loved me, so have I loved you."

But Jesus immediately says, "Now remain in my love. If you obey my commands, you will remain in my love" (Jn 15:9-10).

This could sound conditional – "If you obey, then you can remain" – but it is not. It is not that the Father or Jesus will reject us if we don't do His will; it is *we* who are doing the rejecting. To remain in His love, in the sphere or reality of His love, is to obey His commands, to do His will, to enter into His Kingdom. Why would anyone in their right minds want to be anywhere else?! To step outside His realm of love is spiritual insanity. To obey Jesus' commands keeps us spiritually centered and sane in the realm of God's love.

One of the results of such knowledge and obedience is joy. Jesus continues, "I have told you this so that my joy may be in you and that your joy may be complete" (Jn 15:11).

The night before his death in his farewell discourse to his disciples the Gospel of John has Jesus focused on one command, which is this:

> *Love each other as I have loved you. Greater love has no one than this, that he lay down his life for his friends* (Jn 15:12-13).

We are to love each other as he loved us. Such love is profound, deep, and ultimately impossible for us on our own. But with His love, all things are possible. When His love is combined with ours, making our love perfect, then we can love as He loves. And such love leads to this: the willingness to give our lives for our friends.

In some ways it is easier to think we love all of humanity in the abstract than it is to actually love particular individuals. Jesus died on the cross for all of humanity. But he also laid down his life for his friends, for particular individuals he knew and loved. There is deep feeling mixed in

Global Warming and the Risen LORD

here that comes from living with others, a relational communion born of experience. As the Gospel of John says, Jesus loved the disciple named John. He was not an abstraction. He was described four times in the Gospel as "the disciple whom Jesus loved" (Jn 13:23, 19:26, 21:7, and 21:20). Jesus laid down his life for John, as he did for the rest of the disciples.

For most of us, doing our part to overcome global warming will not involve literally laying down our lives, even for those we know and love. But it will involve tapping the depth of our love that we have for our loved ones, for those within our local Christian community, for our friends.

Everyone will be impacted by global warming, some much more than others. During your lifetime your friends and loved ones will be impacted.

As we saw earlier, here in the United States there will be impacts large and small, from intensifying floods and droughts to becoming separated from pets or losing grandma's handwritten recipe for peach pie because your house burned to the ground.

2. *LOVE YOUR NEIGHBOR*– Mt 22:34-40; Mk 12:28-34; Lk 10:25-28; Rom 13:9-10; Gal 5:13-14; James 2:8; Lev 19:18

Because the word "love" is so associated with feelings, it can have a mushy, sentimental, flighty sense – something not to be carried into dealings with the real world. We can have such feelings for our spouses, children, and "loved ones," as the common phrase puts it. But in dealings with the real world, it is best to store such feelings away – leave them at home.

This sentiment is captured rather strikingly and succinctly in the movie *Cinderella Man*, staring Russell Crowe as the real-life former heavyweight champ Jim Braddock. In the movie, several days before Braddock is to fight the current heavyweight champion Max Baer, the promoter, Mr. Johnston, calls Braddock and his manager in for a meeting to discuss how Baer had previously killed an opponent in the ring. Editorials were calling the upcoming fight premeditated murder, saying that everyone responsible should be hauled into court.

The promoter, Mr. Johnston, says to Braddock, "I'm not getting hung out to dry if something happens to you." Braddock's manager, Joe Gould, (played by Paul Giamatti), replies, sarcastically, "You're all heart." Mr. Johnston quickly and firmly retorts, "My heart's for my family, Joe. My brains and my balls are for business … and this is business." Our hearts are for our families, but business is business.

If "love" is tied to feelings associated with our families, our "hearts," then what can it mean to "love" our neighbors as ourselves? Are we to manufacture such feelings for others outside our families? And can such feelings be relied upon? This puts us in a bit of a quandary.

When I was in my early 20s, I wrote the following words to try to get myself out of this predicament: "Feelings are fleeting. Love is action." There is some truth to those words. But not the whole truth. Feelings are fleeting; they change like the weather. And love must manifest itself in loving actions in order for it to continue to exist and grow. In this way love is like a muscle. It must be exercised, or it will die.

The word "love" also seems out of place in a context of conflict. If there is love, there shouldn't be conflict. The fruits of love are good things. They shouldn't result in opposition. And yet we all know through our own experience that in a sinful world good begets opposition. This can sour us on the whole operation. We become cynical about this fuzzy love stuff. It's fake, and it doesn't work, anyway.

You might hear someone say in exasperation, "Can't we all just love one another?" And at that moment, even if you weren't disgusted before, you just might be ready to throw something at the person calling for love. "Oh, shut up!" might run through your mind and even cross your lips. "I'm not going to manufacture feelings to create some bogus sense of harmony that is really a lie." Well, no, we don't want to do that.

Whether we are comfortable with it or not, "love" is tied to our emotions. We are not called to spontaneously manufacture fake feelings. But each of us has passion or deep emotion inside of ourselves that transcends our feelings of the moment. We don't experience it all of the time, but it's there. We were made as the *imago dei* to be passionate, passionate with one relationship especially.

Because God is passionate in His love for us, as images of Him we are to reflect that passion back to Him. Because God the Father, Son, and Holy Spirit, are passionate in their love for One Another, so too are we to be passionate in our love for our family, our beloved. And the consequences of such passion are not to be contained. They are to radiate outward, just as they do from God the Father, Son, and Holy Spirit.

When you come in contact with those outside of the circle of those you love you may not like them, you may have absolutely no feelings one way or another towards them. But because of your passion for the LORD, who loves them, you must treat them like you would treat Him. You must treat them lovingly like they were His child – because they are! Your passion-filled love for Him must result in good to them. Regardless of your feelings of the moment, you are to do actions that are intended to result in their well-being – because of your passionate love for the LORD.

This is precisely what Jesus teaches us when he lifts up the command to love our neighbors as ourselves. In two of the three Gospels in which it appears – namely Mark and Matthew – this teaching is in the context of conflict.

Jesus is teaching in the Temple during the last week of his life. The religious leaders are looking for a way to discredit Jesus, and are even prepared to have him killed if necessary to get rid of him. So it is in the gritty realism of our sinful world that Jesus talks to those who want to kill him about loving their neighbors.

To prove that this traveling hick preacher, this miracle worker from the sticks, doesn't know what he's talking about, the religious leaders ask him what is the greatest commandment in the Law. In effect, they were asking, "If there is one thing our lives should be about, what is it? What is the most important thing in life?"

Without missing a beat, Jesus quotes Deuteronomy 6:4-5, something that observant Jews of the time recited in the morning and in the evening: "Hear, O Israel, the Lord our God, the Lord is one. Love the Lord your God with all your heart and with all your soul and with all your mind and with all your strength" (Mk 12:29-30).

Jesus confirms that we are to love God with everything that we are. There wasn't anything for the opposition to say in response. They knew he was right.

But, knowing that these religious leaders were missing something, knowing that they knew the right book answer but didn't have a clue to its ramifications, Jesus immediately says, "And the second is like it: 'Love your neighbor as yourself,'" (Mt 22:39, quoting Lev 19:18). To make things perfectly clear, Jesus adds, "All the Law and the Prophets hang on these two commandments" (Mt 22:40).[456]

(And to make things crystal clear, the Holy Spirit ensured that this commandment from the Old Testament was the most frequently cited in the New, found not only in the three Synoptic Gospels, but also in Romans, Galatians, and James.)

Do you want to know what life is all about? What we are called to? What we must do to become who we were made to be? What Jesus teaches us to pray for in the Lord's Prayer, "Thy Kingdom come, Thy will be done"?

Here is Jesus' answer: Love God with everything that you are, and with all the passion you can muster, and love your neighbor as yourself.

What does it mean to love our neighbor as ourselves? It means that you should "Do to others as you would have them do to you," as Jesus elsewhere teaches in his version of the Golden Rule (Lk 6:31). As Paul so famously says in 1 Corinthians 13, love manifests itself in this way:

> *Love is patient, love is kind. It does not envy, it does not boast, it is not proud. It is not rude, it is not self-seeking, it is not easily angered, it keeps no record of wrongs. Love does not delight in evil but rejoices with the truth. It always protects, always trusts, always hopes, always perseveres* (vv. 4-7).

Why does Jesus add the second commandment to love our neighbors as ourselves? He does so because you can't love God unless you love your neighbor. While God loves you, He loves your neighbor, too.

Global Warming and the Risen LORD

These two commandments joined together by Jesus are what the Church has called The Great Commandments, and they are what our lives should be about.

In the Gospel of Luke Jesus has this same discussion in a different context and the dialogue goes a little differently (Lk 10:25-28). An expert in the Law tests Jesus and asks a similar question, "'Teacher,' he asked, 'what must I do to inherit eternal life?'"

Jesus then asks the expert what answer he finds in the Law, and he replies with the Great Commandments. Jesus affirms his reply and then says, "'Do this and you will live'" (v. 28). With this Jesus essentially connects a true life today with life in the Kingdom when it comes in its fullness.

Do you want to know what real life is? Do you want to truly live? Do you want to taste a glimpse of God's future for His people and the rest of His creation right now? Then love God with everything you've got and your neighbor as yourself. Do this and you will live, really live.

But the expert, mister smarty-pants, wants to show off what he knows (and maybe put Jesus in his place). He then asks, "'And who is my neighbor?'" In other words, who am I required to "love" to earn the prize of eternal life, and who are not my neighbors and thus can be treated in less than a loving manner? This weasely question sets up one of the most memorable and loved of Jesus' stories, the parable of the Good Samaritan.

During Jesus' time Jews regarded Samaritans with utter contempt. They understood Samaritans to be the ancestors of those peasant Jews left behind when the ruling elites were taken into exile, who then intermarried with the foreigners brought in by the Assyrians to repopulate the land. Over the ensuing centuries they sometimes aligned themselves with the enemies of the Jews. They were thus considered traitorous half-breeds, and from inferior Jewish stock to boot. Perhaps more importantly, the Samaritans had their own version of the Torah and their own Temple on Mount Gerizim, which the Samaritans claimed was the proper place for the Temple rather than Jerusalem. Religious and ethnic differences that would appear small to outsiders fueled the strong emotions each group had for the other. John 4:9 put it mildly when it says, "Jews do not associate with Samaritans."

Jesus responds to the desire of the expert to know legally where love ends with a story. And in that story it is one of those heretical, traitorous, half-breed Samaritans who shows us what it means to love your neighbor.

> *A man was going down from Jerusalem to Jericho, when he fell into the hands of robbers. They stripped him of his clothes, beat him and went away, leaving him half dead. A priest happened to be going down the same road, and when he saw the man, he passed by on the other side. So too, a Levite, when he came to the place and saw him, passed by on the other side. But a Samaritan, as he traveled, came where the man was; and when he saw him, he took pity on him. He went to him and bandaged his wounds, pouring on oil and wine. Then he put the man on his own donkey, took him to an inn and took care of him. The next day he took out two silver coins and gave them to the innkeeper. "Look after him," he said, "and when I return, I will reimburse you for any extra expense you may have"* (Lk 10:30-35).

The biblical lawyer was looking for the love loophole. "We're just talking about our neighbors, right, and by that I mean the appropriate kinds of neighbors. So I don't have to love those other people, right?"

By having the Samaritan be the one who demonstrated love by his actions, Jesus in effect says that everyone is our neighbor – even or especially others we hold in contempt. And furthermore, those of us who think of ourselves as religious, as doing the right things to appease God and look righteous to others better think again.

Here is where this parable and global warming intersect for us. The priest and the Levite were not the ones who robbed the man, just as in our time we didn't create the conditions of poverty, a situation that makes the poor much more vulnerable to the impacts of global warming. But they did pass by on the other side. Righteousness and love are the presence of good

and loving acts, not simply the absence of bad ones. By not helping the man in the ditch, the priest and the Levite made his plight worse and failed to love God and be who God created them to be.

Today, collectively, we are in fact making the plight of the poor worse through global warming. Knowing their plight and not doing what we can to help to overcome global warming is like passing by on the other side.

We may be highly observant of the outward signs of what it means to be religious in our community. So, too, I'm sure, were the priest and the Levite. That's exactly why Jesus chose them to be characters in his parable. But if we don't help the poor who through no fault of their own find themselves victims in the ditch of global warming's impacts, then all the religiosity in the world won't bring us closer to the LORD.

Why? Matthew 25 lets us in on a little secret about the parable of the Good Samaritan. While the Good Samaritan is Christ-like in his behavior, **it is Jesus Himself who is the man in the ditch**. To get close to the LORD is to tend to him when he's in the ditch. The point of the parable is to be a neighbor, to be a Good Samaritan, to go and do likewise.

In terms of the problem of global warming, right now, as a Christian, it is as if you are approaching a victim of global warming in the ditch. You are just within sight of the person. You don't yet quite know what is going on. Is it risky to go over to this person? You can't quite yet tell anything about who it is – just that there is a lump in the ditch that looks human. Whoever it is could be drunk you think to yourself – maybe not a victim at all! But as you venture closer you come to find that it is a child, not a man. It is a young girl.

She is in distress. Is she sick? Weak from hunger? Both? Maybe she has an infectious disease. Where are her parents? Who is responsible for this young girl? How did she get in this situation? Suddenly you notice that someone else is in the ditch with her. It is the Risen LORD.

It is here that we must ask ourselves some questions in light of what we have learned thus far in this book about the spiritual resources God has given us in our journey with the Risen LORD.

Who are we becoming with the spiritual blood of Christ's saving grace flowing in our veins?

Who are we becoming with the uplifting strength and guidance of Christ's Spirit, the Holy Spirit, residing in our hearts?

Who are we becoming with the Risen LORD walking beside us, with his glorious face reflecting the compassionate fire of the Father's love?

Who are we in light of this?

Are we becoming religious people who will pass by on the other side and ignore the victims of global warming? Or are we becoming those who let our passion for the LORD transform us into those who have the courage and strength to love others as ourselves, whoever and wherever they may be, in whatever situation they find themselves?

What will our response to the impacts of global warming on others, especially the poor, say about our walk with the Risen LORD?

SPOTLIGHT ON AFRICA

The projected impacts of global warming on sub-Saharan Africa in this century would be challenging for a rich continent to cope with. But sub-Saharan Africa's poverty and instability will make it the region hardest hit by climate change.[457]

Sub-Saharan Africa's population is around 1 billion.[458] With a growth rate nearly twice that of the rest of the world, its population will nearly double in 20 years. By 2030 every third person born in the world will be from sub-Saharan Africa, and over half the population will be under age 24. By 2050 every other person born in the world will be from this poor region. Projections are that after 2100 the population will stabilize at around 2.85 billion, about three times the current size.[459]

While parts of sub-Saharan Africa are thriving, it is the world's poorest region, and is the only one in the world that has become poorer in this generation.[460] Forty percent of the population in sub-Saharan Africa are currently undernourished and live on less than $1 a day, and by 2015 over 400 million will fall into this category (up from 315 million in 1999).[461] More than a third of African children under five are stunted,[462] which as we covered earlier has important lifetime effects if they survive.

Because of its poverty, dependence on locally grown food, recurring droughts and floods, civil unrest and political instability of states close to failure, and diseases like malaria and the AIDS pandemic, parts of sub-Saharan Africa are in crisis or live on the edge of crisis. The National Intelligence Council of the CIA doesn't see things getting better anytime soon: "In 2025, Sub-Saharan Africa will remain the most vulnerable region on Earth in terms of economic challenges, population stresses, civil conflict, and political instability." [463]

Global warming will make coping with these problems worse, in some cases much worse. One area of serious concern is how drought intensification due to global warming will impact food security.

In sub-Saharan Africa, over 90% of agriculture is rain-fed, which requires predictable rainfall during the right season in moderate amounts over time. More than half the population lives on food grown locally.[464] This makes them very vulnerable to drought, and currently one-third of Africans live in drought-prone areas.[465] An-

other disturbing trend is that land suitable for agriculture is disappearing at an alarming rate due to desertification and other factors. A staggering two-thirds of such land is projected to be lost by 2025.[466] Thus, any further disruption in Africa's ability to grow rain-fed crops will have serious consequences, especially for the poor. This is exactly what global warming will do.

Crop reductions by 2020 due to global warming could be as much as 50% in certain areas.[467] For the poor with little or no margin of error, such as the 40% who are undernourished, this will be devastating. We think that major impacts from global warming will be way off in the future, but this is not be the case for the poor in Africa. Having done little to nothing to cause the problem, they are nevertheless on the front lines for receiving the initial impacts. Indeed, those impacts have already begun. Looking out to the end of the century, crop revenues could decrease as much as 90%, with small farmers bearing the brunt.[468]

3. *LOVE OUR ENEMIES; OVERCOME EVIL WITH GOOD* – Mt 5:43-48; Lk 6:27-36; cf Rom 12:9-21

The third of the five great loves is loving our enemies. Previously we touched upon Jesus' command that we are to love our enemies, noting the spiritual freedom displayed by Jesus in such words and actions. (This freedom springs from Jesus' relationship to his Father and to his fulfilling his Father's will.)

For Christians, it is hard to admit when we think Jesus sounds, well, nuts. (It's his spiritual freedom talking – he's free enough to sound nuts to us.) But if we are honest with ourselves, loving our enemies sounds crazy and worse – misguided, harmful, and at times flat out wrong. Maybe we aren't to take this literally. Jesus can be hyperbolic, after all. He doesn't literally mean we should pluck out our eyes or chop our hands off, for example (Mt 5:29-30).

When we see where the inspiration comes from for the concept of loving our enemies, it becomes clearer why it sounds so farfetched. We are to love our enemies because God the Father loves His enemies.

But we are not God, and pretending that we are God gets us into all sorts of trouble, right? This loving our enemies stuff sounds like something best left to God, a little too high and mighty for us.

To demonstrate that God the Father loves His enemies Jesus points out, "He causes his sun to rise on the evil and the good, and sends rain on the righteous and the unrighteous" (Mt 5:45).

Really? God even loves evil people? Serial killers? The Hitlers of the world? This feels, frankly, obscene. Is Jesus really saying that we should be nice to Hitler? That led to more evil, didn't it?

Let's get real. There are some for whom sin has so captured their lives that they have crossed over from being regular sinners like the rest of us into being partners of evil. History demonstrates that showing mercy to such partners of evil is folly.

But if you want to know where the line is between garden-variety sinners and those who are in league with evil, well, that simply points out part of our quandary. For a finite sinner to seriously label someone else evil is profoundly, spiritually dangerous. As history has also proved over and over, it can be the justification for killing people – and to do so in the name of good. Indeed, right before Jesus commands us to love our enemies he warns us that Christians will be thought of as evil by others simply because they are Christians (Lk 6:22).

The perception of who is evil is very much in the eye of the beholder. Once Jesus challenged the authority of Israel's religious leaders, they said that he was in league with Satan (Mk 3:22; Mt 12:24; Lk 11:15). When they had Jesus killed, they thought they were doing the right thing and protecting the nation (Jn 11:49-50). Jesus warned the disciples that the same thing could happen to them: "a time is coming when anyone who kills you

 Global Warming and the Risen LORD

will think he is offering a service to God. They will do such things because they have not known the Father or me" (Jn 16:2-3). Such was the case with Saul before he became Paul.

For Christians, because we think we are "good people" by definition, given that we are Christians, our enemies must be on the other side, the wrong side, the side of evil. It's not much of a nudge to simply put our enemies in the category of evil. And evil must be destroyed, right?

Now we have gotten ourselves into some dangerous moral waters. Jesus' command to love our enemies saves us from all kinds of moral hazards given our own sinful finitude. So it is actually part of loving ourselves. But for it to be authentic it will need to be more than that. We will actually have to love our enemies through our actions.

Jesus provides us with very concrete examples of what we are to do towards enemies. "If someone strikes you on the right cheek, turn to him the other also" (Mt 5:39). "If someone forces you to go one mile, go with him two miles" (Mt 5:41). "Love your enemies and pray for those who persecute you, that you may be sons of your Father in heaven" (Mt 5.44-45a).

Oh, boy. Here's where things start to sound a little nuts. Be like God the Father by turning the other cheek. It sounds like we are to become passive wimps who allow bullies to kick the crap out of us. Would that mean that God the Father is a passive wimp, too, who allows Satan to kick the crap out of Him? How can such a God save us? No thanks!

Is the Father really passive in the face of evil? Does He simply allow His enemies the run of the field? Heaven forbid!

In this discussion we may have conveniently forgotten something Paul tells us that as sinners we are "God's enemies" (Rom 5:10). And yet He sent His Son, the Prince of Peace, to bring about reconciliation. Even "when we were still powerless, Christ died for the ungodly" (Rom 5:6). "God was reconciling the world to himself in Christ, not counting men's sins against them" (2 Cor 5:19). Christ reconciled us "by making peace through his blood, shed on the cross" (Col 1:20).

Through this unforeseen – indeed, shocking – transformative act of divine redemption we have been changed from enemies to friends. God was and is doing what He always does: overcoming evil with good.

So if loving our enemies sounds crazy, it is right in line with the Gospel, the good news that even though we were God's enemies, He has acted to transform us. Because of this, Paul says about his own efforts, "If we are out of our mind, it is for the sake of God" (2 Cor 5:13). He adds that we are compelled by Christ's love to truly live, and "that those who live should no longer live for themselves but for him who died for them and was raised again" (2 Cor 5:14-15). And because God has acted to transform us from enemies to friends "from now on we regard no one from a worldly point of view" (2 Cor 5:16). It is that view, the "normal" or worldly view, which leads to revenge, to payback.

Viewed from the idea that we were God's enemies, it may not sound so crazy when Jesus reminds us that God the Father "causes his sun to rise on the evil and the good, and sends rain on the righteous and the unrighteous" (Mt 5:45). Since we ourselves are recipients of divine mercy, we can have a deeper appreciation for Jesus' words when he concludes that in acting lovingly towards our enemies "you will be sons of the Most High, because he is kind to the ungrateful and wicked. Be merciful, just as your Father is merciful" (Lk 6:35-36).

To perceive someone as an enemy is a judgment that we make based on what we believe they have done to us. The Christian way to deal with enemies is to transform them into friends, not in some naïve way that allows them to continue to abuse us (if that is what they are actually doing). For us to become doormats for their harmful deeds, to continue to allow them to participate in evil acts against us, is not loving them. It is enabling their moral descent and deepening our own self-loathing.

So, not revenge. Not payback. But also not a doormat or enabler of evil. What, then?

True transformation.

Let's look again at what Jesus tells us to do. "If someone strikes you on the right cheek, turn to him the other also" (Mt 5:39).

If this still feels wimpy, a bit of historical context will help. To strike someone on the right cheek with the back of the hand was something one did to one's inferiors. To strike someone on the left cheek with one's open hand is something one would do to an equal. To not hit back, and yet at the same time demonstrate that you will not be cowed by unjust violence and will simultaneously claim your true humanity as one created in the image of God, is a way to try to overcome evil with good. It would be a "transformative initiative" as my mentor and friend Glen Stassen puts it.[469]

Jesus also commands us to "pray for those who persecute you, that you may be sons of your Father in heaven" (Mt 5:44b-45). To pray for what is best for our enemies? Well, in the larger sense that means the same thing it means for us, namely, that they would do the Father's will, that they would love God and their neighbors as themselves. If they do God's will, then they will cease to do evil. If we pray for their true well-being, that the love of God would be poured into their hearts, then we are praying for them to be transformed from evil – whatever makes them beyond the moral pale in our eyes – to good. And if so, then they will stop doing evil to us and to others, and will join us in striving to do the LORD's will.

The way to ultimately defeat evil is through love, in part because it starves evil of one of its power sources, the bile that is created in our hearts when we feel we have been wronged. To love our enemies is part of defeating evil in the world and not allowing it a toehold in our own hearts, driving us away from the LORD.

Paul closes his own discussion about loving our enemies with an interpretive key that explains things quite succinctly: "Do not be overcome by evil, but **overcome evil with good**" (Rom 12:21, emphasis added).

So where does loving our enemies come into play in the context of global warming? Viewed from this perspective, all of this talk about mercy is quite good news for those of us in the United States.

Maybe in this instance the command to love our enemies is not meant for us. When it comes to global warming, the folks who primarily need to hear Jesus' commands to love their enemies are those who will be the most severely impacted.

When we let it sink in that historically the United States has been the major contributor to global warming thus far and will remain so for decades,[470] and that per person the United States will probably continue to contribute more to the problem well into this century,[471] then we will be even more grateful that the Father is merciful and "causes His sun to rise on the evil and the good, and sends rain on the righteous and the unrighteous" (Mt 5:45).

God the Father has provided natural blessings like rain to everyone, but because global warming has and will lead to more droughts, we are tinkering with that rain. Since the blessings of rain include water to drink and the growth of crops, then global warming is depriving the poor of these blessings, as we saw earlier.

4. *LOVE THE LEAST OF THESE*– Mt 25:31-46

On August 4, 2008, Andrew Revkin, who at that time was the senior science writer for the *New York Times*, posted a blog with this intriguing title: "Are We Stuck With 'Blah, Blah, Blah, … Bang'?"

In discussing what Revkin described as "the human habit of dawdling in the face of looming risks" such as climate change, one respondent to his blog noted that until we have direct experience of a problem we are prone to tune out those who are warning us to make preparations for a disaster. So for those who choose not to pay attention, the warnings come across as "Blah, Blah, Blah." But then the "Bang" (or disaster) happens.

So, with global warming, Revkin asks, "Are We Stuck With 'Blah, Blah, Blah … Bang'?"

Revkin goes on to quote comments about climate change from David Ropeik, an expert on risk communication and longtime student of how humans deal with risk. Ropeik notes, "The psychological literature on the perception of risk has shown that we fear, and demand protection from,

risks that can happen to us, not to polar bears or ice caps." Ropeik further comments, "Even as potentially catastrophic as climate change might be, if people don't sense climate change as a direct personal threat, reason alone won't convince them that the costs of action are worth it." This is because in such instances our behavior is determined by "our overpowering animal instinct to survive." Unfortunately, "issues like climate change just don't ring our alarm bells."

To demonstrate this, Ropeik suggests the following informal experiment. "Ask a few friends: 'Name one way that climate change will significantly negatively impact you in the next 10 or 20 years.'" He notes that "Most of the people I know struggle with that one."

As an expert in communicating about risk, Ropeik advises, "local risks need to be emphasized, in the concrete terms that will give people more of an idea of what climate change might do to them."[472]

Ropeik, whose job is to help clients know how to communicate to achieve desired outcomes, especially to overcome "the human habit of dawdling in the face of looming risks," as Revkin describes it, has come to some basic conclusions about human nature when faced with major long-term risks like global warming.

First, to overcome "the human habit of dawdling," something needs to "ring our alarm bells." Reasonable-sounding arguments from experts based on the best scientific findings available, even if partially or fully accepted by those of us who are the targets of the arguments, are not enough to motivate us to do what is necessary.

Ropeik's basic understanding of human nature in the context of dealing with long-term risks, especially new ones whose impacts are still largely theoretical like global warming, leads him to conclude that fear for one's personal safety needs to be the primary motivator.

So if we can find a way to scare the crap out of people in terms of their own personal safety, we'll be successful.

Now you may find what I'm going to say next surprising. I don't think there is anything inherently wrong with fear. It is a natural response to danger. It can also awaken us from our lethargy. Fear of judgment from a holy and righteous God is a theme that runs right through the Bible. For sinners like ourselves, to fear a holy and righteous God is quite appropriate. To fear the LORD is a good thing.

But wait a minute. Isn't fear a "bad" feeling? Well, it is certainly an emotional state a normal person doesn't like to be in. But that's the point. Fear motivates us to avoid what makes us fearful. To be fearful of real physical and spiritual (Lk 12:5) dangers is appropriate, and if it motivates us to avoid them, then fear can lead to good results.

But precisely because being in a state of fear is something we want to avoid, we can do so by denying real dangers – especially those that are not right in our face, as Ropeik points out. Fear and dread are appropriate responses to what global warming will do. But that doesn't mean we will let ourselves experience them and face up to the problem.

For arguments' sake let's say we followed Ropeick's advice and tried to scare people about impending local impacts, and they accepted what we said about dangers to themselves rather than finding a way to deny it. Then where would we be?

It might be enough to get us to act, to make certain changes. It might be enough to get us to support our country in making changes. These in turn might be sufficient to have the United States help lead the world in reducing global warming pollution enough to avoid the worst of the impacts to ourselves – and as a collateral benefit, the world's poor. There could be some real truth to this particular version of the trickle-down theory applied to the problem of global warming.

But would fear for our personal safety really sustain us through our overcoming global warming marathon? I have my doubts. Even if it would, Christians must ask ourselves a further question: can that be the sum total of our motivation?

I just don't think fear for my personal survival is going to lead me to care very much for the poor in developing countries who will be hit hard by global warming. Why should it? The connection is profoundly tenuous. In such a context we may throw them some crumbs of compassion after we've taken care of ourselves. But such a token won't be for them, it will be

Global Warming and the Risen LORD

to make us feel better – or less bad – about ourselves and our preoccupation with our own well-being.

For Christians I believe there is something else besides fear, something else besides our own survival, which can motivate us to act in the face of global warming. As followers of the Risen LORD, who in the Garden of Gethsemane overcame the survival instinct, who even overcame the fear of being separated from his Father as he became everything he was not by taking on the sins of the world, we can also be motivated by something else: Christ's love for us and our resulting love for our LORD.

Don't get me wrong. I think we should talk about local impacts of global warming that will impact us directly. Indeed, I did so in chapters 3-4, albeit for the purpose of understanding global warming's impacts on our loved ones and friends.

However, as we will see, here is where the best scientific findings about who will be the most severely impacted meet up with where each of us is in our own spiritual maturity. The chances are quite good that most of us in the United States will escape life-threatening consequences from global warming. (If that's what it takes to set off our "alarm bells," then they will never go off.)

In contrast, the chances are quite likely for those in poor countries, that many of them could face life-threatening – or at a bare minimum, life-diminishing – consequences. As I've already pointed out, individuals in the United States have been and will continue to be major contributors to the problem, whereas the poor in developing countries have contributed next to nothing.

So in the "Blah, Blah, Blah … Bang!" global warming scenario, while we have been tuning out the "Blah, Blah, Blah," it is the poor who first get the "Bang!"

This is where I believe Jesus' teaching in Matthew 25 is more than sufficient to help guide us in our response. Here Jesus teaches us that it is the failure to proactively care for "the least of these" that prevents those judged unrighteous from entering the Kingdom.

But with global warming it's not just a case of failing to do righteous acts, of "benign neglect," to put a charitable spin on what Jesus accuses the unrighteous of. It's more than that.

Jesus says to the unrighteous, "I was hungry, and you gave me nothing to eat" (v. 42). But as we will see, global warming has been and will continue to make the hungry hungrier. Jesus says, "I was thirsty, and you gave me nothing to drink" (v. 42). But global warming has been and will continue to make the thirsty thirstier. "I needed clothes … I was sick" (v. 43). Global warming will make the needy poorer and the unhealthy sicker.

So, when it comes to global warming, we've got a ways to go simply to get back to a situation of "benign neglect," of having done no harm. The first thing we need to do is stop contributing to making the lives of the poor worse through global warming. It's as if in the parable of the Good Samaritan we actually aided the robbers in some way unbeknownst to us – and *then* we also passed by on the other side.

Once we've stopped making their situation worse, the second thing we must do is help them in their need rather than pass by.

As portrayed in Matthew 25, helping the poor in their need is what the righteous do, which Jesus presents as a matter of course for them. Just like the Good Samaritan, they come by, see someone in distress, their hearts are filled with pity, and they feed, clothe, and heal the unfortunate soul. They probably don't even think about it. They just do it.

And unbeknownst to them, they are actually doing these things for Jesus.

When he tells them so, they are surprised. "Lord, when did we see you hungry and feed you, or thirsty and give you something to drink? When did we see you a stranger and invite you in, or needing clothes and clothe you? When did we see you sick or in prison and go to visit you?" (vv. 37-39).

Jesus responds with one of the most memorable lines in Scripture: "I tell you the truth, whatever you did for one of the least of these brothers of mine, you did for me" (v. 40).

The righteous of Matthew 25, the Good Samaritan, they are not motivated by fear for their own survival. As in the Good Samaritan's case, it actually could be dangerous to help the least of these. They are motivated by compassion, by a readiness to suffer alongside those they seek to help. And it is here that they more fully image Christ, the true image of God. It is here that they more fully do the Father's will. Thus by freely yet unselfconsciously losing themselves in service to others do they find their true selves. They become ever increasingly glorious and spiritually beautiful.

But, honestly, are most of us Good Samaritans? The simple fact that the "good" Samaritan is lifted up by the LORD and the Church as a paragon of virtue demonstrates that we are not.

Based on my own experience – the stubbornness of my own sinfulness – I certainly can't claim to be one of the righteous as they are portrayed in Matthew 25.

Once again am I grateful for the mercy showed to Peter, our true companion in personal failure and redemption. Full of love for my LORD and full of myself, I may boast, "Even if all fall away, I never will. Even if I have to die with you, I will never disown you."

And yet how many times have I passed Him by in the ditch?

Thankfully, the LORD ain't through with me yet! I will continue to fail him, but He doesn't fail me. "With fear and trembling" I work out my salvation (Phil 2:12), comforted by what I believe. For I believe that the transformative power of God's grace in the blood of my Savior matters, "so that in him we might become the righteousness of God" (2 Cor 5:21). I believe that the "resurrection effect" matters. I believe that the continual presence of the Risen LORD and the Holy Spirit matters. As they did for Peter and Paul, these spiritual realities more real, actually, than our everyday experience of our sin-bound reality – can transform us here and now. I believe that love His love for us and then our love for Him – I believe that love rather than just fear, can motivate us; indeed, that for this marathon race of overcoming global warming, it is what will sustain us.

Central America is the most vulnerable part of the world to the consequences of hurricanes. According to the Intergovernmental Panel on Climate Change (IPCC) and the insurance industry, "in the next decades climate-related disasters could cost US$300 billion per year," with the possibility of reducing such costs if strong adaptive measures are taken.[473] As a point of comparison, losses for all disasters worldwide in the 1990s were $69 billion annually.[474]

As discussed in chapter 4, global warming will probably intensify hurricanes. A recent study suggested that for every 1°C (1.8°F) rise in sea surface temperatures, the number of Category 4 and 5 hurricanes could increase by 31% globally.[475] The IPCC projects that ocean temperatures worldwide could rise 2.2 °C, with a temperature rise in the Gulf of Mexico of 1°-3°C.[476] So we are looking at the possibility of Category 4s and 5s increasing by 60% or more. And as I pointed out earlier, such jumps in Category belie the increase in destructiveness, because destructiveness increases exponentially. For example, the destructive potential of a Category 4 hurricane is 250 times that of a Category 1.

The story of Hurricane Mitch can give us a sense of just how destructive hurricanes have been to Central America. In October 1998, the second hottest year on record behind 2005, Mitch formed in the Gulf

As we contemplate such a spiritual transformation within ourselves, we must hear the Risen LORD say to us, as he does to Thomas, "Stop your doubting and believe."

Like the disciples during Jesus' ministry before his crucifixion, we are still learning what type of Messiah the Risen LORD is. But those of us who have confessed Him as our LORD and our God have put ourselves under His Lordship, however incomplete and fragmentary is our servanthood. We are willing to be taught. We are willing to have our eyes opened to a more complete understanding of what it means to follow the Risen LORD.

as a Category 5 storm. Thankfully, it made landfall as a Category 1. But it also brought with it torrential rains, with 1-2 feet per day for a total of 75 inches of rain. One area experienced 25 inches of rain in just six hours.

Three million people were severely impacted by Mitch, with 11,000 dead and thousands still missing. It likely killed more people than any other Atlantic hurricane in the last 200 years. Hundreds of thousands of homes were destroyed. There was over $5 billion in economic damages.[477]

The country of Honduras was the hardest hit, with 70% of crops (cost: $900 million), 70%-80% of the transportation infrastructure, and 25% of the schools destroyed. Over 25 rural villages were completely washed away.[478] In the aftermath there were 30,000 cases of malaria,[479] not to mention other health impacts. As for poor households in Honduras, on average they lost 30%-40% of their income from crop production with the poorest losing around one-third of their assets. Honduras' national poverty rate increased by 8%.[480]

One of the families devastated by Hurricane Mitch was that of Olbin and Edgardo Casares, brothers who were 10 and 11 when the hurricane hit. Their father was killed and their house was washed away. A relative who didn't want them took in their mother. So they hitchhiked to San Pedro Sula, Honduras' second largest city, where three months later a reporter found them living behind an old building. At that time they begged or shined shoes, but they had no shoes of their own, and the only clothes they had were the ones they left home with. For such children, hunger, disease, and the threat of predation are constant dangers.

"'If it wasn't for Mitch, I'd be home now,' says Olbin ... Asked whether his mother knows where he is, Olbin just shrugs. 'I don't know where she is anymore. It's just me and my brother.'"[481]

In this spirit, before I conclude this discussion about our calling to love the poor as we follow the Risen LORD in overcoming global warming, let me bring another bit of honesty to the table.

For many of us in the United States who are not poor, the poor are invisible to us, and therefore largely forgotten. As Christians we know that Jesus tells us to care about their welfare, and so every now and then we remember them and probably feel guilty, about what we're not quite sure.

The simple fact is, for the vast majority of our lives we literally don't see them. A primary reason is very concrete – they don't live where we do. Some live in other countries and some in other neighborhoods. When we do see the poor who live here in the United States, we usually ignore them. For a whole host of reasons we are probably afraid of them. It becomes, out of sight, out of mind. It is no accident that Jesus calls these brothers and sisters of ours "the least of these."

When it comes to the problem of global warming, have the poor also been ignored, or, at the most been given lip service because we know that we should somehow care?

I encourage you to do a variation on Ropeick's informal experiment. Ask a few Christian friends the following question: "Name one way that climate change will especially impact the poor in poor countries in the next 10 or 20 years." Will most Christians you know struggle with that one?

It is our propensity to ignore the poor, to make them invisible, to have them disappear from our minds, that has me placing this section on loving the poor after the discussions about loving one another, loving neighbors/others and loving even enemies. This is because, in reality, they are farther out from ourselves.

Maybe my writing of this book and your reading it will give us eyes to see, so that the next time we encounter the Risen LORD in the ditch in form of a poor victim of global warming we won't pass by on the other side. We will instead join the Risen LORD in His fight to overcome global warming and care for the victims who are least among us. In that moment will we experience "life in His name."

5. _LOVE OF GOD_— "We love because he first loved us." (1 Jn 4:19)

In our exploration of the five great loves I have put the love of God last because it surrounds and infuses all the other discussions.

The title of this section, "Love of God," has an intentional double meaning. "Love of God" could mean our love for God. But it could also mean God's love for us. Here it means both; for us to do the former we must have the latter.

This leads us to the first letter of John. Despite its relative obscurity due to its physical placement in our Bibles, the first letter of John contains in

its fourth chapter one of the most famous declarations in Scripture, which it proclaims not once but twice: "God is love" (1 Jn 4:8, 16).

These simple but profound words constitute a deep, deep well of spiritual truth, truth about our God. In one way it is something that even small children intuitively grasp. Yet in another way it is something none of us will ever come close to understanding or experiencing completely.

The spiritual fact that God is love is demonstrated in the love between the Members of the Trinity, an eternal communion of love. In his magisterial prayer in the Gospel of John, Jesus says to the Father: "you loved me before the creation of the world" (17:24).

That God is love is at the heart of what God wants us to know about Himself, as most supremely revealed to us in the life and death of Christ. As 1 John 3:16 puts it, "This is how we know what love is: Jesus Christ laid down his life for us." And as the same numbered verse in the Gospel of John so memorably states, "For God so loved the world, that he gave his only begotten Son, that whosoever believeth in him should not perish, but have everlasting life" (Jn 3:16, KJV). The fourth chapter of 1 John states, "This is how God showed his love among us: He sent his one and only Son into the world that we might live through him. This is love: not that we loved God, but that he loved us and sent his Son as an atoning sacrifice for our sins" (vv. 9-10).

Now, have you merely run your eyes over these words, these Scriptures, or have you taken at least a moment to drink them in?

What we are trying to begin to start to fathom is what makes existence meaningful, where passion and purpose meet, where the Sacred touches our finitude in the incarnation, crucifixion, and resurrection. God's love is where they meet. This is the spiritually deep place where God leads us like a child because that's what we all are in our spiritual helplessness and ignorance. We cry out like hungry babies, starved for real love, for that is ultimately the only thing that can satisfy our spiritual hunger. Only from God can we *receive* such love. He gives His love continuously. But we must receive it.

Here, literally, is the good news. He loves us despite the fact that we are sinners, that we are, indeed, His enemies, that we killed His only begotten Son who freely gave his life for us. The horror of being separated from his Father's love was so terrifying that in the Garden of Gethsemane Jesus asked, "Take this cup from me" (Mk 14:36; cf Mt 26:39; Lk 22:42). And yet his love for the Father and for us was so great that he became everything he was not as he took on the sins of the whole world, as he experienced the absence of the Father's presence, "'My God, my God, why have you forsaken me?'" (Mk 15:34; Mt 27:46). For love's sake the eternal Triune communion was broken. God loves us so much He broke His own heart to save us.

Through such sacrificial love are we spiritually fed, is our spiritual hunger satisfied. "This is my body, which is broken for you" (1 Cor 11:24, KJV).

Is it any wonder, then, that 1 John simply proclaims, "God is love." And yet as our own experience of love tells us, love is ultimately mysterious. It is so because God is love. Just as we cannot know everything about God, nor will we ever know everything about love. But we can know something about love's true character, as revealed to us in Christ. Again, as 1 John 3:16 says, "This is how we know what love is: Jesus Christ laid down his life for us."

Thus, who God is, is revealed in what God does. For Christians, as the New Testament proclaims, God is most fully revealed in the incarnation, crucifixion, and resurrection. Here do we find that life and love are inextricably connected. The Universe doesn't exist simply for the sake of existing – not life for life's sake. It exists because part of the perfection of love amongst the Trinity is its expression for One Another.

Out of such love springs life. Existence, life, flows from love. We exist because of God's love. But as images of God we also exist to freely love Him back – love for love's sake. And so what is the meaning and purpose of life? It is not simply to exist but love for love's sake. Love of God. True life exists because of love, and true life is the expression of that love. As Jesus said about us, "I came that they may have life, and have it

abundantly. I am the good shepherd. The good shepherd lays down his life for the sheep" (Jn 10:10-11, NRSV).

It is the Good Shepherd who most supremely teaches us about the love of God. As the true *imago dei*, Jesus shows us what the ultimate point of our existence, loving God back, is like. To be an *imago dei* is to image or reflect God in our freedom, as Christ did most supremely when he said, "Father, if thou be willing, remove this cup from me: nevertheless not my will, but thine, be done" (Lk 22:42, KJV). Jesus' whole life was one of sacrificial love. "I am among you as one who serves" (Lk 22:27). As it says in Philippians, He "emptied himself, taking the form of a slave" (2:7, NRSV). And his love is a perfect reflection of the Father's love. "As the Father has loved me, so I have loved you" (Jn 15:9).

Jesus' most supreme act of loving God back was to be the Good Shepherd and lay his life down for us, his sheep, to love us with sacrificial love. And so in loving others do we also love God, as Jesus teaches us in the Great Commandments and with his whole life and most fully in his death. It is here that Jesus completely expresses the dual nature of the love of God – God's love for us and how we are to love Him back.

The call to love our fellow image bearers is precisely the reason 1 John 4 proclaims, "God is love." Why have we been told that God is love? To help us understand that to be in loving relationship with Him means that we must love others. As verses 7-8 state, "Dear friends, let us love one another, for love comes from God. Everyone who loves has been born of God and knows God. Whoever does not love does not know God, because God is love." Verse 12 continues, "No one has ever seen God; but if we love one another, God lives in us and his love is made complete in us." And verse 16 states, "God is love. Whoever lives in love lives in God, and God in him."

To love others, then, is how we come to know God, is how we can draw closer to God and have our communion with Him deepened, because God is love. When we love, we live in Him and He in us. But 1 John warns us that the opposite is also true: "Whoever does not love does not know God, because God is love." Further on it states: "anyone who does not love his brother, whom he has seen, cannot love God, whom he has not seen" (v. 20).

To know God is to love God, and those who love God, love others. To not do so is to fail to be a true image in our lack of reflecting God's love, to fail to become what we were created to be, to fail in the purpose of our very existence.

Jesus did not fail. And so for us to become *imago dei* is to become imago Christi. As 1 John 2:6 tells us, "Whoever claims to live in him must walk as Jesus did."

We were created in love for this very purpose: to love Him back. This is what makes our small, transitory, seemingly insignificant lives eternally meaningful. If there is significance to our lives, if there is purpose to our existence, if we want somehow to have our time here matter in the vast scheme of things, it is found here in the love of God.

As I said at the beginning of the chapter 9, we are part of the greatest love story ever. In light of His inestimable love for us, is it no wonder that Jesus identified as the greatest commandment that each of us is to love God "with all your heart and with all your soul and with all your mind and with all your strength" (Mk 12:29-30).

We can never repay Him. But we can have our lives be characterized by a love-filled gratitude in response as we live in and live out the five great loves.

So what, then, does loving God mean when it comes to overcoming global warming? It will mean things great and small, as we shall see. Overcoming global warming will provide us with many opportunities to love our loved ones and friends as Christ loves us, to love others as ourselves, to love our enemies (or have them love us), and to love the least of these as if they were the Risen LORD Himself.

But to do so we must understand what global warming will do as described in Part 1. Emotionally and intellectually we must come to grips with the magnitude of the impacts, especially upon the poor, so that we understand why this is one of the greatest challenges of this century, why each one of us is called in some way to walk with the Risen LORD as He overcomes global warming.

Did reading about what global warming will do make you weary? To the extent that you have allowed the love of God to exist within you, to that extent will you have compassion, which means to suffer with the one who is suffering. And this is also where we will find the Risen LORD, in the ditch suffering with those who are suffering – in this instance because of global warming.

But here's the thing. Many times when I get weary I fail to hope in the LORD to renew my strength. And as such I don't receive the promises of faith so beautifully described in Isaiah 40:31:

> but those who hope in the LORD
> will renew their strength.
> They will soar on wings like eagles;
> they will run and not grow weary,
> they will walk and not be faint.

When I am weary, instead of hoping in the LORD, I end up failing to spiritually fly, I run and get even wearier; I walk and spiritually faint (Isa 40:31). I don't want to love anymore, and so in my weariness I fail.

You know what? I'll let you in on a little secret. The Father knows. The Risen LORD knows. The Holy Spirit knows. God knows that you and I fail in love. God knows that you are weary, that you can frankly lose hope in yourself. But even when we understand that our hope cannot come from ourselves, from our own power, God knows that much of the time we don't have faith in our ability to let the power of His love carry us forward. We must let Him. But so much of the time we don't believe it. Although we lose our faith in Him, He does not lose His faith in us.

Precisely because God loves you He is not going to let you wallow in your failure. God will not leave you there. Why do you think Jesus came? Why do you think he died so that grace can abound? Why do you think the Risen LORD walks besides us? Why do you think the Holy Spirit is within you? Why do you think we have all of these spiritual riches?

As we have seen, the Risen LORD didn't let Peter just stay where he was spiritually. Peter had failed – spectacularly. He had denied Jesus three times. Although it was painful for Peter, the Risen LORD asked him three times whether he loved Him. For each denial Peter had to reaffirm his love. When Peter says yes, the Risen LORD tells him to love Him by serving those He loves.

The Risen LORD didn't let Paul stay where he was spiritually, either. And Paul was a really tough case. The LORD had to go so far as to blind him. Only then did Paul come to a place where he could truly see the LORD and His will, where he could let grace and the Risen LORD's presence and the indwelling of the Holy Spirit and the communion of the Church transform him from Saul the destroyer into Paul the Apostle.

The Risen LORD will not let you stay where you are spiritually either. He loves you just as much as he loved Peter, as he loved Paul.

Maybe, like Paul first opposed the Church, you have actually been on the other side of the global warming fight. But just maybe, like Paul, the LORD can use you to do great things in working with Him to overcome global warming.

Maybe you have failed to grasp the Risen LORD's Lordship over the problem of global warming. Honestly, this failure is child's play compared to so many other spiritual truths related to His Lordship we don't grasp. If the failure of the disciples teaches us anything, it teaches us to constantly entreat the Holy Spirit to test carefully our desires against the Lordship of Christ and the will of the Father. Our hearts are like the parable of the wheat and the tares (Mt 13:24-29). Because God made us and the Holy Spirit lives within us, we have His desires in our hearts. But we are also still sinners, and mixed in with God's desires are our own sinful desires. If we are not careful, our hearts can once again become the devil's playground.

Maybe you are afraid because you have a vague sense that the changes required to overcome global warming will diminish our economic well-being. If we are afraid to make the changes necessary to overcome global warming, if we seek to cling to our current understanding of how our economy is powered and not let the old ways go, then we will in fact help to cause the destruction of a good deal of life; by seeking to "save" or cling to what we believe makes our current lifestyles possible, we will lose this opportunity to become our true selves in faithfulness to the Risen LORD's efforts to overcome global warming. (As will be seen in chapters 16-19, such fear will impede the coming clean energy revolution.)

The simple fact is, we will continue to let things like fear of the wrong thing interfere with loving God. The spiritual treasures we have been given – the transformative power of His grace, His presence – are still contained in "jars of clay" (2 Cor 4:7). My beloved grandparents and spiritual ancestors, Jumo and GranHelen, who lived in "the bombing capital of the world," failed in fear to be completely righteous when it came to civil rights. Their example teaches me that "There is no one righteous, not even one" (Rom 3:10), that grace covers a multitude of sins. It also challenges me – and you – to get this one right.

The disciples, too, let fear lead to spiritual failure. They failed Him; He didn't fail them. They deserted Him; He never deserted them. They denied Him; He never denied them.

It is the same with us and the Risen LORD. We may fail, deny, or desert Him on global warming, or any other challenge that He claims Lordship over, but that doesn't mean the Risen LORD fails or denies or deserts us on overcoming global warming or anything else. He will just keep loving us and picking us up and dusting us off and inviting us to follow Him into the journey of our lives: loving Him back.

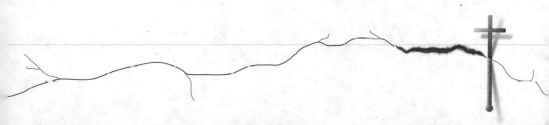

PART 3
Overcoming Global Warming
with
The Risen Lord

15

THE SPIRITUAL GOAL
OF OVERCOMING
GLOBAL WARMING:
TRANSFORMING THE FUTURE

Overcoming global warming is about creating a new future, one better than the one that will come if we fail to act. It's about tomorrow and the next day and fifty years from now. In comparison to all of human history, it is a mere blink of the eye. But it is our time, and for us to not see clearly what must be done and then do it, this blink, this moment, will set in motion harmful consequences that will curse future generations for thousands of years.

How, then, will we warm our hearts and steel our will to run with perseverance the race set before us? On our own we will fail. But we are not alone.

As Christians we look into the future with our hearts filled with the Father's love for us, with Jesus' blood cleansing us from all unrighteousness, with the Holy Spirit encouraging and guiding us, with the gloriously luminescent Risen LORD before us as the true *imago dei*. We look into the future reminded of how God transformed sinners like Peter and Paul and doubting Thomas and the other apostles. We look into the future with Jesus' teachings emblazoned upon our minds as we recall His commands to love one another even as he loved us, to love our neighbors as ourselves, to love our enemies and hope that they love us when we act as enemies, to love the least of these, those we find in the ditch of global warming's impacts.

Above all, we look into the future with grateful hearts for all these blessings as we stumble forward in our childish attempts to fulfill the true meaning and journey of our lives – to love God back. And in so doing our future becomes one of ever-increasing freedom, ever-increasing glory, ever-increasing spiritual beauty. In following the Risen LORD as He leads us in overcoming global warming in this great cause of freedom, that's our future. Are you ready?

THE CHRISTIAN VISION OF THE FUTURE FOR THE PRESENT

If we are going to talk about walking into the future with the Risen LORD, it would be helpful to briefly discuss what the Bible says about the future. Countless books have been written on this subject. Obviously I cannot give such a full treatment in this chapter. But it is important to have a basic understanding of the Christian vision of the future, especially as it relates to our discussion, to help frame our actions in the present and to help us have a sense of where the Risen LORD might be headed.

The Christian vision of the future is meant to help guide – not distract – us in the present. Unfortunately, at times during the course of Christian history certain interpretations of the biblical vision of the future have become distractions, including in our own day. I will try not to repeat that mistake here. For our purposes I will focus this discussion on the Kingdom of God, given that it is at the heart of Jesus' teaching. Even though at times the New Testament's discussion about the Kingdom of God can feel obscure and hard to comprehend, the basic idea is relatively straightforward.

Simply put, the Kingdom of God is where God reigns as King, as Sovereign. The Kingdom of God is where God's will is continuously and completely done. This is told to us succinctly in the Lord's Prayer: "Thy kingdom come, Thy will be done, on earth, as it is in heaven" (Mt 6:10, KJV; cf Lk 11:2-4). Heaven is currently where God's will is continuously and completely done. When God's will is done on Earth as it is in Heaven, when God ushers in His Kingdom in its fullness and this current fragmentary, sin-filled yet redeemed existence is utterly transformed, then God's Kingdom has come to Earth and then will His Kingdom spread from Heaven to Earth. As Revelation 21:3-5 (TNIV) says, after the heavenly city has come down to Earth:

> *Look! God's dwelling place is now among the people, and he will dwell with them. They will be his people, and God himself will be with them and be their God. He will wipe every tear from their eyes. There will be no more death or mourning or crying or pain, for the old order of things has passed away. He who was seated on the throne said, "I am making everything new!"*

We will dwell in God's Kingdom on a "new earth" (Rev 21:1) in our resurrected, transformed, spiritual bodies like the one the Risen LORD had after His resurrection.

If the Kingdom of God is the equivalent of Heaven on Earth, the earthly version of God's reign where His will is continuously and completely done, then during his earthly life Jesus of Nazareth was the embodiment of God's Kingdom. As the true image of God, as a true reflection of his Father's will, Jesus lived *in* and lived *out* the Kingdom. He was the inbreaking of the Kingdom into sinful, rebellious human reality. He was a walking, talking, breathing, laughing, crying, healing manifestation of His Father's reign. In Christ the Kingdom did come, and with His Second Coming, it will come in its fullness.

God's intent from the beginning was for Earth to be His Kingdom also, for us to freely reflect His dominion as His images. Having been made in the image of God, to freely image or reflect His will, we were to enjoy with God a heavenly place on Earth. From the beginning we were made to live – in freedom – in God's Kingdom. But with our freedom we chose death – alienation from God.

Jesus came that we might have life, and have it abundantly, as God intended. Jesus Christ is life. He showed us what true life is. And because he was true life, because he completely did his Father's will, by giving his life could he defeat death – with his death he defeated death. And with our reconciliation to the Father through Jesus, once again He has given us the promise of true life in the Kingdom.

In God's future, at a time of the Father's own choosing, He will fully and completely establish His Kingdom on Earth. He has not abandoned His creation or His intent for it. Far from it. Through His Son he has reconciled all of creation, and when His Son returns we also shall be raised to walk in newness of life in bodies like that of our Risen LORD. We will dwell with the LORD on a transformed New Earth, where His Kingdom will have come, where the Lordship of Christ will reign, "so that at the name of Jesus every knee will bow – in heaven and on earth and under the earth – and every tongue confess that Jesus Christ is Lord to the glory of God the Father" (Phil 2:10-11, NET).

But this Peaceable Kingdom, this Kingdom where the Prince of Peace will have been made King, is not yet here. Our hearts long for it, because God made us to live there. Indeed, as Paul teaches us in Romans 8, the rest of Creation, while it currently groans "as in the pains of childbirth right up to the present time," also "waits in eager expectation" for the fullness of the Kingdom where it "will be liberated from its bondage to decay and brought into the glorious freedom of the children of God" (Rom 8:19-22).

Even though we reside in a place and time where God's will is not continuously and completely done, where His Kingdom has not yet come, where the full flower of freedom has not yet arrived, nevertheless, by virtue of His Son's obedience even unto death, those of us who acknowledge His Lordship are granted citizenship *now* in His Kingdom. You may live where you currently reside, but your true citizenship is in Christ's Kingdom. As Paul says in Colossians, "He has rescued us from the power of darkness and transferred us into the kingdom of his beloved Son, in whom we have redemption, the forgiveness of sins" (Col 1:13-14, NRSV).

As Christians our true citizenship is in the Kingdom of Christ but at present we still live in an existence that is not conformed to the Father's will.

We still live in a land of rebellion that for the most part knows not that it rebels against true life.

And yet, within ourselves, and within society, and within the rest of Creation, pockets of resistance to the rebellion reign. When, however incompletely, we do God's will, when, however fragmentarily, we freely follow the resurrected LORD, to that degree can we catch glimmers of the glory of God's Reign, can we hear ever so faintly the sounds of eternity. In freely doing God's will, we step into God's reality and out of the shadow reality that encompasses so much of our lives and darkens the life of the world. We step into the true light that has been enlightening the world since God said, "Let there be light."

Here's the connection to our present topic: the tyranny of global warming today works against the inbreaking of God's Kingdom. It takes us in the opposite direction of where the Risen LORD is headed. It is part of the shadow reality that darkens not only our lives, but also the lives of those we love, the world's poor, and God's other creatures. As part of our participation in the resistance to the rebellion, it's time to take on this tyranny.

The fullness of the Kingdom will only arrive with the Second Coming of the LORD Jesus Christ. But in our already/not-yet existence, even as our own sinfulness continues to weigh us down and hold us back and have us moving in the wrong direction, we are called to be true images of God as we image Christ, as we follow the Risen LORD, as we do the Father's will. And in our lifetimes, one of the tasks for us in doing His will, in participating in the inbreaking of His Kingdom to the extent that we do His will, is to follow the Risen LORD in overcoming global warming in this great cause of freedom.

THE SPIRITUAL GOAL – BECOMING A CHRISTIAN AGENT OF TRANSFORMATION WITH THE HEART OF A SERVANT

In the next chapters, we will discuss some specifics about how we both as individuals and as a society can work to overcome global warming. Before we do so, some clarification about the actual spiritual goal would be helpful. What should be the major characteristic of our efforts to work with the Risen LORD in overcoming global warming in this great cause of freedom? From my perspective there tends to be a fairly common misperception of what the spiritual goal for Christians should be in this cause, namely that it should be about "personal sacrifice."

Extremes can capture the imagination. Both secular environmentalists and Christians can have extreme elements or strains of asceticism, of self-denial tinged with punishment, which can come to the fore in our minds when we begin to talk about what needs to be done. We might think to ourselves, "Those who sacrifice the most, who deny themselves the benefits of an economy powered by fossil fuels, are the real global warming moral superstars." As a result, one hears, "We must reduce our carbon footprint," and thinks, "That means going without. Less fun. More pain. Well, I guess as Christians we should be ready to sacrifice."

It is quite true that as Christians we should be ready to sacrifice when following the Risen LORD calls for it. We should even be ready to lay down our lives.

However, the current *stereotype* of what "personal sacrifice" means in responding to global warming – a sour, dour, denial framing of our future driven by a guilt-punishment dynamic – is a recipe for failure in overcoming global warming. It's as if all my talk about becoming more glorious and more free and more ourselves was really just some positive spin put on the real situation: we're going to have to endure a time of depravation – and we deserve it, too. Such an emotional and mental crouch is the opposite of what we need: a forward-leaning stance combined with a transformational mind-set.

Again, let me be clear. If at times in our journey of following the Risen LORD we are called to certain types of sacrifice, including of the material blessings of this life, we must be ready to do so with the anticipation that such sacrifices would actually bring freedom and joy.

In addition, the spiritual corrosion of riches and materialism is a profoundly serious issue. Jesus taught a good deal on this subject (e.g., Mt 6:24-33; 19:16-30; Lk 6:24; 18:18-30). But, as Jesus' teachings attest, the danger of materialism is its own issue that has existed long before the problem of global warming.

For individuals who, for their spiritual well-being, need to be set free from materialism, doing so can have positive co-benefits for dealing with global warming.

But hoping that more preaching about materialism/consumerism will solve global warming is unrealistic and misguided. For some it functions like an incantation, evoked as if doing so actually accomplishes something.

The goal of overcoming global warming in this great cause of freedom is not to achieve as individuals a misguided understanding of climate change purity (dare I say, moral superiority?) found in a quixotic attempt to reduce our carbon footprint to zero *by our own efforts* (which is in fact impossible living in the United States). Overcoming global warming is not about donning global warming hairshirts and living in a cave. That won't make you an overcoming-global-warming-spiritual-superstar. It is not really about making sacrifices in the sense of freezing in the dark, of some type of energy flagellation – at least it doesn't have to be if we are smart and committed. The more time and spiritual energy we waste in this misguided lacuna, the less time we will have to devote to the real spiritual goal. It also misses or ignores the fact that overcoming global warming is about addressing both causes *and* consequences, and freezing in the dark with your hairshirt on won't help the poor overcome the consequences.

Our mindset needs to be that we are going to start doing climate-friendly things, not that we are going to stop doing certain "bad" things. We need to be creative and discover common-sense solutions that others can also implement. We can show others how a joyous, love-filled life today includes helping the Risen LORD overcome global warming together in this great cause of freedom.

Furthermore, we can't just think, "What can I do?" We must also think, "What can we do?"

Finally, the focus needs to be on the Risen LORD and on helping those who are in the ditch of global warming's impacts, not on our own spiritual purity. The latter will come when, with servant-hearts, we focus on working together to help others – not how we as individuals look while we're doing it. As in all areas of our lives, we must guard against a spiritual preening, a public flaunting of our righteous acts (Mt 6:1-6, 16-18; 23:5-7; Mk 12:38-40; Lk 18:9-14).

This is not to say that we shouldn't be witnesses through our behavior, including what we can do personally to reduce our carbon footprint. We must. But we must do so in an encouraging, inviting manner that says, "See, you can do this, too!" rather than, "See how good I am compared to you! I'm willing to do the really hard stuff that only spiritual superstars like me can do. Look at how good I am! And I solved it all by myself, too!"

What, then, is the spiritual goal here? It is to do our particular part in following the Risen LORD in overcoming global warming. He wants each of us, using the gifts He has given us, and with hearts like his, hearts of a servant, **to become Christian agents of transformation, to be forward-leaning team-builders as we strive with Him to work with others in overcoming global warming in this great cause of freedom**.

I think some who use the word "sacrifice" in relation to global warming are actually wondering quite appropriately, "Is this going to require something of me?" If you are wondering this, then you would be rightly concerned that what may sound to you like the spiritual version of a free lunch – "Don't worry, there's no sacrifice involved" (which is not what I'm saying, by the way) – will end up being empty spiritual calories. You might be worried that if you are not called to make changes you will not grow; you will not begin to more truly image or reflect Christ; you will not be part of the resistance to the rebellion against the Kingdom; you will not be really following the Risen LORD in overcoming global warming.

So is overcoming global warming going to require something of you? That's a big YES.

In this great cause of freedom, becoming a Christian agent of transformation in overcoming global warming will require each of us to make changes. It will take effort, sometimes quite a lot. It is about being

vigilant, about doing things differently, which can be uncomfortable, even painful. But here's the good news: you will not be alone. You will be part of a movement spiritually spearheaded by the Risen LORD.

With the Risen LORD walking at our side we will make changes as individuals and as families. We will support and help bring about changes in our workplace and in public policy that will restructure our economy in climate-friendly ways. We will support the poor in developing countries in their efforts to overcome the consequences of global warming and achieve climate-friendly economic progress. In this great cause of freedom we will literally be part of a worldwide movement that is working together to overcome global warming.

Here in the United States, will we have some personal and societal "growing pains" as we make the necessary changes in this great cause of freedom? Yes. This will fall harder on some than others. An analysis by the non-partisan Congressional Budget Office (CBO) released in May 2010 found that one version of comprehensive climate change legislation could result in total employment being "slightly lower" during the next several decades.[482] Public policies should help those who might lose their jobs to transition to new ones. For some regions of the country dependent upon coal to create their electricity, their utility prices may go up more than others. Again, public policies should address this and even out the burden.

Here is a principle that we can follow both as individuals and as a society in this great cause of freedom: solve it the smart way – always; solve it the "sacrificial" way – only when necessary. As we will see, if we don't solve it the smart way, we won't solve it at all.

So this movement within the resistance to the rebellion will take effort. It will take creativity. It will take discipline. It will take staying power. It will take working together. This great cause of freedom may end up requiring some sacrifice. And staying faithful to following the Risen LORD as He leads the way in overcoming global warming will take all of the spiritual resources that God has so graciously provided.

But this is also why those of us who choose to follow the Risen LORD in this cause and become Christian agents of transformation can make a lasting contribution. With the LORD's help we can be vigilant, we can be disciplined, we can have staying power in overcoming global warming. This great cause of freedom, this movement within the resistance to the rebellion, needs people whose relationship with the LORD will sustain them for the long haul; it needs people with servant-hearts who will patiently and consistently look for ways they can contribute as we work together to overcome global warming.

The spiritual challenge, therefore, is making and keeping our commitment to change, our commitment to follow the Risen LORD in overcoming global warming. The challenge is to be saints in the world, forward-leaning team-builders, continuously transforming it with our LORD, to become, in this great cause of freedom, servant-heart-global warming-transformation-champions.

Of course, just like the disciples, at times we will fail. We will not always be faithful. Our sinful nature is a stubborn, stubborn reality. We are, as Martin Luther recognized, justified by Christ's imputed righteousness while simultaneously still sinners – which is why Christ our Savior must forever accompany our walk in life. The "goodest" of good news is that God in Christ has accomplished our salvation. It is a spiritual fact in the mind and heart of God. As Luther taught, Christ's righteousness has been imputed to us. At the moment of our salvation, while still in our sins, Christ's righteousness covers our sins in an act of sheer grace, sheer unmerited favor. He has imputed to us citizenship in His Kingdom. But, as Augustine taught, Christ's righteousness also has been imparted to us. It is a spiritual power within us, and this grace-filled, imparted righteousness matters in the here and now. Through further spiritually transformative acts of grace, God is working to sanctify us, to transform us into His righteousness, into what we were made to be, His true image, into *imago Christi*, into true citizens of His Kingdom, into those who fully and consistently and naturally and freely do His will. In the here and now this means following the Risen LORD on the road to His Kingdom.

But to the degree we truly follow our Risen LORD, to that degree we become more glorious, more spiritually beautiful, more free, more our true selves, more citizens of His Kingdom. This already/not-yet quality of our personal history is similar to the already/not-yet nature of human history when encountering God's reality, when in fleeting moments human history glimmers with God's reality as the mist of our sinful finitude clears ever so slightly and we catch a glimpse of the Kingdom of God.

Today the harmful consequences of climate change, the tyranny of global warming, is one of the forces arrayed against the inbreaking of the Kingdom. It is part of the rebellion. It works against the creating and sustaining and reconciling activities of Christ. It is part of the "not-yet" of human history – a freedom denier, a freedom destroyer.

And yet we know this: "For freedom Christ has set us free" – free to be free, not fake selves, not hiding in the shadows of our sinfulness. But free to be true images of Christ. We are being called not to be like Peter before the crucifixion – a wannabe hero infatuated with the false hope of his own glory. We need to be like Peter after the resurrection: true disciples, at peace with living or dying if it be the LORD's will, tempered by the crucifixion in terms of our own sinfulness but buoyed by the resurrection in terms of the power of God.

The truth is, I believe the Risen LORD is asking much more of us than to feel bad about not becoming the stereotype of a "global warming spiritual superstar." The One with the nail-scarred hands is actually asking us to change and to keep changing as we work with Him and others in overcoming global warming. In this great cause of freedom He is asking us to become in every area of our lives a Christian agent of transformation with a servant-heart. He's inviting us to join Him and others in transforming our society into a climate-friendly one that leads the world in generosity for the sake of those in the ditch, for His sake. For freedom Christ has set you free, free to become who He created you to be.

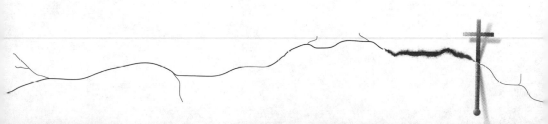

16

OVERCOMING THE CAUSES –
IS IT POSSIBLE?

But is it possible?

Is it possible to overcome global warming?

All things are possible for the LORD, of course. But in overcoming the major problems we face as individuals, as a society, as the world, the LORD has decided to work with us and through us and the wonderful and terrible gift of freedom He has given us. Jesus told us, "with God all things are possible" (Mt 19:26) and Paul assures us, "I can do everything through him [Christ] who gives me strength" (Phil 4:13). We know that the Risen LORD walks beside us, "And remember, I am with you always, to the end of the age" (Mt 28:20, NRSV). As Christians we can rest in these promises.
Nevertheless some of you might be wondering, "I appreciate all of these inspiring words. But since God is going to work through us – a dicey proposition, I must say, frankly – has anyone scoped this out? Done the X's and O's? You know, actually tried to see if this might be possible in the real world?"

Good questions.

When it comes to overcoming the *causes* of global warming we can also take some hope from major studies that have looked into how we can begin. There are, thankfully, many ways the LORD can work through us and others to overcome the causes of global warming in this great cause of freedom. Because here in the United States and in many other countries the burning of fossil fuels is so entwined with everything we do, because, frankly, we have not been optimal stewards of the energy available to us, because dealing with the *consequences* includes the enhancement of the economic and democratic dimensions of freedom, opportunities to do better abound!

In overcoming global warming we must do two basic things:

1) address the sources, or *causes,* of the problem by slowing and eventually reversing the accumulation of greenhouse gases and black carbon in the atmosphere due to human activities; and

2) adapt to the *consequences* of global warming that have occurred and will occur even if we make a serious effort now to address the sources.

Causes and consequences – addressing those are what we're about in overcoming global warming. This chapter and the next three chapters (17-19) deal with how we can overcome the causes with the Risen LORD by our side. Chapters 20-21 will explore how we can overcome the consequences.

Addressing the sources or causes of the problem is called *mitigation.* We want to reduce or mitigate the sources. As the old saying goes, when you're in a hole, you need to stop digging. Right now we are still adding a huge amount of greenhouse gases and black carbon to the atmosphere. We're still digging the hole. We need to slow, stop, and reverse that trend. If we don't, we could face much more devastating impacts than those I have described in previous chapters. As Christians, we must do everything we can to not let this happen.

This chapter addresses the question, "Is it possible to overcome the causes?" Thankfully, as will be explained, the answer is yes.

Chapter 17 demonstrates how it is possible to overcome the causes of global warming *worldwide*. This will give us a "bird's-eye view," so to speak, of the worldwide potential to overcome the causes.

Chapters 18 and 19 provide an "on the ground" perspective, describing what mitigation will look like here in the United States (chapter 18) and in developing countries (chapter 19). Concrete examples will illustrate how it is indeed possible to overcome the causes of global warming.

IS IT POSSIBLE TO OVERCOME THE CAUSES?

Major studies looking at mitigation have found that it is quite possible for us to make the transformations necessary to keep the global temperature from rising higher than two degrees Celsius, or 3.6°F, above preindustrial levels,[483] a goal agreed to by most countries in the *Copenhagen Accord* in December 2009.[484] Keeping the global temperature from rising no more than this would provide us a decent chance of avoiding dire consequences.

For example, in the United States it would make future heat waves 55% less intense, reduce sea level rise from thermal expansion by 57%, and cut in half significant changes in precipitation, lessening droughts and floods.[485] More fundamentally, it would also reduce the possibility that an irreversible momentum would be created, leading to abrupt and devastating impacts worldwide.[486]

Keeping the temperature rise to 2°C above preindustrial levels (or about 1°F above 2009 levels) is possible with today's technologies at a modest cost or even a net economic benefit to society as a whole. The IPCC's 2007 report suggests an annual investment of between 0.2% and 3.0% of global GDP would be required.[487] A more recent study by the respected business consulting firm McKinsey & Co. (in cooperation with Shell Oil, the World Wildlife Fund, and prominent academic experts) projects that the cost would be less than 1% of global GDP annually.[488] Another McKinsey & Co. study gives an annual cost range of 0.6%-1.4% of global GDP.[489] These figures don't include the economic and other benefits we would gain by making these investments.

So for an investment of about 1%-2% of global GDP we could set the world on the path to overcoming global warming. In comparison, in 2005 the world spent 3.3% of global GDP on insurance (excluding life insurance)[490] and the United States spent 4.8% of GDP on national defense. (We spent about 6% on national defense during the Cold War.[491])

Four things need to be pointed out about these cost estimates for addressing climate change.

First, two examples from history suggest that cost projections for reducing pollution can be much higher than the actual costs. The costs of the Montreal Protocol, the international treaty to phase out chemicals that deplete the ozone layer in the atmosphere, have been 87% less than the estimates. Similarly, costs to implement the 1990 Clean Air Act have been between 53% and 94% less than originally projected.[492] While dealing with greenhouse gases will be more complicated, McKinsey & Co. suggests that the same principle will apply to overcoming global warming: there will be many cost-saving measures in the future that we don't forsee today." [493] As these two cases demonstrated, once industry was required to find solutions, they responded with "product substitutions, technological innovations, and changed business practices."[494] Production prices dropped as companies learned how to do new practices and procedures better ("learning curves") and as the products become mass-produced ("economies of scale"). Even though we know this is how things will work, they are not included in cost projections today precisely because many of the factors that will reduce costs are unforeseeable now.

Second, these cost estimates do not include the savings from health benefits, reducing our dependence of foreign oil, and other benefits, which are significant. For example, a George W. Bush Administration analysis showed that cleaning up our air via the 1990 Clean Air Act led to benefits exceeding costs by 12 times[495] – an investment with an incredible rate of return. A recent report from the National Academy of Sciences found that

Global Warming and the Risen LORD

using fossil fuels to create electricity and power our vehicles cost our society over $120 billion in hidden expenses, mainly in health costs.[496]

An earthy example of tackling water pollution and global warming pollution simultaneously concerns hog waste from large hog farms. A recently devised approach in North Carolina called the "Super Soil System" simultaneously eliminated 95%-99% of the various water pollutants while reducing its global warming pollution by 97%, earning an additional $1.80 or more per hog by selling the carbon credits it subsequently generates.[497]

As for our oil dependence, in 2006 the total of all costs related to our oil imports – including trade impacts, loss of jobs, and defense outlays – was estimated to be $825 billion per year. This is like adding $8.35 to the price of a gallon of gasoline.[498]

Thus, while cost estimates for investments to overcome global warming don't incorporate such savings, these investments will lead to savings from cleaning up our air and water, and reducing our dependence on foreign oil from unstable regions and countries that deny religious freedom to their citizens.

Third, the economic costs of overcoming global warming are not simply economic losses. These costs would be **investments**. We would simply be making different investment decisions. For example, instead of building a traditional coal-fired power plant, we might build a wind farm. Or instead of buying a regular SUV, we would buy a hybrid car.

One area where we absolutely must invest in is energy efficiency – doing more with the same amount of energy. This must be the foundation upon which our mitigation or greenhouse pollutant reduction efforts will be built. Worldwide, more than half of the needed reductions are projected to come from efficiency improvements.[499] Purchasing a hybrid car as opposed to an SUV is one example. If you bought a 2009 Toyota Prius instead of a 2009 Chevrolet Tahoe SUV, you could save $1,255 per year in fuel costs, reduce global warming pollution by over 7 tons per year, and reduce our dependence on foreign oil.[500] Another specific example is LED lighting, which are "three to four times more energy efficient than incandescent bulbs and last up to five times longer than compact fluorescents." This makes them the longest lasting lighting available.[501]

Many companies that have already been working to reduce their carbon footprint through efficiency gains have achieved significant savings. "Johnson & Johnson, for example, cut its CO_2 emissions by 22 percent from 2003 to 2006 due largely to energy efficiency measures, while growing annual revenues by 27 percent – an improvement in carbon productivity of 64 percent."[502]

Throughout the world's economy there is a great deal of potential for such savings via efficiency gains. McKinsey & Co. estimates that 35%-45% of the mitigation potential of developed countries is in the form of efficiency gains that would save money.[503] However, some of these efficiency gains cost more up front even though they save money over their lifetimes – an investment opportunity that is often not taken. To overcome global warming, this must change.

Climate-friendly investments can also produce a better rate of return on job creation when compared to the old way of doing business. One analysis found that "on average, three to five times as many jobs were created by a similar investment in renewable energy versus than when the same investment was made in fossil-fuel energy systems."[504] Climate-friendly jobs can also be created in areas that have recently lost jobs or where traditional production has slowed down. For example, a company that manufactures a particular type of solar power called Concentrating Solar Power (CSP), has designed such a system that can be mass manufactured "on existing auto production lines and shipped as a kit that can be installed by the most basic construction crew."[505] As the CEO explains, it can be "stamped out like a Chevy and installed like a Maytag."[506] Once in operation, CSP power plants create 94 jobs to run the plant compared to between 10 and 60 jobs for coal and natural gas power plants, yet their operating costs are 30% lower.[507]

Finally, the costs of inaction, of doing nothing, are almost unthinkable in terms of human suffering and that of the rest of God's

Global Warming and the Risen LORD

creation. But even in economic terms, the costs of inaction are substantial. Although currently hard to estimate, a major report by Sir Nicholas Stern for the British government puts the costs at between 5%-20% of global GDP, and it considers these costs conservative and permanent.[508] By comparison, during the Great Depression GDP in the United States fell by 25%, whereas during the Great Recession of 2008-09, the loss of GDP in the United States was 2% (through February 2009).[509] Thus, the permanent minimal cost of not addressing global warming would be more than double the 2008-09 recession, and the upper end of its economic cost would be close to experiencing a permanent global Great Depression.

Comparing these estimated costs of inaction to the estimates of the investments needed to overcome global warming of a little more than 1% of global GDP, there's every reason to move forward quickly.

Global warming can and must be overcome. But we must be clear-eyed in understanding that the breadth and depth and speed of the change required in addressing the causes are enormous. To keep the rise in temperature to 2°C above preindustrial levels will require a tenfold increase in carbon productivity.[510] Has something like this ever been achieved? Yes. "This is comparable in magnitude to the labor productivity increases of the Industrial Revolution."[511] But, in comparison, we will have to achieve this "carbon revolution" in one-third the time. We need to do it in forty years, by 2050.[512]

If we don't improve our carbon productivity in an aggressive but manageable way, increasing it by 5.6% per year, then at a certain point to reach our goal we would have to take drastic measures, a crash-carbon-diet if you will. If we were required to live on the per-person carbon budget we will need to achieve by 2050, at our current low level of carbon productivity, we would be forced to make some choices. For example, on a given day you could choose to do *just one* of these four things: (1) a 12-25 mile car ride; or (2) a day of air conditioning; or (3) buy two new T-shirts without driving to the store; or (4) eat two meals. As McKinsey &Co. puts it, "without a major boost in carbon productivity, stabilizing GHG emissions would require a major drop in lifestyle for developed countries and the loss of hope in developing economies for greater prosperity through economic growth."[513]

So, harkening back to the discussion about "personal sacrifice," and putting it harshly, if we are lazy and self-centered and dumb we *will* have to sacrifice, and we will force the poor in developing countries to sacrifice even more. If we are smart and vigilant and want to help others, we can enhance our quality of life (e.g., a cleaner environment) while simultaneously helping the poor to develop cleanly and sustainably.

Given all of this, if we tackle this challenge with gusto the world could end up making money while avoiding tremendous suffering and a permanent recession-or-depression-size reduction in the economy.

So is it possible? The results from major studies from McKinsey & Co. and others lead to this basic conclusion: **we can do it**.

But to do it, those of us who are Christians will have to play our part and become Christian agents of transformation in overcoming global warming. Working with others, we will have to keep at it. We will have to keep pushing. We must choose to be smart, to make and support climate-friendly decisions. In this great cause of freedom we have to choose to invest, to support good policies and good leadership. Just as righteousness is the presence of good acts, not simply the absence of bad ones, we must be proactive in overcoming global warming; we must be forward-leaning team-builders. The good news is this: **if we do, we can**. We won't if we don't.

We also must begin to act NOW. Starting in 2010, the world needs to begin overcoming the causes of global warming in a major way. To keep us to a rise of 2°C, emissions of CO_2 must peak between 2015-2020.[514] The International Energy Agency (IEA) has warned that after 2010 each year of delay in getting started would add approximately $500 billion to the worldwide cost, and that "a delay of just a few years" would put the 2°C goal "completely out of reach."[515] This, in turn, would put us in greater danger of crossing thresholds better left uncrossed.

In overcoming the causes, the bottom line is this: **it's time to get real, get smart, and get busy**.

For those of us ready to follow the Risen LORD as he leads the way in overcoming global warming, we have the spiritual resources necessary to take on this tremendous challenge. Here before us are incredible opportunities to love God back.

A KEY INGREDIENT: PUBLIC POLICIES TO UNLEASH THE PRIVATE SECTOR

We have seen that it is indeed possible to keep warming to 2°C above preindustrial levels (or about 1°F above 2009 levels) if the world achieves a tenfold increase in carbon productivity by 2050 – a doable but daunting challenge. Even the International Energy Agency (IEA), a quite conventional organization known in the past for discounting the potential of alternative energy sources, has recently stated, "A global revolution is needed in ways that energy is supplied and used."[516]

For such a revolution to take place in the time required, public policies are needed to catalyze and focus private sector implementation.

A GLOBAL EFFORT REQUIRED

We must remember that global warming is a *global* problem. The countries of the world must work together to overcome the causes or it won't work. The participation of major emitters like the United States and China is especially important. We need an international treaty with verifiable global warming pollution reduction targets and timetables for every nation – leaving each country to determine how to meet its emissions reductions. In addition, it is essential for the rich countries to assist the poor nations in overcoming both the causes and the consequences.

NATIONAL POLICIES THAT HARNESS PRIVATE SECTOR IMPLEMENTATION

The needed tenfold increase in worldwide carbon productivity by 2050 should be implemented primarily by the private sector to ensure efficiency, innovation, cost-effectiveness, and choice. But in this marathon of overcoming the causes of global warming, governments must create and enforce the "rules of the road." They also must lay out a safe course and the ultimate destination in order for the private sector to play its role in the timeframe required.

The myriad of decisions that go into training for and running the race must be left up to each runner, so to speak. Governments must create the rules, set the conditions, and then get out of the way and let the runners run the race. Put another way, while we must avoid an ideological aversion to the use of government power in the marketplace, we need also to strive for as little government intrusion in the markets as possible to allow them to thrive and play their part in bringing about the common good. But as we have seen in the United States with the Great Recession of 2008-09, without proper government oversight, huge problems can occur that then require massive government intervention to fix. Better to strike the right balance from the beginning.

What this means in the United States is that we need a comprehensive national climate change policy that will move our economy away from carbon-intensive ways of doing business towards climate-friendly ways, and do so by using the power of markets as much as we can to spur innovation, efficiency, and choice.

Given this, two of the most important functions of government policies are to

      (1)   send a price signal to the marketplace strong enough to shift investment decisions away from carbon intensive investments towards climate-friendly ones; and

      (2)   provide businesses the certainty they need about what will be required of them in the future so they can plan accordingly.

As for sending a price signal, until quite recently in the United States, releasing CO_2, other greenhouse gases, and black carbon into the atmosphere has been "free." Well, not really free. It has been what economists refer to as a "negative externality," meaning that the costs to society have not been borne by the producer or the consumer of the products that cause the global warming pollution. Instead, the costs are borne by others in the form of the damages that have been and will be caused by global warming. These damages have not been included in the price, thereby sending the wrong price signal and preventing the market from optimizing the common good.

To put it starkly, we have been freeloading off our children's future. There has been a cost we haven't been paying. We may not have understood that until recently, but now it's time to send the right price signal by having our goods and services reflect their true costs when it comes to global warming. The fact that this hasn't been the case means that we have been denied the information we need to make the right choices for the common good. Once the prices are right and they begin to reflect the costs of global warming, then the markets can play their constructive role.

The good news is that when the price of goods and services reflect their true cost, we can use the power of the market and the choices of businesses and consumers rather than onerous and inefficient government regulations and bureaucrats to decide how we are going to overcome global warming. It's actually about *increasing* freedom: more freedom in how we overcome the causes, and more freedom by overcoming the tyranny of global warming's consequences.

A specific policy mechanism that accomplishes both the price signal and the certainty, and does so in a manner that takes advantage of market dynamics to drive innovation and cost-effectiveness, is called a cap-and-trade system.[517] It was first used in a significant way by President George H. W. Bush with the 1990 Clean Air Act, which, as we have seen, has been highly successful in reducing air pollution cost-effectively, doing so between 53%-94% less than the original cost projections.

Here's how a cap-and-trade system works. A limit, or "cap," is placed on how much total global warming pollution will be allowed each year (either across the economy or across a sector, such as utilities). To achieve the gradual glide path of emissions reductions you tighten the limit or cap each year on a long-term timetable that everyone is made aware of. The timetable provides the certainty – you know what's coming. Each entity (such as a power plant) receives or buys a certain number of "carbon credits," with the total credits equaling the amount of global warming pollution allowed by the cap. The entity can either use its credits or, if it reduces its carbon pollution below its allowed amount, sell its excess credits to another entity for a price. The gradual tightening of the cap reduces the total number of credits, making each credit more valuable over time, which in turn provides the incentive (or price signal) for entities to move away from carbon-intensive products and services towards climate-friendly ones.

It is this "buy and sell" or "trade" part of the cap-and-trade system that creates much of the positive economic benefits arising out of tapping into market dynamics – e.g., choice and cost-efficiency. Since global warming is a problem created in the atmosphere, emissions reductions anywhere on the planet achieve the same reduction benefits. So trading allows us to find the cheapest ways to achieve the desired level of reduction.

If, for example, it is cheaper for one plant in Texas to reduce its emissions compared to another in Vermont, then a "cap-and-trade" system allows the Texas plant to reduce its emissions and get paid for it, while the Vermont plant is given the flexibility to either reduce its emissions or pay the price for continuing to pollute.

Here in the United States there has been a great deal of controversy surrounding both an international agreement and national climate change legislation that would implement a cap-and-trade system as one of the policy instruments. The most vehement opposition has come primarily from two sources: (1) some of the companies and businesses that are very carbon intensive and don't want the price of their products to reflect their true cost when it comes to global warming pollution; and (2) those who are opposed to nearly all government intrusion into the marketplace. Whether due to self-interest or ideology, the opposition has only delayed our inevitable transition to a clean energy economy, increased our overall risk, made the poor more vulnerable, and driven up the costs of implementation.

On the other side, for those in the United States who have been focused primarily on the passage of international and national climate policies, there has been a tendency to view those as ends in themselves. However, by themselves they do absolutely nothing; they are simply words. In this marathon of overcoming the causes of global warming, the approval of such policies represents the crack of the gun to *start* the race. Everything before that point was establishing the rules, laying out the course and destination, and training to run the overcoming-global-warming-marathon. **We still actually have to run the race**, which we will be doing for the rest of our lives. It's not just pass a law, ratify a treaty, and we're done. At that point we've only just begun.

Once the policies start to be implemented and the divisive politics have burned themselves out, that's when our country will come together and really take to heart that our efforts are part of a much larger plan to overcome global warming.

CONCLUSION

In this overcoming-global-warming-marathon, once we have the rules established and enforced and we've got the course and destination determined via the right government policies, how are we actually going to run the race? If the private sector should be the primary implementers of a tenfold increase in worldwide carbon productivity by 2050 to ensure efficiency, innovation, cost-effectiveness, and choice, how could this possibly be done? The next three chapters will address this question.

Here are three points to keep in mind as we enter into the discussion of how we will overcome the causes of global warming.

First, we know that we can. We also know it is possible and easily affordable using (1) off-the-shelf technologies, (2) technologies that have a realistic chance of being deployed in the near future, and (3) proven changes in forestry and agricultural practices – the latter primarily in developing countries. We also know from past experience that creativity and innovation will bring about new and better solutions that we cannot foresee today.

Second, overcoming the causes of global warming will require efforts by a majority of the people on the planet. Mitigation potential worldwide is diffuse, not concentrated. It is "scattered across scores of industry sectors and geographies."[518] This is not something a few powerful people can solve. Nor are the solutions located in one or two sectors of the economy or regions of the world, so that we could simply have the United States switch to wind power to create electricity, for example, and the problem would be solved.

There are some important concentrations of mitigation potential, to be sure. The largest is actually the forestry sector, where we find 25% of the world's greenhouse gas mitigation potential, followed by the power sector at 22%.[519] But achieving the potential in these two sectors, especially in forestry and agriculture, simply proves the point: it will require the participation of billions of people all over the globe, and in a good number of cases will involve solutions or practices that will need to be customized and implemented locally.

This diffusion of the potential to reduce global warming pollution is a major challenge for the world, meaning it is **a tremendous opportunity for Christians**. There are over 2 billion of us all over the globe who can be Christian agents of transformation in overcoming global warming.

The third point highlights another incredible opportunity for Christians and others who want to enhance the lives of the poor. According to McKinsey's analysis, 67% of the greenhouse gas mitigation opportunities required to keep the world below 2°C above preindustrial levels (or about 1°F above 2009 levels) are found in developing countries.[520] Let that sink in: **two-thirds of the greenhouse gas mitigation potential resides in the developing countries**. This means there will need to be huge investments in developing countries in clean energy, energy efficiency, and in carbon sequestration via the protection and growth of forests and agricultural practices that store carbon. Thus, many positive co-benefits could result for developing countries in order for the world to overcome the causes of global warming. However, to help ensure that the poorest will benefit, Christians and others concerned for their well-being will need to be proactive in looking out for their interests.

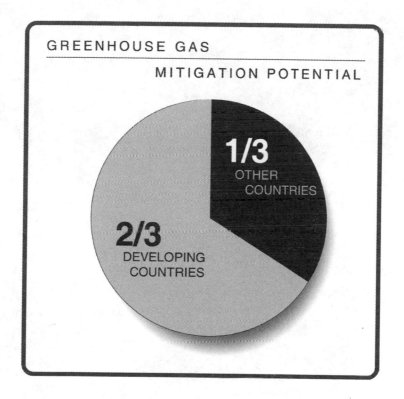

GREENHOUSE GAS

MITIGATION POTENTIAL

1/3
OTHER
COUNTRIES

2/3
DEVELOPING
COUNTRIES

17

OVERCOMING THE CAUSES –
MITIGATION WORLDWIDE

This chapter will provide an overview of how global warming needs to be overcome worldwide. The next two chapters will provide concrete examples of how it can be done in the United States (chapter 18) and developing countries (chapter 19).

Experts from McKinsey & Co. have identified five areas where our efforts need to be turbocharged if we are to keep the temperature rise below 2°C by achieving a *worldwide* tenfold increase in carbon productivity by 2050, improving it 5.6% per year:

1.   Energy efficiency;
2.   Climate-friendly energy sources;
3.   Development and deployment of new technologies;
4.   Changing the attitudes and behaviors of business managers and consumers; and
5.   Forests.[521]

Each will be considered in turn.

1. ENERGY EFFICIENCY

Energy efficiency is the foundation upon which overcoming the causes of global warming will be built. It is the largest way in the next several decades that we will overcome the causes of global warming, with the International Energy Agency (IEA) estimating that over 50% of the solutions worldwide will come from efficiency through an investment of $6 trillion.[522]

Energy efficiency is also a tremendous money-saving opportunity for both developed and developing countries. McKinsey & Co. estimates that 35%-45% of the greenhouse gas mitigation potential of developed countries is in the form of efficiency gains that would save money.[523] In the United States, a National Academy of Sciences report found that efficiency gains throughout the economy could reduce energy use by 15% by 2020 and 30% by 2030, with additional reductions possible if more aggressive efforts are taken.[524]

McKinsey and Co. found that a $520 billion investment in the United States through 2020 would cut energy use by 23% and save $1.2 trillion.[525] As for developing countries, if they were to achieve economic progress using off-the-shelf energy efficient technologies and by building efficient buildings they could cut their energy demand 25% by 2020, saving $600 billion a year.[526]

Where do we find the efficiency opportunities we need to capture *worldwide*? Here's where:[527]

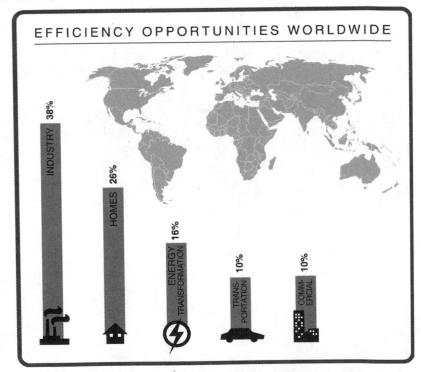

EFFICIENCY OPPORTUNITIES WORLDWIDE

INDUSTRY 38%
HOMES 26%
ENERGY TRANSFORMATION 16%
TRANS-PORTATION 10%
COMM-ERCIAL 10%

Global Warming and the Risen LORD

According to one estimate, if between 2009-2020 the world were to capture all of its energy efficiency opportunities at an annual cost of $170 billion, it would yield $900 billion a year in savings.[528]

2. CLIMATE-FRIENDLY ENERGY SOURCES

Currently, 81% of the world's energy needs are met by fossil fuels.[529] Projections are that approximately $20 trillion will be invested in the area of energy by 2035, much of it in developing countries.[530] The world cannot successfully achieve mitigation if collectively these investments are mostly in carbon-intensive energy sources, like they are at present. We need to "decarbonize" the world's energy system through efficiency, carbon capture, and especially by switching from oil and coal to alternatives. Just as there was an Industrial Revolution, now we need what McKinsey & Co. describes as a "carbon revolution" – but one that happens much more quickly.[531]

So what are the alternatives to traditional coal-fired power plants, which currently still produce over 50% of electricity in the United States and over 40% worldwide, or to vehicles powered by gasoline from oil?

McKinsey & Co.'s analysis concludes that by 2030, 70% of global electricity could be "decarbonized" or supplied by a mixture of wind, solar, hydro, nuclear, and coal with carbon capture and storage (CCS).[532] This would be an increase of 40% of these sources compared to their 2005 levels. They also project that biofuels – or fuels to power vehicles that are made from plants – could supply 25% of global transportation fuel needs.[533]

Others are more bullish on the potential of renewable sources like wind, solar, hydro, and geothermal, in combination with strong efficiency gains, to supply the world's electricity needs without adding more nuclear or coal-with-CCS.[534] The potential for renewables is indeed quite large. For example, an area covering less than 4% of the Sahara Desert "could produce an amount of solar electricity equal to current global electricity demand."[535]

A recent mapping of worldwide wind potential suggests that even if only approximately 20% of wind potential were captured, it could supply "over seven times the world's electricity needs" and 100% of all the world's energy needs.[536] Wind power from three states, Texas, North Dakota, and Kansas, could theoretically supply all US electricity needs, even excluding large areas within these states for environmental reasons.[537]

Likewise, in the United States there is enough offshore wind potential to satisfy the electricity needs of the entire country.[538] China also has more wind potential than their current energy needs.[539] In the United States, land in seven Southwestern states suitable for Concentrating Solar Power (CSP) plants could generate "nearly seven times the nation's existing [electricity] capacity from all sources." Indeed, one-fifth of current US electricity could be produced on a plot of land the size of Phoenix.[540]

However, capturing such potential renewable energy and getting it into homes and buildings in the form of competitively priced electricity available at the flip of a switch, is another matter altogether. When these practicalities are considered, the potential for renewables is still quite large. Wind, for example, could provide 20% of US electricity by 2030 while creating 500,000 jobs.[541]

While nuclear and coal-with-CCS (carbon capture and storage) are less desirable in numerous respects, a realistic appraisal at this time suggests they should not be ruled out in terms of their mitigation potential. In the case of coal, the top five countries in terms of coal reserves are the United States, Russia, China, Australia, and India.[542] None of these countries, including the United States, are going to give up using coal anytime soon. At current consumption rates, the United States has enough coal to last for 100 years.[543]

The National Academy of Sciences concludes that "the entire existing coal power fleet could be replaced by CCS coal power by 2035" by retrofitting old plants and building new plants with CCS.[544] Currently in China about 70% of its power comes from coal-fired plants, and in a recent five-year time span they "built the equivalent of America's entire coal power generation system."[545] At its present rate China builds a coal-burning power plant about once a week. China's demand for energy is expected to grow 3.9% per year through 2030, and much of this will be met by coal.[546]

The IEA suggests that worldwide 14% of the solution will come from CCS.[547] McKinsey & Co.'s analysis concludes that even if conventional coal (i.e., coal burned *without* CCS) continues to be an important part of the energy mix, we can still keep the temperature rise to below 2°C.

3. DEVELOPMENT AND DEPLOYMENT OF NEW TECHNOLOGIES

The US National Academy of Sciences concluded in a 2009 report that "extensive" research and development will be required in this decade "to enable accelerated deployments of new energy technologies starting around 2020." It advises, "A portfolio that supports a broad range of initiatives from basic research through demonstration will likely be more effective than targeted efforts to identify and select technology winners and losers."

However, the report does recommend some key areas for further research – e.g., advanced batteries, fuel cells, advanced solar technologies – as well as some technologies that have already gone through the basic research and development (R&D) stage and now require demonstration projects. These include new nuclear technologies, cellulosic ethanol, and carbon capture and storage (CCS) to be used with coal-burning power plants.[548]

How much of an R&D investment will be needed to put the world on a trajectory to overcome the causes of global warming via a tenfold increase in carbon productivity? McKinsey & Co. estimates that the world will need to spend $20 billion a year by 2020 and $80 billion by 2050.[549] Honestly, such sums are incredibly modest. Others have suggested upwards of $700 billion a year.[550] For comparison's sake, if we were simply to take the $450 billion the world spends annually on subsidizing energy and petroleum products[551] – the causes of global warming – and shifted it to decarbonized energy production, we'd be well on our way.

Unfortunately, over the last several decades, energy R&D funding worldwide has been heading in the opposite direction, falling by nearly half from 1980 to 2007. While private R&D spending, which currently dwarfs public spending, has been increasing, it still only represents 0.5% of revenue. This pales in comparison to the 8% in the electronics industry and 15% in the pharmaceutical industry. Not surprisingly, this anemic investment has produced only a small amount of technological advancement, as demonstrated by the fact that in 2005 patents for renewable energy were a miniscule 0.4% of the total.[552]

It is time to get real, get smart, and get busy – and nowhere is this truer than in investing in energy R&D. The good news is that investments in R&D yield rates of return anywhere from 20% to 50%, much higher than investments in capital.[553]

4. CHANGING THE ATTITUDES AND BEHAVIORS OF BUSINESS MANAGERS AND CONSUMERS

Changing attitudes and behaviors of business managers and consumers is the fourth area identified by McKinsey & Co. as playing a major factor in achieving a tenfold increase in carbon productivity by 2050 in order to keep the warming at or below 2°C. Their point here is that while price signals resulting from climate policies will help drive constructive changes, achieving a tenfold increase in carbon productivity by 2050 will require more than putting a price on carbon.[554] Business managers and consumers must also be educated and motivated to change above and beyond responding to price signals.

There are times in both our personal and professional lives where we don't change, even if something costs more. Reasons include: (1) we may not know why we should, either because the information is not readily available or because we have not actively sought out the necessary information; (2) it may not be worth the bother for us to change; we are willing to pay more to keep on doing things the way we've always done them, either because we are too busy, too lazy or both; (3) we may be ideologically opposed to making the changes the price signals are pushing us towards. Thus, in many cases both large and small we will need to choose to

Global Warming and the Risen LORD

change as we play our part in overcoming global warming, including our consumer choices.

One interesting example where the decision of a business manager helped change consumer behavior is former Walmart CEO Lee Scott's decision in September 2007 to sell only concentrated laundry detergents. This decision was made due to Walmart's commitment to sustainability, which they fulfilled in May 2008.[555]

Because as consumers we tend to equate volume with value – e.g., a laundry detergent bottle with more liquid in it must wash more loads than one with less – detergent manufacturers add "fillers" that increase the volume but do nothing to clean clothes. This also takes up scarce retail shelf space, leaving less for competitors. Detergent fillers also have a significant environmental and climate cost in terms of their additional packaging and added weight during transportation, given that concentrated detergents have 20% fewer lifecycle emissions.

So something that falsely appears to provide more bang for the buck – a big bottle of detergent – ends up giving a real bang to the climate. Lee Scott's decision to have Walmart sell only concentrated laundry detergents will save during a three-year period more than 400 million gallons of water, 80 million pounds of plastic resin, and 125 million pounds of cardboard, as well as the energy saved by not producing and transporting all this unnecessary stuff.[556]

Another wrinkle to this detergent story is the fact that Walmart's desire to address climate emissions helped inspire one of its suppliers, Proctor and Gamble, to produce a detergent that gets clothes as clean in cold water, reducing emissions from heating wash water. Both Walmart and Proctor and Gamble have also invested in educational efforts to help customers understand the advantages of these concentrated, cold water detergents.

McKinsey & Co. points out that this effort at overcoming global warming "involved no new technology, no significant capital investments, no loss of functionality for consumers and, in economic terms, no loss of GDP." What were required were intentionality and a commitment to change. The leaders at Walmart and Proctor and Gamble chose to take reducing global warming pollution "into consideration as an objective in the design for the product and its supporting value chain. This in turn, required managers and consumers to change their mindset."[557] They also chose to help their customers understand that a smaller bottle of detergent cleaned the same amount of clothes as a bigger one, and that they could get their clothes just as clean in cold water.

McKinsey & Co. further point out that even if there were a price on carbon provided by something like a cap-and-trade system, it probably would not have sent enough of a price signal alone (concentrated detergents would have been 15¢ cheaper a bottle) to shift consumer behavior.[558]

As a consumer, let's say you have been buying big bottles of detergent for years and you are dubious of such concentrated, cold-water detergents. You're also not crazy about adjusting how you have done laundry for years. You'd rather just continue on autopilot, buy the big bottle, wash in warm or hot water.

But as a follower of the Risen LORD, the One who is leading the efforts to overcome global warming, the one who taught us that to be faithful with the small decisions helps us prepare for the big ones (Lk 16:10), well, this is a no-brainer. You buy the concentrated cold water detergent and are thankful for such an easy way to make a difference that pretty much fell right in your lap.

Whether you are a business leader like Lee Scott, or an average consumer, it is the willingness to change things both large and small that will add up to overcoming global warming. In this century, overcoming global warming will involve literally billions of decisions by people all over the world. And in the United States, we don't have to be Lee Scott to make a difference, given that households account for 38% of our CO_2 emissions, which is more than the total of any other country except China.[559]

Household actions can indeed make an important contribution. One recent study published in the journal of the National Academy of Sciences analyzed 17 such actions that would involve "readily available technology, with low or zero cost or attractive returns on investment, and without appreciable changes in lifestyle." From past studies on the most effective ways to get people to change their behavior in these 17 areas, they estimated "the proportion of current nonadopters that could be induced to take action." The results? A reduction of 7.4% of all US emissions within 10 years, which is greater that all emissions from iron, steel, and aluminum production and petroleum refining combined.[560]

The 17 actions this study chose to analyze that individual households could take were as follows:

> (1) weatherizing one's home;
>
> (2) upgrading the heating and cooling (HVAC) system;
>
> (3) installing low-flow showerheads;
>
> (4) using an efficient water heater;
>
> (5) purchasing efficient appliances;
>
> (6) driving on low rolling resistance tires;
>
> (7) purchasing a fuel-efficient vehicle;
>
> (8) regularly changing HVAC air filters;
>
> (9) tuning up air-conditioning units;
>
> (10) maintaining vehicles routinely;
>
> (11) washing clothes in cold water;
>
> (12) lowering water heater temperature;
>
> (13) eliminating standby electricity;
>
> (14) setting thermostat setbacks;
>
> (15) line drying of clothes;
>
> (16) adjusting driving behavior to save gas; and
>
> (17) carpooling and combining trips.[561]

While, as we have just seen, we can make important changes through our buying decisions that help overcome global warming, I'd like to make an important point. Our decisions as individuals involve much more than what we choose to buy. As followers of the Risen LORD we are much more than our consumer choices. In overcoming global warming there will be many other types of individual decisions that have nothing to do with consumer choices, such as whether to donate money to a Christian organization helping the poor overcome the consequences of global warming, or to join a group at your church that wants to overcome global warming together, or to support government policies that help address the causes and consequences of climate change, or to become a Christian climate entrepreneur with the heart of a servant. Our consumer choices are important, even vital, to overcoming the causes of global warming. As mentioned earlier, no righteous act is too small to the one who bore our sins on the cross. But following the Risen LORD in this great cause of freedom will involve much, much more than our consumer choices.

5. FORESTS

Because trees absorb carbon and help hold carbon in soil, forests provide 25% of the world's mitigation potential, the largest of any sector.[562] The vast majority of this potential resides in tropical countries such as Brazil and Indonesia, where 50% of their projected business-as-usual emissions to 2030 would come from the destruction of forests if significant changes are not made.[563]

Forests have value in their own right, of course, independent of climate considerations or how they benefit humanity. But as Haiti, the poorest country in the Western hemisphere, unfortunately demonstrates, the absence of them helps illustrate some of the value they do have for human well-being.

In Haiti there is a saying, "Either this tree must die, or I must die in its place."[564] While desperate circumstances may dictate that a father cut down a tree to feed his starving family, such desperation can contribute to long-term devastation. In the case of Haiti, the country was once nearly all forested, but now there is less than 3% of forest cover left. Due to the resulting soil erosion between 1950 and 1990, the amount of land suitable for farming has been reduced by two-fifths. Deforestation has also reduced evaporation back to the atmosphere over Haiti, and rainfall has decreased by as much a 40% in many areas. When the rains do come, even moderate storms can produce devastating floods and landslides. Streams become choked with sediment and pollution, which also degrade estuaries and coastal ecosystems.

Forests help purify water systems. The absence of this service combined with the presence of excess sediment and pollution result in unclean water in Haiti, which in turn leads to over 90% of the children being "chronically infected with intestinal parasites."[565]

Recently there have been some attempts to monetize the "ecosystem services" that forests provide to humans. One estimate suggests, "Halting deforestation could save the global economy between $2 and $5 trillion per year in lost ecosystem services."[566] The Masoala National Park in Madagascar (a country where 70% of the population lives on less than $2 a day) can serve as an example. Some of the ecosystem services provided by its 888 square miles have been estimated to be worth the following:[567]

Medicines (e.g., those derived from the rosy periwinkle)	$1,577,800
Erosion Control	$380,000
Carbon Sequestration (in trees, other plants, and soil)	$105,111,000
Recreation/Tourism (e.g., to see the red ruffed lemur)	$5,160,000
Forest Products for Local Residents	$4,270,000
TOTAL =	$116,498,800

Climate mitigation provides the vast majority of the monetary benefits in this example. Not only can people make money through carbon sequestration in the form of trees, it will cost the world 50% more if we don't capture the opportunities offered by the world's forests.[568]

While at its heart the solution is relatively straightforward – don't cut down trees and do plant more of them – achieving this becomes complicated in a hurry. McKinsey & Co. noted that such efforts are difficult "due to the diffuse nature of the opportunity, the fragmentation of the potential actors, the complexity of implementing effective land-use policies in developing countries, and the need for substantial capacity-building."[569] Once again, it presents tremendous opportunities for Christians around the world to make a difference.

Given the incredible economic pressures to cut down forests and turn them into products such as paper, beef, soy (80% of which goes to feed livestock), and palm oil,[570] one of the most important ways we will be able to have forests be part of overcoming global warming is to make a tree more economically valuable alive rather than dead.[571]

One Christian organization that knows a great deal about forestry in poor countries is Plant with Purpose (formerly Floresta), founded in 1984. Its Executive Director, Scott Sabin, puts it plainly: "Any real solution must take into account both environmental and economic considerations." Scott once walked deep into a national park in Indonesia that was riddled with illegal logging activity and observed, "The efforts to protect the forest are well intentioned. But without corresponding changes in the incentives for the people who rely on the land, nothing will change."[572]

Having both national and international policies that pay to store carbon in the form of live trees, whether currently standing or planted for this purpose, will help provide the financial incentive to keep them alive. In this way, overcoming global warming can also contribute to maintaining all of the other benefits forests provide to people, not to mention maintaining the habitat of thousands of God's other creatures who are threatened with extinction due to human activities.

In addition to the right financial signals being sent, those of us who follow the Risen LORD and are called to image Christ the Creator must become much more involved than we have been. When Scott Sabin first started at Plant with Purpose years ago as a volunteer, his father said to him, "Planting trees for Jesus? That's about as marginal as you can get."[573] It should not surprise us to find Jesus at the margins, of course. And while Plant with Purpose's work has grown significantly since then, overcoming global warming will require much, much more from the Church as a whole than our current efforts produce.

A recent example of a reforestation project by the Christian organizations World Vision Australia and World Vision Ethiopia that is earning carbon credits through an international program called the Clean Development Mechanism (CDM) is the community-based Humbo Farmers

Managed Natural Regeneration Project in Ethiopia. Before the project began in 2006, the area had been denuded and degraded for over two decades, causing floods, mudslides, and erosion, all of which contributed to hunger, malnutrition, and decreased water availability.

In 2007-08 over 540,000 trees were planted on the site of the project, and another 90,000 were planted in the surrounding area.[574] Besides the generation of carbon credits, other benefits include: protection from soil erosion, flooding, and mudslides; sustainable supplies of clean water due to the recharge and filtration of water supplies; habitat for wildlife; and creation-and-climate-friendly economic progress through the "managed harvesting of forest products, including firewood, fodder, edible fruits and leaves, medicines, and dyes."[575]

However, as an independent analysis of the project concluded, "The process the Humbo project had to undertake to secure CDM carbon credits was highly complex and likely would have been unsuccessful without considerable resources of an organization such as World Vision."[576] The legal and financial complexities alone were quite daunting, such as: understanding Ethiopian land-use and property-rights; organizing the local communities into seven incorporated cooperatives; reaching an agreement with the World Bank on forestry management practices; the "development of a sophisticated financial forecast that built elements of commodities futures trading into the project model";[577] and the completion of the CDM project paperwork, which in this instance took 96 pages, filled with technical information and analysis.[578]

As well, for the project to be successful, World Vision needed the trust of the local communities and government that they had earned by working in the area for over 20 years. "Difficulties experienced during implementation could have derailed the project and it not been for the strong relationships in place."[579]

The CDM has been criticized for its overly burdensome process. What economists call the "transaction costs" of a deal – in this case the legal work, the financial and scientific analyses, the paperwork – must be significantly reduced if forests in developing countries are going to provide

25% of the solution as projected by McKinsey & Co. Even with these improvements, such projects will still require various types of technical expertise – e.g., scientific, legal, financial – as well as up-front financial resources. The poor will need help – whether from fellow Christians or others of good will in their own country, from their government, from Christian organizations like World Vision, from other organizations ready to lend a hand, or from a combination of all three.

Having the proper financial incentives is critical to having forests play their role in climate mitigation. However, the poor must also be protected from being taken advantage of or cheated out of the benefits of these projects. Government policies – from the international level on down – must protect the lives of the poor and vulnerable by

(1)   respecting the land tenure and resource rights of those dependent upon forest resources;

(2)   consulting local communities regarding projects;

(3)   strengthening the ability of local and community organizations to successfully negotiate and benefit from such carbon sequestration projects; and

(4)   making certain that the poor share equitably in the revenue generated from these activities.[580]

To meet these goals, Christians will need to work with others to ensure that the right policies are enacted and enforced, thereby allowing the poor and their allies to take advantage of these new opportunities.

Overcoming global warming makes many of the world's trees more valuable alive than dead. When at least some of the incredible benefits they provide to us and God's other creatures are properly incentivized in our economic system, and when the right policies are enacted and enforced to protect the vulnerable, then new opportunities will be created that allow the poor to lift themselves out of poverty by keeping trees alive. Then we will be able to reverse the Haitian saying quoted earlier – "Either this tree must die, or I must die in its place" – where desperation leads to devastation.

Instead, by keeping trees alive the lives of the poor will be enhanced so that such desperate circumstances can be transformed.

In the future, the right incentives and policies can lead to more situations like those in the Haitian village of Moren. Since 1997 Plant with Purpose has worked in Moren, and is currently assisting over 70 families there to reforest and utilize other soil conservation techniques. In a single year over 2,000 trees were planted. One Haitian farmer who has participated from the beginning is Jean Marie Forvil. He has been able to increase his savings 20 times – "an exceptional accomplishment in a place where saving money had once seemed virtually impossible."[581] As a result of such improvements, many families like Mr. Forvil's have been able to send their children to school.

Like in the Haitian village of Moren, the Mosquitia region of Honduras provides examples of success stories that, with the right incentives and policies, could be multiplied all around the world. In this region a Honduran Christian organization named MOPAWI has been helping the indigenous population for over two decades to achieve sustainable economic progress through conservation practices. One of their approaches is to replace slash-and-burn agricultural practices – which destroys rainforest and depletes the soils within a few years, forcing poor farmers to repeat the cycle – with what is called an "alley cropping" approach using the Inga tree.

Alley cropping involves planting crops between rows of trees (in this instance corn and beans). The Inga tree, native to Central America, is utilized for several reasons. First, it replenishes the soil's productive capacity because it helps put nitrogen and phosphorus back into the soil. Second, it grows quickly; creating a canopy in two years whose shade helps kill weeds. Third, the trees can be carefully pruned year after year, with the larger branches used for firewood and the smaller ones left with the leaves to create mulch that retains moisture and feeds the crops.[582]

MOPAWI's Executive Director, Osvaldo Munguía, reported to me that as of the fall of 2009 they had worked with 20 households to teach them the Inga alley cropping approach, which has improved their yield of corn and beans by 30%. In addition, one of the farmers, Florentino Garcia, from the village of Los Flores, Capapan, has "harvested enough firewood from his Inga plot to provide three families with the necessary firewood for the entire season."[583] Thus, Inga alley cropping avoids deforestation by providing an alternative to both slash-and-burn agriculture and to cutting down the rainforest for fuel wood. It replaces harmful practices with ones that are even more productive, and sustainably so.

Another example from MOPAWI's work, in this case with an indigenous people called the Tawira – whose name means "people of the beautiful hair" – involves benefits the Tawira receive from a native species of palm tree called *Elaeis oleifera*. The Tawira call *Elaeis oleifera* "Dawan Yamnika Kum sa," which means "A blessing from God." The blessing comes in the form of an oil that is extracted year round from the fruit bunches of the trees. For centuries the Tawira have used the oil for a variety of purposes, including as a moisturizer for their hair – hence their name, "people of the beautiful hair."

Its use as a hair product led to the founding of an entire company called Ojon (which was purchased by Estée Lauder in 2007 while retaining its name, leadership, and approach). Ojon markets its products by stating, "From the remote tropical rainforests of Central America, an ancient native secret to naturally beautiful hair has been discovered."[584] In 2003 representatives from MOPAWI helped the company's founder, Denis Simioni, find the Tawira and their "blessing from God," or *Elaeis oleifera*.[585] This mutually beneficial relationship now supports 2,000 local jobs. MOPAWI helps the Tawira ensure that the oil continues to be harvested sustainably by them in the traditional manner and that the Tawira's rights are protected and the community develops in a healthy way. Ojon highlights all of this in its marketing.[586]

One of those who have benefited, Arturo Blucha from Kalpu, in the Rio Kruta, summarizes the outcomes this way: "We are producers from 40 villages; we have improved our food supply because we can buy food from the stores now; we can recover quicker from Tropical storms and flooding; we are able to send our children to school and do repairing of our houses; we can buy tools for working the land."[587]

When international climate policies help to make rainforests more valuable alive than dead, and when such incentives are combined with sustainable practices like the ones just described, then will "the people of the beautiful hair" and others like them have both their lives enhanced and their ways of life protected.

Here is a summary of the discussion thus far of how the causes of global warming can be overcome. It will require

(1)   an international treaty, including mitigation requirements for each nation, with the rich countries helping the poor nations achieve climate-friendly economic progress;

(2)   national policies that harness private sector implementation;

(3)   major investments in energy efficiency, climate-friendly energy sources, the development and deployment of new technologies, climate-friendly decisions by business managers and consumers; and

(4)   massive efforts in the forestry sector.

Such investments need to begin now, in a major way. Our calling as followers of the Risen LORD is to be Christian agents of transformation and to use the gifts and opportunities we have been given to lead the charge in the mitigation revolution.

Having done a "fly-over" to get the big picture worldwide, the next two chapters will provide "on the ground" examples of how mitigation is already happening in the United States and in developing countries.

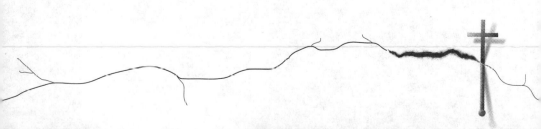

18

OVERCOMING THE CAUSES –
MITIGATION IN THE UNITED STATES

This chapter is focused on concrete examples of what overcoming the causes of global warming already looks like in the United States, including ways we can all participate. We will explore examples of energy efficiency in homes, commercial buildings, industry, and transportation. Then we will discuss why this hasn't happened more often. Next will be examples of how clean energy technologies will be a major driver of jobs in this decade and beyond. Finally, through the example of zero energy homes, the possibilities from combining efficiency and renewable energy will be highlighted.

ENERGY EFFICIENCY

Before getting into specifics, let me do two things: (1) clarify the difference between efficiency and conservation; and (2) recap why efficiency is essential. Both efficiency and conservation are good stewardship practices, but they are also different.

Energy efficiency is getting more out of the same amount of energy, more bang for your buck, so to speak. It is cutting out the wasted energy, the energy that is not producing anything.

Conservation is using less energy, like turning off the lights or turning down the thermostat to 67°F in the winter and up to 73°F in the summer. Conservation activities don't necessarily mean that you have improved your efficiency.

However, with efficiency improvements, such as weather stripping or regularly replacing your home furnace filter or by buying a more efficient furnace you could keep the temperature the same as before, say 70° year-round, but use less energy to achieve the same result. Depending upon how much your efficiency gain was, you could even save or conserve energy and still have the thermostat at 70°. Or, on top of these efficiency improvements you could go the extra mile and continue to keep it at 67°F in the winter and up to 73°F in the summer – saving even more by combining efficiency and conservation.

In general, the only way that efficiency doesn't help to overcome global warming, or its contribution is diminished, is if instead of producing the same result with less energy, we decide to take the financial savings and produce even more with it. (This is known as the "rebound effect.") Continuing with our earlier example, after weather stripping and buying a more efficient furnace we decide *not* to keep the thermostat at 70 degrees. Rather, we increase it to 72° in the winter and to 68° in the summer. In doing so, depending upon our efficiency gains, we wipe out or diminish its effectiveness in overcoming global warming. Thus, we must choose to have our efficiency gains make their proper contribution and not increase our consumption simply because it has become cheaper.

Now let me summarize the threefold connection between efficiency, conservation, and overcoming the causes of global warming. First, as we transition away from carbon-intensive energy sources such as oil and coal, we must use less of such sources to meet our energy needs in order to reduce global warming pollution. Second, without such efficiency and conservation gains it will take much longer for alternative energy sources, such as wind and solar, to be able to supply our energy needs – time we don't have if we want to avoid the worst consequences of global warming. Third, the financial savings from efficiency is a crucial element in making mitigation affordable. In summary, there is no realistic scenario for overcoming global

Global Warming and the Risen LORD

warming that doesn't include significant efficiency gains. It is foundational. Now to the specifics.

HOMES

In our homes we're talking about such things as weatherization by adding insulation and caulking and sealing the house. Such efforts can reduce energy use by 10% to 25%. Yet a recent survey found that nearly three-fourths of those polled underestimated the benefits, assuming weatherization would only save 10% or less.[588] Another way to save energy is by installing a programmable thermostat – no need to cool or heat your house when you are away, for example. This can save $100 annually and reduce global warming pollution by 10 tons. Other options include upgrading anything that uses power to a more efficient one (e.g., an Energy Star approved appliance) at the end of its life. This includes the water heater, furnace, A/C unit or heat pump, TV, washer, dryer, refrigerator, stove, dishwasher, and changing all lights to LEDs, which are over 50% more efficient, contain no mercury, and last up to five times as long as compact fluorescents (CFLs).[589] New refrigerators, for example, are on average twice as efficient as ones built before 1993. A final home improvement option is window replacement. New energy-efficient windows, with low-emissivity (or low-e) coated glass, gas fills, spacers, and improved frames, can save up to 16% on heating costs and up to 23% on cooling costs.[590]

Home improvements that reduce energy use and money should not be available only to those who can afford them. Helping low-income families in the United States make their homes more efficient represents a win-win. For over 30 years the federal government's Weatherization Assistance Program (WAP) has helped families earning approximately $15,000 or less to do just that, reducing their heating and cooling bills by 32%.[591] The average household spends 5% of income on energy, whereas the average low-income household spends 15%, with some on fixed incomes as high as 35%. After help from WAP, this is reduced by 5% for the average low-income household and by 14% for those on fixed incomes.[592]

Much bigger savings can be achieved by building new homes with efficiency in mind from the beginning. For example, an Energy Star qualified new home will be up to 30% more efficient than a standard home, and is required to have (1) effective insulation in attics, walls, and floors; (2) highly efficient windows; (3) tightly sealed construction (especially ducts); (4) efficient heating and cooling equipment; and (5) efficient appliances, lighting fixtures, and ventilation fans. Those purchasing such homes could be eligible for an energy efficient mortgage at a favorable interest rate. And such homes will be quite attractive when it comes time to sell.[593]

COMMERCIAL BUILDINGS

Many of the same ideas that apply to homes apply also to commercial buildings, resulting in significant savings. For example, a local grocery store in Sacramento, Vic's Market, has saved $48,000 a year by upgrading their lighting and freezers. The customers are happier, too. Store owner, Vic D'Stefani explains: "People complained that it was too cold in the freezer aisle before, so with the new equipment they shop longer."[594]

Church buildings also fall into this commercial buildings category, and significant savings can be had in our church facilities as well. For example, Prestonwood Baptist Church in Plano, Texas, a church with over 25,000 members, has instituted an energy efficiency program that has reduced their energy use by over 30%, saving them approximately $725,000 a year. The program has included switching to efficient lighting, installing occupancy sensors, reducing hot water settings, and reconfiguring facility usage patterns.[595]

A final example comes from the giant retailer Walmart. Earlier I highlighted their decision to only sell concentrated laundry detergents. But their efforts at "sustainability" have involved even more. In 2005 they launched a major initiative with three overarching aspirational goals: (1) to be supplied by 100% renewable energy; (2) to create zero waste; and (3) to sell products that are produced sustainably. Their efforts on all three will continue to make important contributions to reducing global warming pollution.

One of their specific near-term goals is to reduce the greenhouse gases created to power their stores and warehouses by 20% from 2005 levels by 2012. This will be achieved in large measure by improving the energy efficiency of their buildings. They have begun to open stores that use 20%-45% less energy, a major part of which is achieved with more efficient lighting. This includes what they term "daylight harvesting," which involves approximately one skylight for every 1000 square feet paired with computer-controlled daylight sensors that dim the lights when not needed. This system can reduce the electricity used for lighting by up to 75% during daylight hours. Each system saves enough energy annually to power the equivalent of over 70 single-family homes for a year. As of September 2007, Walmart had installed this system in nearly 2,500 stores.[596]

INDUSTRY

Tremendous opportunities for energy efficiency gains are found in the industrial sector, which includes such things as the manufacturing of paper, chemicals, cement, steel, plastics, glass, and the products made from them. In the United States, by 2020 over half of our energy will be consumed by industry. An investment of $113 billion from 2009-2020 would yield a return of $442 billion and reduce total US energy use by 18%.[597]

One major approach to increasing efficiency is called cogeneration or combined heat and power (CHP). CHP has actually been utilized in the industrial sector for a century (and is starting to be utilized in commercial buildings). Standard electricity generation involves using a power source like the burning of coal to create heat to boil water to create steam that then powers a generator that creates electricity. So if you have heat and water you can create electricity.

The manufacturing process creates heat as a by-product. By capturing this heat you can (1) create electricity, (2) use it to heat buildings in close proximity, (3) cycle it back into powering the manufacturing process, or (4) do a combination of these. In industrial manufacturing a CHP system can nearly double the standard 45% efficiency to 70%-80%.[598] But even greater efficiencies are possible, as demonstrated by the CHP system at ExxonMobil's oil refinery in Beaumont, Texas, which has an operating efficiency of 88% – thereby reducing global warming pollution by the equivalent of 397,000 passenger vehicles per year.[599]

Other specific examples of increasing industrial efficiency include: (1) upgrading to more efficient boilers at the MinnTac plant of US Steel Co, the largest producer of steel in the country; this resulted in $790,000 in annual savings for an upfront cost of $1.2 million, with a payback in less than 1.5 years;[600] (2) fixing pipe leaks at Dow Chemical's St. Charles Operations petrochemical plant in Hahnville, Louisiana, resulting in $1.9 million in yearly savings for an upfront cost of $225,000 and a payback in six weeks;[601] and (3) a company-wide efficiency improvement program at CEMEX USA, a large manufacturer of cement and other building materials, which through 2008 has resulted in a 16% reduction in global warming pollution or the equivalent of 21,000 passenger vehicles per year.[602]

TRANSPORTATION

While transportation represents 10% of the worldwide opportunity to increase carbon productivity, here in the United States it represents 34% of our total.[603] (We own 30% of the world's passenger vehicles, over 220 million of them.[604])

As individuals, we can make an important difference with our transportation choices. Indeed, driving represents over half the global warming pollution created by the average household.[605] It's one of the reasons back in 2002 I led an educational campaign that captured a great deal of media attention using the question "What Would Jesus Drive?"

To walk the walk and "drive the talk," my wife and I only have one vehicle, a hybrid Prius, which averages around 45 mpg – and therefore produces two-thirds less global warming pollution than a vehicle that gets 15 mpg.[606] We purposely purchased a home that was a short walk to public transportation, which we use for commuting and other purposes. Making such major decisions as home and vehicle purchases with overcoming global warming in mind allows one to build such contributions into the structure of everyday life. It simply becomes the new normal.

Currently, there is no pollution control device that will reduce CO_2 emitted by vehicles; so policies that increase fuel economy (or fuel efficiency) are an important part of overcoming global warming in the United States.

While driving a more fuel-efficient vehicle is a critical component, we must not fall prey to the "rebound effect" mentioned earlier – in this case driving more because our efficiency gains make it cheaper to do so. Studies show that when the government raises fuel economy standards, vehicle miles traveled (VMT) go up, diminishing the gains.

Consistently using public transportation is one of the biggest ways an individual can reduce global warming pollution. One study found that "Reducing the daily use of one low occupancy vehicle and using public transit can reduce a household's carbon footprint between 25-30%."[607] It also reduces congestion (both annoying and inefficient) and can save over $6,000 a year for a household that goes from two cars to one.[608]

Policies that increase fuel economy and public transportation options provide us more freedom of choice in how we contribute to overcoming the causes of global warming.

Even if public transportation isn't currently available to you, or if you are not currently in a position to purchase a more fuel-efficient vehicle, there are at least 10 choices you can make in three basic areas.

☑ **Vehicle Maintenance –**

You can help to maximize the energy efficiency of your vehicle by

1.   keeping your tires inflated (improves mileage by about 3%);

2.   regularly replacing your air filter (up to 10% improvement); and

3.   keeping your vehicle tuned up (from 4% to 40% improvement).

These vehicle maintenance efforts could reduce up to one metric ton of CO_2 and save you up to $250 per year.

☑ **Reducing Your Vehicle Miles Traveled (VMT) –**

You can reduce your VMT by

4.   carpooling to work;

5.   combining errands to make fewer trips;

6.   walking;

7.   biking; and

8.   using public transportation when possible.

☑ **Driving Habits –**

Finally, you can increase your fuel economy by

9.   removing excessive weight from your vehicle; and

10.   driving modestly – avoid aggressive acceleration and braking, and maintain a proper speed (saving 5% in town and 33% on the highway).[609]

BARRIERS TO IMPLEMENTATION

I've highlighted numerous examples from a variety of sectors that would suggest increasing energy efficiency is a no-brainer. So why aren't we doing it? The reasons for not taking advantage of energy efficiency opportunities are called "barriers" by those who study such things.[610]

Part of the reason is one we have already encountered: decision-making is about as decentralized as it can possibly be. Just about every adult in the country (and a good number of the children and teenagers) are decision-makers when it comes to energy efficiency. As one report puts it, "efficiency potential is highly fragmented, spread across more than 100 million locations and billions of devices used in residential, commercial, and industrial settings. This dispersion ensures that efficiency is the highest priority for virtually no one."[611]

So since you the reader are an energy efficiency decision-maker, ask yourself the same question. "If it's such a no-brainer, why aren't I doing it?"

First you have to actually know the facts. Not thinking about it or not being aware of the possibilities or having incorrect assumptions is part of the problem. Did you know that weatherizing your home could decrease your household energy use anywhere from 10% to 25 percent? Or were you part of the majority that thinks it saves 10% or less?

Second, you have to make it a priority. When you purchase things that use energy, for example, efficiency must now be a priority (made easier in part by being able to look for the Energy Star label). This is part of being spiritually vigilant in playing our part to overcome the consequences of global warming.

Third, the upfront costs of more efficient products may have thrown you off. They can be more expensive than the inefficient ones. LED lights are an example, so are hybrid vehicles. They are investments with some of the best rates of return around. But that higher sticker price can still be an impediment, especially when one doesn't think through the long-term savings – or overcoming global warming.

Fourth, the opportunities may not have been available to you. For example, why didn't your house come ready-made as an energy-efficient one? Wouldn't that make sense as a good selling point? It will. But until recently it hasn't, leading to a lack of incentive for builders to make them that way since it costs more to do so. We need minimum government standards to level the playing field and provide a floor for efficiency improvements; but building codes are local affairs, meaning there is not an easily implemented policy fix.

The fact that the decision-making on energy efficiency is highly dispersed highlights again an important theme about mitigation: it represents a tremendous opportunity for Christians to make a difference. Odds are that the vast majority of energy efficiency decisions made in this country are either made by Christians or are made by someone who can be significantly influenced by Christians.

CLEAN ENERGY CREATES JOBS

The clean energy industry (which can include jobs in energy efficiency) has already been a major creator of sustainable jobs in the United States. Between 1998 and 2007 such jobs grew "by 9.1 percent, while total jobs grew by just 3.7 percent."[612] As mentioned earlier, three to five times as many jobs are created by investments in renewables compared to fossil fuels. One estimate suggests, "new investments in cleaner energy could help create 2.5 million new jobs in the United States by 2018."[613]

A recent example of the explosive growth of jobs in the renewable energy industry in the United States is Hemlock Semiconductor based in Hemlock, Michigan, the world's largest producer of polysilicon, a key element in solar panels. They recently doubled their employees in Hemlock from 600 to 1,200, and in 2012 will open a plant in Clarksville, Tennessee, worth $1.2 billion that will employ 900.[614]

A small but growing start-up company is SunRise Solar Inc. in St. John, Indiana, which in its first year in 2003 had $95,000 in earnings, but grew to approximately $3 million in 2009.[615] The founder, Bill Keith, a former roofer, invented a solar attic fan and took out a second mortgage to start the company. As he told ABC News in February 2009, "For the last six years it's like a 45-degree angle straight up [in sales], that's without incentives. If an incentive got put into place, we would grow like gangbusters. We'd have to put a lot more people on. We'd be buying a lot more raw goods."[616]

SunRise Solar has been growing because they have a great product: a fan that is easy to install, costs nothing to power, yet saves customers up to 30% on their utility bills. As Keith puts it, "By putting in a solar-powered attic fan, you're helping customers reducing their energy bills, extending the life of their roof, cutting down on mold and mildew, and doing it all by using the power of the sun."[617] He adds, "My fan runs the fastest when it's the hottest outside. When the sun is the most intense, this thing is running the fastest to remove [heat]. So it's the perfect marriage to shave the energy."[618]

Keith started out assembling the solar attic fans in his barn. Soon it became too much, but he needed more capital to expand. At the same time, one of his suppliers, Indiana Vac-Form, was facing layoffs due to slowdowns in the manufacturing of RVs, cars, and trucks. So SunRise and Vac-Form decided to join forces, with Vac-Form doing the assembly, warehousing, and shipping. Bret Wolf from Vac-Form explains, "They've helped level off the peaks and valleys of our business, allowing us to better plan for people, resources, and machine time. That stability also benefits the companies supplying our raw materials … SunRise Solar has helped us maintain our business in a very difficult economy."

From Wolf's perspective, good jobs and a healthy economy "will come from small companies like ours putting our heads together to innovate and incubate new ideas. It's like the farm. You plant seeds and at first don't see the fruits of your labor, but eventually things crack open and begin to grow and develop." What is exciting for Bret Wolf is that such innovation is happening all across the United States: "If a thousand small start-up companies succeed and they all employ a few dozen people, and each of their suppliers also grows, doesn't that make a significant difference to America?"[619]

In 2009, even amidst tough times, SunRise had four full-time employees with plans to add four to eight more. Vac-Form was able to maintain their employment level at 24, with the possibility of increasing two to four more in order to keep up with the demand for SunRise's solar attic fan.

Keith has been encouraged to relocate to save money and to shift his production overseas to China. But he decided to stay in Indiana because "there's something about being where your roots are."[620] And he has an additional motivation: "I want to employ my neighbors."[621]

SunRise Solar also tries to work with local or national suppliers. Their raw plastic comes from a local company called Spartec, while their stainless steel brackets are made by a locally owned family shop; the bolts and screws are from Hammond, Indiana, the fan blades and wire from Illinois, and the motors come from Pennsylvania and Ohio.[622]

As Bill Keith travels the country promoting his solar attic fan he loves to sit in the window seats on flights so he can look out the window at takeoff and landing. "I'm looking down on all those rooftops and thinking, 'there's my field of dreams right there. That's my future.'"[623]

624

Global Warming and the Risen LORD

The benefits for employment don't end just with commonly understood "clean energy jobs." Indeed, most of the job benefits accrue to many other "regular" jobs in every region and state of the country. For example, the construction, installation, and maintenance of wind power involves environmental engineers, iron and steel workers, sheet metal workers, machinists, electrical equipment assemblers, construction equipment operators, and industrial truck drivers, among others. In like vein, mass transit involves civil engineers, rail track layers, electricians, welders, metal fabricators, and engine assemblers.[625] When the United States economy fulfills its clean, green promise, there will be one word that will describe the jobs created: *American.*

INTEGRATING EFFICIENCY AND CLIMATE-FRIENDLY SOURCES

While we can achieve great gains by either energy efficiency or alternative energy sources, the combination of both working together in integrated systems is what will wean us off fossil fuels.

An example of such an integrated system is zero emission or zero energy houses. While this may sound far-fetched, it's not only doable, it's done-able. We've actually been there and done that. Now it needs to become the norm in our country and around the world.

In 1998 the US Department of Energy (DOE), working with a local homebuilder, had two 2,425-square-foot homes built around the corner from one another in a Lakeland, Florida, subdivision. One house was to be the "control" house. The second was the zero energy house with an identical floor plan. The zero energy house incorporated a variety of energy efficiency measures and included a solar array on the roof for producing electricity. "The objective was to test the feasibility of constructing a new single-family residence that was engineered to reduce the home's energy loads to an absolute minimum so that most of the cooling, water heating, and other daytime electrical needs could be met by the solar systems."[626]

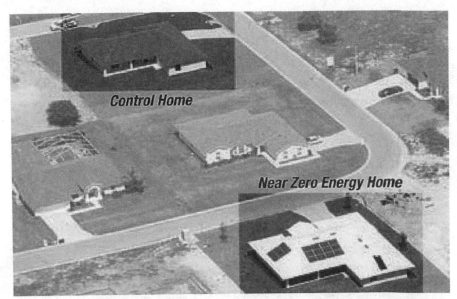

A bird's-eye view of both homes: the completed control (above left) and near zero energy (below right) homes in the Windwood Hills development in Lakeland, Florida.

This early effort exceeded expectations. The near zero energy home uses 90% less energy than the conventional one. On the hottest day recorded in 1998, it used 93% less energy. During peak demand daytime hours, it actually produced a significant amount of excess energy that was sold to the power company.

Such homes can be built in various shapes and sizes and for all income levels, as demonstrated by a Habitat for Humanity home built in Denver in 2005. This 1,280-square-foot, three-bedroom home actually produces nearly 25% more energy than it consumes throughout the year – making it not only a zero energy home, but also a positive energy home. As Habitat said about their efforts, "This home has proven it is possible to build efficient, affordable zero energy homes in cold climates with standard building techniques and materials, simple mechanical systems, and off-the-shelf equipment."[627]

Global Warming and the Risen LORD

628

As of 2009 over 40,000 homes across America, in all climate zones, have been built as part of this DOE program, achieving anywhere from 40% to 100% energy reductions.[629]

I can't think of any reason why starting in 2020 all new homes built in the United States shouldn't be zero energy homes.

19

OVERCOMING THE CAUSES –
MITIGATION IN DEVELOPING COUNTRIES

Two-thirds of greenhouse gas mitigation potential is found in the developing countries. At the end of chapter 17 we covered the potential in forests. This chapter focuses on energy use.

To overcome the causes of global warming, approximately 33% of the world's efforts need to involve the installation of climate-friendly energy-related technologies in developing countries, including technologies that use energy efficiently and those that produce low-carbon energy or successfully store what carbon they produce.[630] Developing countries, especially China, India, and Brazil, have begun to ramp up such investments, accounting for 23% (or $26 billion) of worldwide investments in 2007, up from 13% in 2004.[631] But given that half of the total capital stock that will exist in 2020 in developing countries will be built between now and then, climate friendly investments must increase much more quickly to avoid locking in inefficient, carbon-intensive investments. Doing so in the area of energy efficiency "could slow the growth of their energy demand by more than half … a reduction larger than total energy consumption in China today."[632] With an outlay of only half of what they would otherwise spend if they locked themselves into inefficient systems, such investments would pay for themselves and even turn a hefty profit through energy savings – $600 billion a year by 2020.[633]

Almost half of the energy efficiency potential in developing countries is in China,[634] a country poised to lead the world in greenhouse gas reductions by 2020.[635] This is the case even while they simultaneously have to build the equivalent of a city of 1.25 million every month to accommodate their rural migration.[636] One study suggests that by 2030 there is the potential for China to reduce (1) imported oil by up to 40%, (2) energy produced by coal from 80% down to 34%, and (3) greenhouse gas emissions by 50% from a business-as-usual scenario.[637]

So while there is tremendous potential in developing countries for climate-friendly economic progress, windows of opportunity will not stay open forever. An international climate change treaty can help accelerate climate-friendly investments and ensure the clean energy revolution takes place.

Climate-friendly economic progress in developing countries not only helps to overcome the causes of global warming; it simultaneously improves the lives of the poor. It is these opportunities that I want to focus our attention on as Christians.

Approximately 1.5 billion people in poor countries still lack access to electricity (85% in rural areas, mainly South Asia and sub-Saharan Africa). By 2030, on a business-as-usual path, this number is projected to decrease by only 200 million, while Africa is actually projected to see an increase. For cooking purposes, 2.5 billion people will still burn firewood, charcoal, agricultural waste, and other traditional fuels in inefficient and unsustainable systems that in many cases lead to health concerns. For example, approximately 1.5 million women and children die each year from respiratory problems caused by the resulting indoor air pollution.[638] Finally, because of scarcity due to unsustainable practices, the collection of fuel wood, usually by women, can take hours, exposing them to harm and robbing them of time that could be used for other purposes.

It doesn't have to be this way. We'll first examine how cookstoves can be improved, followed by climate-friendly ways to create electricity for those who lack it.

 Global Warming and the Risen LORD

EFFICIENT COOKSTOVES

What may come as a surprise is that inefficient cookstoves and the fuel used in them by 2.5 billion people are important contributors to global warming in two ways. First, if wood is the fuel, it may contribute to deforestation and other unsustainable practices that contribute to greenhouse gases. (As a reminder, worldwide deforestation contributes approximately 17% of greenhouse gas emissions.)

Second, the emissions from inefficient cookstoves create black carbon, which, as discussed in chapter 2, is probably the second leading cause of global warming. Unlike greenhouse gases, black carbon and other pollutants associated with it can also have regional impacts affecting about 3 billion people,[639] including 20%-50% more warming,[640] the melting of snowpacks and glaciers in the Himalayan region,[641] and regional drought.[642]

Given these global and regional consequences, better cookstoves could play a major role in overcoming global warming. According to one study, black carbon emissions could be reduced 70%-80% in South Asia and 20%-40% in East Asia by using stoves that don't produce black carbon (e.g., solar or biogas stoves). Significant health benefits would also result. In these areas, indoor air pollution from cooking leads to over 400,000 premature deaths.[643]

Another study estimated costs and benefits if a program in India were undertaken to replace "100,000 simple 'unimproved' stove[s]... with improved biomass-gasifier stoves."[644] The health benefit/cost ratio was 13-to-1, meaning every $1 spent would result in $13 in health benefits from breathing cleaner air. The estimated climate benefit/cost ratio ranged from 1.7-to-1 if climate impact costs are low to a 14-to-1 ratio if they are high. Combining health and climate benefits, the benefit/cost ratios ranged from 15-to-1 to 27-to-1. Thus, these efficient cookstoves are investments with an astounding rate of return.[645]

Additional financial benefits could come from the sale of carbon credits. A cookstove effort in India known as Project Surya estimates that even at a very modest price of $6 per ton of carbon, it could be equal to approximately 20% of a participant's salary.[646]

All such potential benefits mean nothing, however, unless successful programs can be implemented. The last 40-plus years are littered with well-intentioned cookstove projects that have failed because the recipients didn't use them, they didn't reduce emissions, or they were either built too cheaply or cost too much.[647]

Recently, however, new efforts have incorporated lessons from the past and a formula for success has emerged. Listed in order of importance, to be successful a project must have cookstoves that

    1)   fit the preferred cooking practices and customary dishes of the customers;

    2)   are affordable, reliable, and desirable to offer for sale to the customer;

    3)   have a successful marketing strategy; and

    4)   are much more efficient and produce significantly less pollution.

Many of you might think this formula is rather self-evident. But numerous well-intentioned efforts in the past have not been focused on the needs and desires of the end users. Rather, they have been focused on the last criteria – producing efficient, less polluting stoves. If such stoves could be offered for free or at nominal cost, then problem solved, right? Well, no. Such stoves – not necessarily designed to fit into the lives of the recipients, possibly designed to be cheap to keep costs down because they are being given away – are not valued by the end users and are therefore not used.

One such project in Chibau Khera, a village in northern India, can serve as an example. Most meals there include *chapatis*, a round flat bread that in this village is "partially cooked on a steel plate heated on a stove" and then baked in the oven to complete the cooking process.[648] The first round of stoves was more efficient and produced less smoke. But they didn't have an opening that allowed for *chapatis* to be baked. This is an essential step

for the taste in this village and therefore the stoves were never used. A second attempt at an improved stove corrected this problem and met with success.[649]

Another revealing aspect of this story involves the effort to make an important part of the stove, the grate, more inexpensively. The grates are made out of steel and cost the equivalent of a full day's pay. So the project managers worked with a local potter to design clay rods that could successfully function as a grate but only cost 20% of what the steel grate cost. While the clay grate has met with some limited success, cultural and social factors have likely prohibited its wide-scale adoption. The materials themselves – clay versus steel – are signifiers of status as well as permanence and well-being. The ability to afford the steel grate is a sign that one is doing well, while a clay grate signifies the opposite. Things made out of clay can be referred to in Hindi (the local language) as *katcha*, meaning something that is temporary or of poor quality – cheap. The opposite of *katcha* is *pucka*, which means something that is permanent or of good quality. For many in Chibau Khera, "The clay grate is *katcha*, while the steel grate is *pucka*." [650]

Hundreds of millions of stoves need to be replaced in fairly short order. It will take **a business model, rather than a well-intentioned charity mentality**, to help focus efforts on the end users, or *customers*. Their needs and desires must to be taken into account in order to produce a stove that is affordable, reliable, yet desirable. A business approach will result in cookstoves that are *pucka*, not *katcha* – and in more ways than one. We must "learn the language" to know what type of stove will sell well in each local area. One poor villager in India named Chetram Jatrav was asked whether she would try an improved stove. "I'm sure they'd look nice, but I'd have to see them, to try them," she replied, with her three children coughing nearby.[651]

Pucka-type stoves will fit the local context. Does the local population cook indoors and standing up? Outside with a stove close to the ground? What are the local food customs, and will that impact what type of stove can be successfully marketed to them?

To keep *pucka*-type stoves affordable for the customer base but also reliable and of high quality, they will need to be mass produced with quality control systems in place. Another approach to making them affordable is the availability of credit, in this instance, microcredit.

This may sound like wishful thinking, but it's being done successfully. A major example is a non-profit in Colorado with a business mentality called Envirofit. They describe their approach as an "enterprise-based business model" utilizing a "global supply chain supporting centralized quality-controlled mass-manufacturing, multi-tiered distribution & sales networks, [and] location-specific marketing strategies." They go on to proudly state, "In creating products for developing world customers, Envirofit utilizes the same disciplined, mature product-development methodologies used by modern industry."[652]

Envirofit's latest stove, the G Series[653], comes with a five-year warranty. It is the result of over five years of basic research, product development, and market analysis done by Envirofit in partnership with Colorado State University's Engines & Energy Conversion Laboratory and the Oak Ridge National Lab. The G Series includes a basic stove, but also offers a variety of accessories that help to accommodate local cooking practices and allow for venting of emissions outside the

home.[654] The stove itself retails for $25, and is made affordable through the extension of microcredit and the fact that the stoves typically pay for themselves in about six months due to the fuel savings.[655] In cities the stoves are sold in stores and shops, and in the rural areas there are local sales representatives (in many cases women from the area) who demonstrate the stoves and take orders.[656]

Initially for Envirofit's customers the most important characteristic of the stove is that it saves money and labor because its efficiency reduces fuel use by up to 60% and cooking time by up to 50%. As one woman villager who had become a local sales representative put it, "The stove saves time and money. If we don't explain this to the people, they won't understand."[657]

But over time, as the customers notice that their eyes burn less from the smoke and their kitchens and homes are cleaner due to less black soot, they come to realize the health benefits as well. "I like it. It makes less smoke than my other stove," said a satisfied customer.[658] And such benefits are considerable, reducing emissions by up to 80%.[659] Field studies utilizing over 100 examples in India of an earlier ceramic stove distributed by Envirofit found that fuel use was reduced by 40% to 50% and emissions by 50% to 70%.[660] Thus, such improved cookstoves cut global warming pollution, including both greenhouse gases and black carbon, by 50% to 80%.

Recently, several Christian relief and development organizations, Food for the Hungry, World Vision, and Plant With Purpose, have begun cookstove efforts as part of something called *The Paradigm Project*. This partnership is guided by the same philosophy as that of Envirofit. Not only are they offering efficient cookstoves in countries like Kenya and Tanzania, they are selling carbon credits and using the funds to expand the program and help the local communities.[661]

To conclude, in many poor countries around the world, where 2.5 billion people still cook using inefficient methods, improved cookstoves – *pucka*-type stoves – represent one way to

(1)   use energy more efficiently;

(2)   reduce deforestation and its greenhouse gas emissions;

(3)   reduce or eliminate black carbon emissions;

(4)   improve human health;

(5)   reduce time spent collecting fuel and cooking, thereby providing women and girls more opportunities;

(6)   increase the family budget; and

(7)   provide local employment.

One word can help to sum this all up: *freedom*.

Pucka-type stoves are *freedom stoves*. They can become one of the stepping stones out of poverty, thereby enhancing the economic dimensions of freedom. More efficient cookstoves can create numerous opportunities for 2.5 billion people to make better lives for themselves and their families by improving their economic well-being. This of course can have positive ripple effects throughout their lives, including making them more resilient to climate impacts. Such stoves can free many of them from indoor air pollution, which kills 1.5 million women and children a year.

Since improved stoves are more efficient, they will free up time spent collecting fuel wood, the majority of which is usually done by women. Worldwide, the average time collecting fuel wood in poor rural areas is more than two and a half hours, with the worst case being eight hours in African countries like Guinea and Ghana.[662] If we were to assume a 50% reduction in fuel use would translate roughly into cutting the collection time in half, that would provide many poor women around the world an extra hour or so a day – for some, up to four hours. This could be used to generate additional family income or in other productive ways.

BIOGAS

Another way to enhance the economic dimensions of freedom through climate-friendly economic progress is for people to be able to produce something called "biogas," a fuel created from organic wastes, including animal and human wastes and crop residues. For nearly all biogas systems, the breakdown of such wastes occurs in conditions without oxygen present, i.e., anaerobic conditions, and is done by microorganisms in a process called "anaerobic digestion," (which occurs naturally in swamps and landfills). The result is a gas that is made up of 50%-75% methane, with the rest primarily carbon dioxide. Since methane has over 20 times the global warming potential of CO_2, it is far preferable to capture such methane and use it. Basically all that is needed to produce biogas are wet organic wastes, an anaerobic container called a biodigester, and the anaerobic microbes. The resulting biogas can then be used for cooking or to power a generator that produces electricity or run a motor that uses natural gas.

If biogas replaces the use of wood or coal, it not only creates clean energy, it also reduces global warming pollution. It would not only be simply climate-neutral, it would be a positive contributor to overcoming global warming because it would reduce global warming pollution that would have otherwise been produced.

Other benefits to switching to biogas include:

- a reduction in indoor air pollution;
- a reduction in deforestation and all of its damaging impacts;
- elimination of the time needed to collect firewood;
- the creation of a safe organic fertilizer as a by-product to boost crop reduction and eliminate use of chemical fertilizers (and their expense); and
- the possibility of also safely utilizing and treating human waste, thereby creating a water sanitation system.

Significant investments in small, household-based biogas power plants for the rural poor have occurred in Nepal, a country of 27 million where 80% of the population lives in rural areas with no electricity, relying on wood for cooking and heating. Over 200,000 such mini-power plants have been installed, providing energy to over a million people. Three-fourths of the leftover by-product (called "bio-slurry") is being used as an organic fertilizer, and 65% of these systems have the household toilet connected, helping to solve sanitation issues.

These households have also saved approximately three hours of work a day by avoiding the need to collect firewood, and reducing time spent both cooking and cleaning off the black carbon from their pots and pans and inside their homes. These biogas systems cost about $350, with the government covering a third of the price and microcredit[663] helping the poor pay their up-front cost.[664]

Before her family bought a biogas mini-power plant, one mother named Khinu Darai from the southern village of Badrahani had to walk three miles every day to collect firewood. As she put it, "Biogas is a blessing for my family. These days I don't have to go into the jungle to collect wood." She added, "It is clean and safe, and we are healthier now as we are not breathing in smoke all the time."[665]

Because of the avoided emissions from reducing deforestation, some of the biogas projects in Nepal have received carbon credits equivalent to over $600,000 annually through an international program called the Clean Development Mechanism (CDM), which the government is using to help pay for the program.[666]

Another way to transform agricultural wastes into electricity has recently been invented by two engineers from India. Manoj Sinha and Gyanesh Pandey, now in their early 30s, grew up without electricity in rural Bihar, the state in India with the lowest per capita GDP. After Sinha received a Masters in Electrical Engineering at the University of Massachusetts and was working for Intel, he began thinking about how he could help his village back in India. "We grew up in those areas. Our

relatives still do not have electricity. We wanted to give back to those areas," Sinha said.[667]

While working on his MBA at the University of Virginia, Sinha shared with fellow student Chip Ransler that he wanted to create electricity for his local village back in Bihar using rice husks – maybe donate a few redesigned generators to do this. Ransler, also now in his early 30s, "did a bit of research and soon suggested that the generators could be a financially viable business."[668] In 2007 they created their own company, Husk Power Systems, which according to the *New York Times*, "now has a proprietary generator that runs on a methane-like gas released by heating rice husks a certain way."[669]

Essentially, Sinha, Pandey, and Ransler have created a private electric utility company catering to poor customers that builds small power plants and local electric grid systems village by village. The grid simply consists of wires starting at the generator that are strung over bamboo poles to the homes or businesses of customers. In a recent interview Ransler said that "The grids aren't fancy – but they work, and they're built by and serve local people, so they are well-maintained." Ransler adds, "We go where the inputs are cheap and where electricity is most needed and valued."[670]

The cheap input behind it all is the rice husks, which normally are left in the fields either to rot or to be burned, which releases methane, a powerful greenhouse gas. Husk Power Systems serves villages that produce about 500 tons of rice husks each year, which the company buys from local farmers, providing them additional income.

Each village system is profitable to Husk Power Systems even at 40% capacity. And once they reach 95% capacity, a village system generates a net profit of $22,500 per year from the sale of the electricity.[671] However, the profit opportunities don't end there. A by-product is silica, a valuable mineral for the production of cement. The company is also applying to have the avoided greenhouse gas emissions certified as carbon credits. As Ransler puts it, "we take agricultural waste and turn it into electricity, minerals and carbon credits."[672]

Currently in five villages serving 12,000, they plan to be in over 100 in a few years. The opportunity for expansion is tremendous according to Ransler. "If you took a map of the world's energy poor areas and compare it to a map of rice producing areas, these two maps would look nearly identical." He says that the system sells itself. "Our marketing is easy. We light up a village – then the surrounding villages see it and they come asking, how can we get this too?"[673]

One villager aptly summed up their situation this way: "We earned our independence from England 60 years ago, but today – when you came into our village – we got independence from poverty."[674]

SOLAR

Those of us living in the United States take electricity for granted precisely because it invisibly powers most things in our homes and much of contemporary society. It is only when the power goes out that we realize how dependent we are on electricity. A Tennessee farmer in 1940 strikingly summed up the difference electricity can make to our lives: "The most important thing in the world is to have the love of God in your heart. The next most important thing is to have electricity in your house."[675] And yet the goal of having electricity available in nearly all of rural America was only achieved less than 50 years ago.[676] For Americans today, electricity is like the air we breathe, taken for granted and yet indispensible for how we live our lives. When the power goes out, it's as if our lives are left gasping for air.

Yet, as mentioned earlier, 1.5 billion people around the world currently lack access to electricity. One of the best ways to quickly bring electricity to the poor who lack it, and to do so in a climate-friendly and cost-effective manner, is by solar energy.

Solar is really starting to take off in developing countries, as evidenced by the skyrocketing success in Bangladesh of Grameen Shakti (*shakti* means energy), a spin-off from the Grameen Bank. Between 1996 and 2003 they installed 10,000 solar systems for homes and small businesses. Microcredit customers with a typical three-year payback purchased them utilizing the money they would have spent on kerosene or candles. In 2009 alone Grameen Shakti installed approximately 250,000 systems, and its founder, Dipal Chandra Barua, has a goal of 7.5 million by 2015.[677]

An innovative non-profit that has been "lighting the way" (sorry, couldn't resist) in the poorest countries since 1990 is the Solar Electric Light Fund (SELF). SELF has an excellent track record of successfully integrating solar power into impoverished areas that have had minimal experience with modern technology or electricity. Their pilot projects are the functional equivalent of R&D (research and development) in this area.

One of the most important uses for electricity is to power health services. In May 2009 SELF installed an electrification system in a health clinic in Kigutu, a remote village in southern Burundi, the second poorest country in the world[678] with an infant mortality rate of nearly 20%. Even though Burundi has a population of approximately 8 million, a 2005 survey showed there were only 156 doctors working in public hospitals.[679]

Providing power 24/7, this particular system receives over 90% of its electricity from solar panels, with a diesel generator as a backup. To help ensure reliability, efficiency, and reduce long-term maintenance costs in this remote setting, a monitoring system was installed "allowing remote technicians to monitor performance via the Internet."[680] Kigutu's health clinic, which serves a total population of 60,000 from the village and the surrounding area, now has electricity for lights, power for refrigerators storing life-saving medicines, and computers with Internet access. Doctors and nurses have the electricity they need to treat thousands of patients suffering from AIDS, malaria, tuberculosis, and other serious ailments.

SELF has also installed such systems in five rural health clinics in eastern Rwanda, which have recently added "solar powered microscopes, blood analysis machines, centrifuges, portable X-ray machines, and sterilization devices. There is, as well, new LED lighting in patient wards."[681]

Having reliable electricity has made a tremendous difference. As Harelimana Assoumpta, Senior Nurse at the Kamabuye Health Center in Rwanda remarked, "Without power, we had to deliver babies and perform other procedures with candles, kerosene lamps, or in the dark. We never knew if this equipment was properly sterilized, we couldn't see if a baby was in distress."[682]

Solar power can also help with food security, as demonstrated by a SELF project in the Kalale District of northern Benin, a least-developed country in western Africa. In the Kalale District people live on about a dollar a day, with 60 cents of that dollar going towards food. During the six-month dry season, many in this area suffer from hunger and malnutrition. The solution? A completely solar-powered drip irrigation system for high value crops on small plots, which has eliminated hunger for the participants, even providing $7.50 a week from the sale of additional crops at the local market.[683]

684

A final example from SELF is their joint project with the Jigawa state government in northern Nigeria to supply all of a village's power from solar technology. Life has changed little over the centuries for this area of northern Nigeria near the Sahara desert. People still live in mud huts, use carts pulled by donkeys, and eke out a subsistence living via agriculture.

685

But with the introduction of solar-powered electricity, the pilot villages now have electricity for

(1)   pumps to distribute water from deep wells to both the village and to agricultural fields for irrigation;
(2)   health clinics;
(3)   schools, where they power lights and computers with broadband Internet connections;
(4)   places of worship;
(5)   streetlights, providing places to congregate safely in the cool of the evening where little food shops have opened up;

(6)   a micro-enterprise building for six small
        businesses, providing shared lighting and
        electricity for sewing machines and other
        commercial electrical devices; and

(7)   home lighting systems purchased with micro-
        credit. [686]

SELF's projects, the R&D of solar integration into impoverished contexts, help demonstrate what is possible, paving the way for taking such efforts to scale.

However, it is my belief that achieving the potential of solar and other alternative energy technologies to help transform the lives of those in poor countries will only take place via a business approach and for-profit ventures. I've already highlighted the examples of Envirofit with cookstoves and Husk Power Systems with biogas.

In the area of solar power I want to shine a light on Selco, an Indian company that currently supplies solar electricity to 105,000 households and plans to expand to another 200,000 over the next several years.[687] One of their early customers was a church in a border village between Kerala and Karnataka where the priest was tired of the power going out during Mass. As Selco's founder and CEO, Harish Hande, recalls, "We installed a single solar light for him. During the next mass, power went off as usual but there was light on Jesus! The whole congregation asked how Jesus was lit up."[688]

Founded in 1995, Selco seeks to dispel three myths: (1) that the poor cannot afford sustainable technologies; (2) that they cannot maintain such technologies; and (3) that "social ventures cannot be run as commercial entities."[689]

Hande, born and raised in India, wrote his Ph.D. dissertation at the University of Massachusetts on the possibilities of solar power in rural India. With Selco he has put theory into practice. The early years were difficult, as Hande survived on his graduate scholarships, credit from suppliers, and social venture capital in the form of loans from several foundations. But with a powerful focus on the needs of his customers, the extension of microcredit to help them buy his products designed for them, partnerships with community-based organizations and academic institutions like Massachusetts Institute of Technology (MIT), and the hiring of local employees, Hande and Selco have come up with a winning formula.[690] Hande states that Selco begins "with the precept that the solution must not be paid from the existing income of the person. It must extend it."[691] As examples he cites "this auto-rickshaw guy who bought our panels, charged batteries during the day and rented them to roadside vegetable sellers in the evening"; this allowed them to sell more produce instead of having it remain overnight and begin to spoil.[692]

Hande tells his sales representatives that if a customer cannot afford their product, they have built the wrong product. "If our product caters to a client's need, there's no way he won't buy it."[693] Such a mentality has led Selco to create such products as solar-powered headlamps for workers who need both hands and could benefit from a light source at night to extend their productive hours, such as construction workers, rubber-tappers, flower-pluckers, and midwives.

A typical Selco customer is a widow named Channama. She works making products at home while her teenage son, Lokesh, works as a laborer. As hard as she and her son tried to get ahead, their combined incomes could only pay for daily living expenses, her daughter's education, and the debt her late husband left. Then she heard about solar lighting from Selco and how it could extend her work hours making products at home. While the local bank was willing to loan her the money to buy the system, she couldn't afford the 25% down payment. Selco came through with microcredit for her to be able to cover this. Now with the extra earnings from working at night, Channama has been able to save a little bit each month.

As was the case with biogas, if uses of solar power are replacing energy generated from wood or fossil fuels, then you have gone from climate-neutral energy to actually reducing global warming pollution that would have otherwise been produced.

These examples of the creation and utilization of biogas and solar power simply give us a hint of what we, as human beings, can do to simultaneously lift people out of poverty and overcome global warming if we tap into our God-given creativity and entrepreneurial spirit. They help us see that poor countries don't have to develop the way "developed" countries did in the twentieth century.

Such sources of energy like biogas and solar are literally homegrown, decentralized power. The availability of such power can free the users from becoming dependent on expensive, imported inputs such as diesel fuel or coal-generated electricity from a big, centralized, expensive power plant and grid system that doesn't even exist in many situations. These decentralized, alternative sources of energy, literally giving "power to the people," via power sources they themselves own, will become less and less expensive as the rich countries continue to invest in such technologies and economies of scale, increased numbers of suppliers, and advancing learning curves drive prices down.

Reliable sources of affordable, creation-and-climate-friendly energy are the foundation upon which sustainable economic progress will be built in this century. We can help pull the 1.5 billion people without electricity out of poverty and allow them to create better lives for their families while simultaneously overcoming the causes of global warming. As we enhance the poverty-reducing dimension of freedom, we will also increase their resilience and reduce their vulnerability to the consequences of global warming.

OVERCOMING THE CAUSES: NOW IS THE TIME

Bill Keith's solar attic fan ... Efficient cookstoves that reduce fuel use by up to 60% and global warming pollution by up to 80% ... Lee Scott and concentrated laundry detergent ... A couple of young entrepreneurs starting Husk Power Systems to sell electricity produced from rice husks to the poor ... Vic's Market saving $48,000 a year by upgrading lighting and freezers ... The indigenous "people of the beautiful hair" having the oil from a palm tree in their rainforest lead to the establishment of an international company now owned by Estee Lauder … A Habitat for Humanity home in Denver that produces more energy than it uses … Selco's solar-powered headlamps for midwives and flower-pluckers.

These stories help us to see the exciting possibilities. But here's the deal: it is time not simply to let a thousand flowers bloom, but to help a billion flowers bloom. Billions of decisions by billions of people will be needed. It's time to wake up and not waste any more opportunities. The scale of transformation needed is massive, but it's been done before, during the Industrial Revolution. However, we must have this overcoming global warming revolution take place in one-third the time. So scale and speed – that's what we need.

Given that CO_2 emissions must peak between 2015-2020, our window for setting the world on the right course is **this decade, with the next few years being crucial**. This is our time to be counted, our time to shine forth, and our time to get real, get smart, and get busy. The days of training and preparation are over. It's time to start this overcoming-global-warming-marathon.

Now is the time.

In the New Testament the word used for this type of time is *kairos*. It means right or opportune moment, and can be of indefinite duration. It is contrasted with *chronos*, or chronological time as measured in seconds, days, months, and years. In the New Testament *kairos* is usually associated with decisive action bringing about deliverance or salvation (e.g., Mk 1:15).[694]

The world has been plodding along in chronological time on the problem of climate change for decades. No more. We are now in *kairos* time when it comes to overcoming global warming. How long will it endure? Until the opportunity for decisive action has passed us by.

Today is the day for us to say, "I will follow you, LORD, as you lead the world in overcoming global warming."

Just as Christ the Creator mysteriously sustains all things by His powerful word, so too is the Risen LORD spiritually leading this great cause of freedom. He is not the author of our acts in this great cause. That is far too crude a description to fathom the delicate dance of our freedom and His will, the mixture of human motives He works with to paint the future, to achieve His purposes. To our finite minds precisely how He leads the way in overcoming global warming is a mystery to fathom but not master.

What we can comprehend is that we are not alone. What we can remind ourselves of over and over again is that the Risen LORD is right there beside us in this marathon of overcoming global warming. When we tire He is there to strengthen us; when we stumble He is there to lift us up; when we veer off course He is there to lead us back onto the right path.

The tremendous task set before us is literally a glorious one that shimmers with the Father's will. As we strive to be Christian agents of transformation in playing our part in achieving a tenfold increase in carbon productivity worldwide by 2050, we can call upon God and all of the gifts He has given us: our freedom, which is only truly real when we image the Risen LORD; the love He has poured into our hearts that must be shared; the creative potential each and every one of us possesses but only exists when expressed; the particular individual talents He has given to each one of us that only become real when we use them; the financial resources He has entrusted to us to be good stewards of that find their true worth in righteousness. And so we see that overcoming global warming takes place in a spiritual context of gratitude. In living out such thankfulness as we tackle the challenges before us, in loving God back, we are presented with new opportunities to become our true selves, the glorious, spiritually beautiful true images Christ created us to be.

20

OVERCOMING THE CONSEQUENCES – *ADAPTATION*

> In earlier days, water used to come
> with low power; now it comes with
> heavy force that sometimes brings
> fishes from the sea to our rooms.
>
> Isatu Fofanah from Kroo Bay,
> Freetown, Sierra Leone.[695]

Ethiopian Kaseyitu Agumas is a poor farmer whose family goes hungry if the rains fail. Intsar Husain from northwest Bangladesh has lost his land because of increased flooding. "My land is in the river. I have nothing now."[696] The young boys Olbin and Edgardo Casares from Honduras were forced to live on the streets after Hurricane Mitch, which killed their father and destroyed their home. A relative who didn't want the boys took in their mother. So they hitchhiked to San Pedro Sula, Honduras' second largest city. "'If it wasn't for Mitch, I'd be home now,' says Olbin … Asked whether his mother knows where he is, Olbin just shrugs. 'I don't know where she is anymore. It's just me and my brother.'"[697]

Recalling these stories told previously can remind us of something obvious. Those of us in the United States are better positioned to overcome the consequences of global warming than a poor person in a developing country. Even if you have done nothing in particular to adapt, you are much less vulnerable because our country is much more resilient. The more solid a society's foundations are, the better the members of that society will be able to weather the storms of global warming. Good education, good healthcare, a strong economy, and vibrant democratic rights, democratic institutions, and democratic governance – these provide a strong platform upon which to overcome big challenges like the impacts of global warming. These provide a deep *resilience reserve* from which to draw.

Many Christian relief and development organizations work to fight hunger, disease, and natural disasters and help those in developing countries build solid foundations upon which to escape from poverty. If you have supported these organizations in the past, such efforts have not been in vain simply because of global warming. These efforts have helped to enhance resilience and reduce vulnerability, creating reserves that can be drawn upon, although nothing like ours. While this will be helpful, to create such a *resilience and reserve* was not our primary intent. We wanted to help them create a better life, not just slow the global warming-intensified downward development spiral.

But all is not lost. Indeed, opportunities abound! By choosing to walk with our Risen LORD and working to overcome both the causes and consequences of global warming, we still can help the poor create better lives for their families. In this great cause of freedom against the tyranny of global warming we can still let freedom ring by helping to create the conditions for successful adaptation.

All of us will have to overcome the consequences of global warming, to adapt to the changes that it will bring. However, this chapter and the next will focus on the poor in developing countries. Here we will discuss whether they will be able to adapt, and what it will take to make this possible.

IS IT POSSIBLE TO OVERCOME THE CONSEQUENCES OF GLOBAL WARMING?

The impacts of global warming that the world has already experienced have answered the question of whether we will need to adapt.

But is it possible to overcome the consequences of global warming through adaptation?

The short answer is YES. But it is only yes if we do two basic things: (1) sufficiently address the causes through mitigation as was described in the previous chapters; and (2) make the necessary investments of time and money to overcome the consequences.

If we don't address the causes as we should, then at some point we will not be able to adapt to the consequences in a meaningful way. The impacts will overwhelm our capacity to adapt. And then we shall *not* overcome. This is especially true for the poor in developing countries, the first forced to face such a situation.

Even if we mitigate or address the causes, can we adapt to the consequences that will still come our way? More specifically, will those of us in the rich countries play our appropriate role, including investing enough in the adaptation efforts of those in developing countries so that they have the resources necessary to adapt?

Of course, the poor have been adapting to such things as floods and droughts for years with varying degrees of success. However, in many cases such coping strategies have been and will be completely overwhelmed by global warming.

A poor family in a slum in Ghana serves as an example. Their home and furniture were made to withstand a certain amount of flooding. The mother explains, "When the rain starts falling abruptly, we turn off the electricity meter in the house. We climb on top of our wardrobes and stay awake till morning … our tables are very high and so also are our wardrobes, they are made in such a way that we can climb and sit on top of them." Unfortunately, these adaptive strategies have reached their limits due to more frequent and more intense flooding, leading to a partial break-up of the family. "I have two children, but because of the floods my first child has been taken to Kumasi to live with my sister in-law."[698]

While what has helped in the past may simply need to be modified over time, relying solely on past strategies could in fact prove dangerous and become what experts call *maladaptive*, given that some of the impacts of global warming will fall outside of historical experience.

JOSEPH: ADAPTATION'S PATRON SAINT

An illustrative situation occurs in the biblical story of Joseph found in the 41st chapter of Genesis, where an unusually severe and prolonged drought required a massive response outside of normal practice in order to avoid dire consequences. Like so much of what will be needed to successfully adapt to global warming, Joseph's story is an example of planning for hard times to come.

Because Joseph accurately predicted the dreams of others, the Pharaoh believed Joseph's interpretation of his dreams: there would be seven years of plenty followed by seven years of famine, a famine so severe that "the abundance in the land will not be remembered" (v. 31). Then Joseph recommended a plan of action:

> *Let Pharaoh appoint commissioners over the land to take a fifth of the harvest of Egypt during the seven years of abundance. They should collect all the food of these good years that are coming and store up the grain under the authority of Pharaoh, to be kept in the cities for food. This food should be held in reserve for the country, to be used during the seven years of famine that will come upon Egypt, so that the country may not be ruined by the famine* (vv. 34-36).

Planning in the present to survive major problems in the future – this is a vital part of what global warming adaptation is all about

Under Joseph's direction and authority, the government took steps in the present to invest in the future, a time when "the abundance in the land will not be remembered." This required a great deal of organization, from the appointment of commissioners to the storage of grain to its proper distribution when conditions called for it. Additional storage facilities probably had to be built, distribution centers created, people trained, the populace educated.

I'm sure there were some doubters. I'm sure a good number didn't like a fifth of their grain being taken by the government for some future

threat they didn't understand or believe in. But I bet they were glad when the famine came and they had food to eat because of Joseph's leadership.

Today, in light of global warming, we see Joseph in a new light. He is the Patriarch of adaptation; he is adaptation's "patron Saint."

THE TWO TYPES OF ADAPTATION

Before going further, it will be helpful to have a working definition of adaptation. If we are going to ask "Is it possible?" we should have an understanding of what "it" is.

According to the IPCC, adaptation actions are those that "**enhance resilience or reduce vulnerability** to observed or expected changes in climate"[699] (emphasis added). This is exactly what Joseph did.

But Joseph was working in an economically well-off country with a well-functioning government and a leader (the Pharaoh) who, with Joseph's help, understood the vision; he "got it" – at least enough to know to put Joseph in charge.

Our focus here is the poor in developing countries. As discussed previously, poverty makes people more vulnerable to the impacts of global warming. But the most vulnerable are the poor in situations that lack good governance and the ability to hold their government leaders accountable. While a wise and benevolent leader in an autocratic system can accomplish great things as the story of Joseph demonstrates, we have concluded in our own day and age that the best way to ensure good governance is to have the leaders of that government accountable to the people. This is in keeping with one of God's greatest gifts: freedom.

There are two complementary and sometimes overlapping ways to achieve adaptation by enhancing resilience and reducing vulnerability. The first is broader, the second more targeted. The first is achieved by realizing the poverty-reducing and democracy-increasing dimensions of freedom.

The second is achieved through projects, processes, and mechanisms designed in whole or in part to address climate impacts.

Both are needed. Neither can be neglected. Each will be described briefly in turn. I will then conclude this chapter with some examples that suggest we can do adaptation at the scale required. The next chapter will describe seven keys to success for targeted adaptation.

LET FREEDOM RING: REALIZING THE POVERTY-REDUCING AND DEMOCRACY-INCREASING DIMENSIONS OF FREEDOM TO ACHIEVE CLIMATE ADAPTATION

Adapting to the consequences of global warming provides Christians and others of good will another reason to enhance the poverty-reducing dimension of freedom. This involves helping people climb out of poverty and stay out through education, proper nutrition and healthcare, and sustainable economic progress that is compatible with or enhances both global warming mitigation and adaptation.[700] It also involves fostering the conditions for democratic freedom and the constructive exercise of that freedom, thereby helping to ensure that governments are responsive to the calls by their people for enhanced resilience and reduced vulnerability to global warming impacts.

Let me illustrate with examples from (a) the area of child nutrition and healthcare, (b) creation-and-climate-friendly economic progress via a plant called sisal, and (c) a constructive exercise of democratic freedom by slum dwellers in a place called "The Big Ditch."

CHILD NUTRITION AND HEALTHCARE

As would be expected, evidence demonstrates that a poor child who has been provided proper nutrition, healthcare, and a cleaner, healthier environment will be more resilient and less vulnerable to the impacts of floods, droughts, and disease.

Let me illustrate with three examples:

- A study in Bangladesh after the 1998 floods found that children who had received proper nutrition and healthcare beforehand fared far better than children who were provided aid after the floods. The after-the-fact aid was "relatively ineffective."[701]
- Another study of drought in Mozambique found that supplementary nutrition programs targeted to drought-

Global Warming and the Risen LORD

prone areas helped reduce the under-five mortality rate "to 41 percent below the national average, even in the context of severe drought."[702]

- Improvements in sanitation infrastructure in Salvador, Brazil (population 2.4 million) "reduced the prevalence of diarrheal diseases by 22 percent across the city in 2003-04 and by 43 percent in high-risk communities" (mainly children).[703]

As one analysis concluded, "When children's health is already compromised by illness and malnutrition, they are far more likely to sustain long-term damage to their development in the wake of a crisis."[704] The bottom line is pretty simple: healthy, well-nourished children will be better positioned to overcome the consequences of climate change. As the old saying goes, an ounce of prevention is worth a pound of cure.

SISAL, THE WONDER-PLANT

The second illustration comes from the area of sustainable economic progress and is captured in one word, just one … sisal. That's right: sisal.

Ok, what's sisal? Glad you asked.

Sisal is a drought-resistant plant that thrives in the tropics and subtropics and is planted and harvested year round, providing income even if food crops fail. The fiber in sisal has made it a cash crop. Sisal fiber products include: ropes, twines, yarn, paper, dartboards, mattresses, rugs, carpets, wall-coverings, cat scratching posts, slippers, and a creation-friendly replacement for asbestos. Given all of these and other uses found thus far, it should come as no surprise that sisal is projected to be a growth market for years to come.[705]

While its drought-resistant, year-round cash crop status alone makes it something to consider for climate adaptation, its benefits don't end there. Not only does it not have to be in competition with food crops, yields can actually increase through intercropping and the use of one of its by-products as an organic fertilizer. In one example maize yields grew by 200%.[706]

But wait! There's more!

Only about 4% of the sisal plant ends up being used to make sisal products. Typically, the rest is allowed to either rot in the fields, releasing methane, a potent greenhouse gas, or burned, releasing carbon dioxide, the most prominent form of global warming pollution.

Understanding that such waste is inefficient, Katani Ltd, an enterprising and socially responsible sisal processing company in the Tanga region of Tanzania, has built the world's first plant that turns sisal biomass into biogas. The biogas is currently used to create electricity for the company and the community, and plans include using it for vehicles and piping it into homes.[707]

Additional income for the local farmers comes from selling the sisal biomass to Katani and from their increased yields from their other crops from the intercropping and the organic fertilizer. The availability of electricity has also allowed for increased income by providing lighting for work in non-daylight hours and to run small-scale industries and businesses. Katani also provides electricity to the local schools and hospitals.

All of this has led to a *wealth* of benefits for local families and the community, including the ability to build better houses, buy bicycles and mobile phones, afford and receive better health care, a reduction of those leaving the community to find work, and an 80% increase in the number of children attending school. In addition, replacing wood with biogas for cooking reduces (1) the health impacts of indoor air pollution – a top killer of children under five,[708] (2) deforestation (a serious problem for the area), and (3) both black carbon and greenhouse gas emissions.[709]

Even apart from overcoming global warming, Katani's production of sisal biogas is a win-win-win: good for the company, good for the community, and good for God's creation. Quite a combination.

Strictly from an adaptation perspective, Katani's sisal biogas production not only increases the wealth and well-being of families and the community, thereby building resilience and reducing vulnerability. But it does so with a drought-resistant year-round crop that could just as easily have been chosen as an example of a targeted adaptation project. As an added bonus it reduces deforestation, thereby helping with mitigation.

DEMOCRATIC FREEDOM AND "THE BIG DITCH"

The final example, one that highlights the importance of democratic freedom, involves a slum outside Buenos Aires in Argentina called El Zanjon or "The Big Ditch." In 2004 the slum dwellers of The Big Ditch became aware that the government knew when the chronic floods that impacted their slum would occur, that climate change was increasing such flooding, but that the government was failing to provide them with this information. They came up with a simple early warning system utilizing whistles, and organized a campaign to petition the local government to help them with resources and information to create this early warning system. As a result of the slum dwellers exercising their freedom to influence their government officials, a telephone line was installed so that when floods are expected a government employee can call and the community can be alerted through the whistle system.[710] Even though they are poor and at risk, the slum dwellers of The Big Ditch found a way to reduce their vulnerability by organizing themselves into a political force that the government had to pay attention to.

TARGETED ADAPTATION: PROJECTS, PROCESSES, AND MECHANISMS

The second approach to adapting to the consequences of global warming is much more targeted, based on what the likely major impacts of global warming will be in a particular area. Is there going to be more flooding? More drought? Higher temperatures? How do we prepare? It is this more targeted approach that is usually meant when the term adaptation is used in the context of climate change.

Four types of small-scale projects – floating gardens, inexpensive flood-resistant housing, several examples of rainwater harvesting, and drought-resistant crops – will serve to illustrate how targeted adaptation can occur.

FLOATING GARDENS

As we have seen, global warming will increase both the frequency and intensity of inland flooding. One consequence will be a diminished ability of poor people living on increasingly flood-prone lands to grow crops. A simple, practical solution made from resources readily at hand is a floating garden. Water hyacinth (a free-floating perennial plant that grows in water) is collected and formed into a raft, upon which soil and cow dung are placed. Seeds of suitable crops are then planted in the soil.

Such floating gardens have been successfully demonstrated in one of the poorest, most remote, and flood-prone areas of Bangladesh, the Gaibandha district, located at the confluence of two major rivers, the Tista and the Brahmaputra. The local populations live below the subsistence level, and out of necessity many fathers leave the area in search of work, leaving behind their families.[711]

Some of the women have been trained regarding the creation and upkeep of floating gardens. They are now planting them to see them through the lean times. One such mother is Tara Begum, who was able to grow such crops as red onion, pumpkin, and okra. "This has made a great difference to my life. Now I have enough food in the floods and I can give some to help my relatives as well."[712]

FLOOD-RESISTANT HOUSING

Another consequence of flooding is damage or destruction of housing. One successful solution being implemented in Bangladesh, using locally available resources, involves creating a two-foot-high foundation upon which to erect one's home. This simple foundation is made of earth with an outer protective layer of cement and stones. The walls of the home are constructed of easily replaceable panels made of jute (a readily available plant in the area). Water-thirsty plants, such as bamboo and banana, are planted around the structure to soak up water and retain the soil. As one father said, "Before, when the rain came, we wouldn't sleep. We were terrified. But now at last we can live our lives in peace."[713]

Flood resistant housing[714]

Before, when the rain came, we wouldn't sleep. We were terrified. But now at last we can live our lives in peace.
- a father in Bangladesh

RAINWATER HARVESTING

As the old saying goes, necessity is the mother of invention. For thousands of years, when the need arose, people found various ways to capture rainwater. Because of increased water scarcity brought on by global warming, many will need to discover anew how to "harvest" rainwater.

One way to capture rainwater for crops in time of drought is by constructing ridges of soil along the contours of fields so that the rainwater doesn't simply run off the hard-baked soils. Before using this technique, Tias Sibanda, a local farmer from the Humbane village of Gwanda, Zimbabwe, frequently harvested nothing during times of drought and would then have to sell some of his livestock to survive – a downward disaster spiral.

But using this rain harvesting technique has made a tremendous difference for Tias Sibanda and his family. In the first year he had two crops, which he calculates saved him from having to sell 12 goats (worth about $320). Tias states, "I am confident of further improvements in the future and, if the drought eases, would soon be able to sell some of my maize crop."[715]

Another rainwater harvesting technique is to capture rainwater that flows off rooftops by a system of gutters and pipes that channel the water into a storage tank. Efforts in Muthukandiya, a drought-stricken village in Sri Lanka, serve as an example, not only of effective use of this technology, but of how intentional efforts at community involvement increased the success rate.

Previous top-down efforts in Muthukandiya by the government proved ineffective. So a relief and development group working in the area, Practical Action, called a meeting of people in the village to ask their views. As a result, this particular rooftop-to-tank storage system of rainwater harvesting was chosen and a plan was developed to make it a reality. A village committee was set up to run the project. Nearly 40 families agreed to participate. Two local masons were trained in how to construct the 1,300-gallon storage tanks. Participating households were trained in how to maintain the system. The entire system cost $195 (equal to a month's income for a family), but the community, in the form of materials and unskilled labor, covered over half of the cost.

The results? During the driest times, participating households have nearly twice as much water as non-participating ones – and the water is much cleaner.

A widow in the village, Nandawathie, has capitalized on the opportunities provided by increased water by growing and selling vegetables at her doorstep. With this additional revenue stream she applied for a loan to install solar power in her house, and she is thinking of building another storage tank to grow more produce. Nandawathie also feels safer not having to fetch water. Her children have less diarrhea and her daughter, Sandamalee, has more time for schoolwork.

Global Warming and the Risen LORD

The benefits from this project are clear and compelling. However, Practical Action reminds us that "a lot of effort and patience are needed to generate the interest, develop the skills, and organize the management structures needed to implement sustainable community-based projects" like this one.[716]

A final rainwater harvesting example involves a community-based project in a poor village in the drought-stricken Kitui district of eastern Kenya. Such rainwater harvesting projects are desperately needed in the country, given that only 4% of its rainwater collection potential is being tapped even though it is chronically water-scarce.[717] The particular technique utilized for this project was a "rock catchment," which requires an

impermeable rock outcrop of sufficient size. In the area where the rock slopes down a wall is built, essentially creating a dam. This particular project provided nearly a gallon of clean water per day within walking distance for each village resident during the dry season.[718]

Drought-resistant Potatoes

DROUGHT-RESISTANT CROPS

One example of switching to more drought-resistant food crops comes from farmers in northern Ethiopia. Because rainfall patterns have become less and less reliable, these farmers have shifted from wheat, which takes nine months to grow, to potatoes, which take three months. This allows them to potentially get three yields of potatoes for every one of wheat. If the rains fail during the year, with potatoes they stand a much better chance of bringing in a harvest or two. Furthermore, working with the Christian relief and development organization Food for the Hungry and the Ministry of Agriculture, these farmers are planting new varieties of potato that "are resistant to blight and yield up to 3 times more than traditional potatoes."[719]

Just as in the patriarch Joseph's case, these examples sound deceptively simple. But as anyone who has managed a project before knows, they are usually much more complicated than meets the eye; they are just as much about process as they are about product. Furthermore, when the project means major changes in the way people do things, part of that process includes education and persuasion and buy-in of those who need to approve and participate in the changes. In many cases, adaptation projects will need to involve both the private and the public sector. Governments, businesses, non-profits, community groups, churches, families, and individuals will have to participate and play their respective roles.

ADAPTATION AT THE SCALE REQUIRED

Again, is it possible? Can we overcome the consequences of global warming? The case studies above suggest the possibilities. This section closes with several examples demonstrating that it is indeed **possible at the magnitude or scale necessary – if we exercise our freedom to make it so**. These examples come from the successes achieved in (1) arresting the downward disaster spiral, (2) reducing risk from flooding, and (2) the Millennium Development Goals.

ARRESTING THE DOWNWARD DISASTER SPIRAL

As discussed previously, successive natural disasters create a downward spiral that thwarts the efforts of the poor to try to create a better life for themselves. Climate-intensified disasters could make this dynamic even worse. But recent efforts in Malawi, one of the world's poorest, most densely populated countries, demonstrate that concerted efforts can thwart this downward disaster spiral. Because of successive floods and droughts, the 2005 harvest was one of the worst ever recorded, declining 29%.[720] Given that 85% of the country lives in rural areas and one third of GDP comes from agriculture,[721] such impacts are particularly devastating. Over 5 million people faced food shortages.

But just as Joseph planned for hard times to come, the government of Malawi worked with relief and development organizations and development financing institutions to help ensure that the population was better positioned to withstand the next round of natural disasters. The resources of many families had been depleted by successive calamities, leaving them unable to buy fertilizers and other inputs. Therefore these items were heavily subsidized and government and non-government entities distributed them in order to raise production. The result was an additional 600,000 tons of maize worth at least $100 million from an investment of $70 million.[722] So not only did this effort thwart the downward disaster spiral, allowing millions to continue creating a better life, it made money for the entire economy in the process.

REDUCING FLOOD RISK

Mozambique is the sixth poorest country in the world and one that will be hit hard by climate change. Both coastal and inland flooding are constant threats with tropical cyclones (hurricanes) roaring in from the Indian Ocean and nine major rivers flowing through the country to the ocean. Heavy rains in 1999 had swollen the rivers and in February and March of 2000, Mozambique was hit with two major cyclones. Seven hundred people died and 650,000 were displaced. But when a similar situation occurred in 2007 only 80 died, an 89% reduction.

What made this dramatic difference? The government worked with relief and development organizations to conduct a detailed analysis identifying the 40 most vulnerable areas, home to nearly 6 million. At the community level disaster plans were developed and training exercises conducted. Early warning systems were created. In 2007 the activation of these plans and systems helped with the evacuation of those most at risk.[723] Even though the citizens are poor, concerted efforts by the government, relief and development organizations, and their local communities made them less vulnerable.

If success can be achieved in two of the poorest countries in the world, Malawi and Mozambique, then success can be achieved anywhere. And such successes in disaster risk reduction not only save lives, they can be highly cost-effective economically, with benefits exceeding costs anywhere from 1 to 38 times, depending upon the project.[724]

THE MILLENIUM DEVELOPMENT GOALS

Finally, let me offer two examples of efforts endorsed by the entire international community. Recent progress on several of the Millennium Development Goals (MDGs) demonstrates that the world has made significant strides in enhancing the poverty-reducing dimensions of freedom.

In the year 2000, world leaders came together and adopted the *Millennium Declaration*, from which the Millennium Development Goals (MDGs) came. The 192 nations of the world have affirmed the *Declaration*, which begins by stating there are "certain fundamental values" that are "essential to international relations in the twenty-first century." The first value listed is freedom, which it describes thusly: "Men and women have the right to live their lives and raise their children in dignity, free from hunger and from the fear of violence, oppression or injustice. Democratic and participatory governance based on the will of the people best assures these rights."[725] The attainment of the MDGs is intended to help achieve freedom for those in poverty. In our broader understanding of adaptation, achieving the MDGs helps enhance resilience and reduce vulnerability to the consequences of global warming. Conversely, the impacts of global warming – if not diminished in their potential intensity through mitigation and blunted or averted through adaptation – will slow, stop, and even reverse the advance of freedom in this century.

Enhancing freedom today includes overcoming both poverty and global warming and their interrelationships. We can take hope from the progress that has been made thus far in achieving the MDGs. While that progress is admittedly mixed,[726] there has been progress nonetheless. As of 2008, the world is well on its way to achieving the first goal of reducing poverty by half by 2015.[727]

This already is an incredible accomplishment, and much of it has been achieved by economic growth. Since the world has yet to ensure such growth is sustainable economic progress consistent with efforts at mitigation and adaptation, we must do precisely this moving forward so that such progress is not reversed. As we saw with Katani biogas production, and will discuss more fully in the next section, overcoming poverty through economic progress can dovetail well with mitigation and adaptation.

Global Warming and the Risen LORD

Indeed, helping poor countries contribute to mitigation through climate-friendly sustainable economic progress is essential if we are to achieve our mitigation goal of keeping global temperatures from rising beyond 2°C above pre-industrial levels. And if we don't help them adapt, then our efforts at economic progress and poverty reduction will be stymied and undone.

The second MDG is achieving universal primary education by 2015. As with poverty reduction, there has been significant progress, with all but two regions having at least 90% enrollment as of 2008.[728] Given that education is essential to the foundation and perpetuation of the economic and democratic dimensions of freedom, this progress is critical to the success of both mitigation and adaptation efforts to overcome global warming.

The progress on the first two MDGs demonstrates that the nations of the world can come together and accomplish major goals that help the poor climb out of poverty, which, if done in a manner consistent with both mitigation and adaptation, is a key ingredient in the poor being able to overcome the consequences of global warming.

CONCLUSION

Overcoming the consequences of global warming is possible if we (1) successfully mitigate or address the causes in a timely manner; (2) continue to enhance the poverty-reducing and democracy-increasing dimensions of freedom; and (3) do targeted adaptation at the scale required. These tasks are challenging, but not impossible. The real question is whether we will choose to do so.

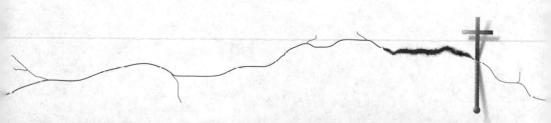

21

TARGETED ADAPTATION –
SEVEN KEYS TO SUCCESS

Projects designed to target the impacts of climate change will need these seven vital ingredients to be successful:

1) Commitment: Moral and Political Will
2) Adequate Funding
3) Good Governance
4) The Right Policies
5) Accurate, Understandable Information for All Decision-Makers
6) An Integrated, Coordinated Response
7) Community Engagement

We will discuss each in turn.

1. COMMITMENT: MORAL AND POLITICAL WILL

The previous chapter made the case that it is indeed *possible* to adapt to the consequences of climate change. However, it is not going to be easy. It presents us with a major challenge, a challenge that some would say is ultimately insurmountable.

Precisely because adaptation is a major challenge, it also presents Christians with an incredible **opportunity** to follow our Risen LORD, with whom all things are possible. He invites us to join Him in His love for the poor and vulnerable. He is there with them in the ditch of global warming's impacts. He beckons us to become more spiritually free, beautiful, and glorious as we play our part in helping the poor overcome the impacts of global warming. Adaptation, like mitigation, will take commitment and staying power – the type that Christ's love provides, a love inside of us, which cannot be hoarded but must be expressed. As it says in 1 John 4:19, "We love because he first loved us." We must be involved regardless of whether anyone else is. Because of our love for Him, we must be involved whether we will be ultimately successful or not.

But that is not an excuse for failure! In order for there to be major adaptation successes, Christians must work with others. In many if not most cases, adaptation will require an integrated societal response that in important respects will involve the international community. It is not something that individual Christians or local churches or even the Church universal can accomplish. Along with our own explicitly Christian efforts we must work with others and through secular institutions. We must be leaven.

The question is not whether humanity has the capacity or the ability. The question is whether we will have the moral and political will. We can do it. But will we? A key part of our contribution as Christians is to help provide this moral and political will, to fight the good fight, to exercise our freedom to follow the Risen LORD as He works to overcome global warming.

2. ADEQUATE FUNDING

A major determining factor in whether the poor in developing countries will be able to adapt to the consequences of global warming is whether or not the rich countries commit sufficient financial resources. The United States in particular, as we have done with AIDS, malaria, and in times of natural disasters, should lead the world with our generosity in helping the poor adapt to consequences they did not cause.

While we may not have understood that our actions in burning fossil fuels would contribute to harmful impacts being visited upon them, this is in fact the case. Our country has a strong value of fairness, and it is only fair that we help those we have unintentionally harmed. Recent contributions from the rich countries, unfortunately, have been woefully inadequate; by one estimate, less than 0.2% of what is required.[729]

This begs the question: what amount of financial resources will be required?[730] It all depends on how broadly we define enhancing resilience and reducing vulnerability. No one has put forward an estimate that corresponds to how broadly I cast some adaptation activities in the preceding section, i.e., the poverty-reducing and democracy-increasing dimensions of freedom.

A 2009 study by the World Bank estimated the cost of *targeted* adaptation to range from $75 billion to $100 billion a year between 2010 and 2050.[731] An earlier study by the UN Development Program estimated the cost of adaptation at approximately $86 billion per year, which would represent a mere 0.2% of developed country GDP, or roughly one-tenth what developed countries spend on their militaries.[732]

A report by the UN Framework Convention on Climate Change (UNFCCC) regarding some of the major areas that will require targeted adaptation provides a range of $28 billion to $67 billion per year. The upper and lower ranges are based upon how severe one assumes the impacts will be.[733] Other respected experts have collectively criticized the UNFCCC funding levels as underestimating the costs "by a factor of between 2 and 3" for the areas estimated. A key sector not included by the UNFCCC, the protection of ecosystems, "could add a further $65-300 billion per year in costs."[734]

What should be the contribution from the US government for targeted adaptation? Given that historically, our generous spirit as a country has led the United States to contribute 20%-30% of the funds for major natural disasters and for such health problems like AIDS, let's assume 25%.[735] Let's also assume that the range of funding needs is somewhere between $28 billion and $100 billion per year. We'll split the difference and call it $64 billion. Twenty-five percent of $64 billion is $16 billion per year. Christians in the United States should not only support this minimum level of funding, we should contribute our own time, talent, and financial resources to adaptation efforts. In addition, as the need increases over time we should support the increase of our country's financial commitments in keeping with our past generosity.

While such funding can be justified in a variety of ways, one important way for the United States *as a country* to understand financing for targeted adaptation is as **a strategic investment**.

First, investments in targeted adaptation will in most cases generate a healthy rate of return (assuming the money is spent as intended). Some analogous examples from our own lives can help to illustrate this. Preventative maintenance of one's house or vehicle, such as cleaning out the rain gutters or keeping the car tuned up and the tires inflated, saves money when compared to the costly repairs resulting from negligence. And as mentioned previously, in disaster risk reduction, benefits can exceed costs from 1 to 38 times depending on the project.

A recent major study on climate adaptation found that between 40% and 68% of projected damages "could be averted through adaptation measures whose economic benefits outweigh their costs," with even greater savings possible "in highly targeted geographies."[736] As this report puts it, "well-targeted, early investment to improve climate resilience – whether in infrastructure development, technology advances, capacity improvement, shifts in systems and behaviors, or risk transfer measures – is likely to be cheaper and more effective for the world community than complex disaster relief efforts after the event."[737]

In other words, much better to avoid a big mess than have to clean one up. Much better to do things right the first time.

As we have already discussed, proper preventative healthcare for children – poor or otherwise – generates a far greater rate of return given that health impacts on their developing bodies can last a lifetime. Keeping a

child healthy and well nourished allows them to grow up to become a productive member of society and avoids costly healthcare in the future while helping them be more resilient to climate impacts.

Second, a stable world is in our national interest. Diminishing the ways global warming functions as a "threat multiplier" (see chapter 6's discussion) helps to keep our military personnel out of harm's way and forestalls situations that can become breeding grounds for terrorists. A stable world also enhances our economic security by facilitating the free flow of commerce.

Finally, remaining true to our character and our values of fairness, compassion, generosity, and freedom keeps us strong as a country.

While adequate levels of funding are essential, money alone will not overcome the consequences. As with any major problem, we can't simply throw money at adaptation. Successful adaptation also will require the other keys to success to which we now turn.

3. GOOD GOVERNANCE

A recent study looking at responses to climate-intensified disasters in developing countries concluded that the most important ingredient for an adequate response was effective governance, whether as a result from (1) good leaders, or from (2) encouragement/pressure from citizens for officials to act like good leaders, or (3) a combination of the two. The qualities of such effective leadership included a willingness to adapt to changing circumstances, the integration of adaptation into other related policies, and a commitment to invest in the long term.[738] While this particular study looked at responses to intensified natural disasters, these same qualities of good governance are needed for most other climate adaptation efforts, such as reducing the spread of infectious diseases due to temperature changes. (Studies have shown that the higher a government's overall effectiveness, the higher the country's environmental performance.[739])

One example of good leadership is Ilo, a city of 70,000 in Peru. Contrary to many other cities in the developing world, Ilo has no informal settlements – despite a fivefold increase in population from 1960 to 2000. Such settlements usually take place on undesirable land, land that in many instances is flood-prone, thereby increasing vulnerability to climate-intensified flooding. But this is not the case in Ilo. Why? Because the local government has worked cooperatively with the poor and their local housing association to provide appropriate housing plots, land titles, drinking water, and waste disposal.[740]

Sometimes we are blessed with good leaders like Joseph who see the hard times coming and ensure that we plan ahead. If we are not so blessed, best to have democratic freedom to encourage them towards good governance. This is what the slum dwellers from The Big Ditch did, as we saw earlier.

A recent increase in the responsiveness of local governments in developing countries is due in good measure to the fact that more and more are moving towards elected mayors. Combine this with local civic associations, NGOs, and local churches that have the freedom to actively look out for the poor, and the chances increase for appropriate adaptation measures to be implemented.[741]

Finally, good governance does not simply involve local authorities. If local officials are not given the tools and resources they need by state and national governments, and if the international community does not provide some of these tools and resources, then successful adaptation could be thwarted.[742] Adaptation will require an *integrated response*, as will be discussed shortly.

4. THE RIGHT POLICIES

Global warming necessitates that all types of policies around the world will have to change – be they government policies, policies of businesses, or policies of nongovernmental organizations. No doubt there will be many new or adapted policies that arise to deal with adaptation in poor countries. Here I will highlight two that may not immediately spring to mind: microfinance and crop insurance.

MICROFINANCE AND MICROCREDIT

Microfinance has been incredibly successful over the last several decades in enhancing resilience by increasing the poverty-reducing dimension of freedom. Microfinance institutions arose because normal banks were not serving the poor, lacking such things as a stable address, steady employment, collateral, or a verifiable credit history. When they began, microfinance institutions had a "double bottom line" of both fiscal and social sustainability. Recently, however, some are beginning to explore how to include environmental sustainability, creating a "triple bottom line."[743]

Microfinance "now reaches more than 93 million poor clients and has helped households in risk prone communities around the world to strengthen their livelihoods and increase their resilience."[744] A key service of microfinance institutions is "microcredit," or extending small loans to spur entreprencurship. (Microcredit was pioneered by Muhammad Yunus and the Grameen Bank in Bangladesh; Yunus and the Bank won the Nobel Peace Prize in 2006 for their work.)

One microfinance institution dedicated to helping the rural poor in Nicaragua, the Fundación Denis Ernesto González López, has seen the necessity of a "triple bottom line" – with environmental sustainability essential to ultimately achieving fiscal and social sustainability. As an inducement to the poor rural farmers, small loans with a reduced interest rate are offered to those willing to farm and forest in ways that reduce water scarcity, increase water quality, and reduce soil erosion.[745] While this effort was not originally designed with climate adaptation as a primary concern, it nonetheless is helping to reduce global warming pollution and helping its recipients adapt to the consequences by reducing poverty, addressing water scarcity, and absorbing carbon in trees through reforestation.

But microcredit could also provide a more intentional and targeted response to adaptation by providing loans to help people purchase land and build or buy homes out of flood-prone areas, for example, or to help them enhance the resilience of their current homes.

INDEX-BASED CROP INSURANCE

One of the normal ways that most of us in the United States reduce our risk to accidents or disasters is through insurance. However, less than 5% of homes and businesses in poor countries have such insurance.[746] Instead, they rely on help from family and friends, or in the case when the whole social network is also devastated, on relief agencies or the government, if such help is available. Unfortunately, when a poor country is struck, outside assistance often covers less than 10% of the losses. Funding for rebuilding can take up to a year or more to organize, and such funds are often diverted from other development programs.[747] All of this helps to keep the poor, poor – and climate-intensified disasters will only worsen this situation unless changes are made.

For those whose livelihoods depend on agriculture, one type of insurance that would be useful is crop insurance. To attempt to meet this need, a new approach called index-based or weather-based crop insurance is currently being tried in poor rural areas in around 15 countries, including China, India, Bangladesh, Ethiopia, Malawi, Nicaragua, and Peru. Instead of receiving a payout if they demonstrate that they have lost their crop, participants receive a payout if a certain type of destructive weather event has occurred normally associated with crop failure, such as a drought, severe rainfall, or flood – occurrences that can be measured with a standardized index. This reduces costs associated with running the program, since an agent doesn't need to visit a farmer's field to determine whether or how much a payout should be. It also avoids a perverse incentive structure created by crop failure policies where at a certain point a farmer may quit trying and allow the crop to fail to get the payout. Instead, now the farmer has an incentive to try to save the crop even if the agreed-upon weather index that triggers a payout has been exceeded.[748]

How else could financial and insurance policies and offerings be changed to help with climate adaptation? Well, this question is actually the point. One of the ways we will we overcome the consequences of global warming is by being creative in adapting successful approaches in such areas as finance and insurance. We will overcome global warming by freely choosing to apply our creativity and our gifts and abilities in the areas where we personally can make a difference. So for those of you whom God has gifted in these areas, this question is actually meant for you.

5. *ACCURATE, UNDERSTANDABLE INFORMATION FOR ALL DECISION-MAKERS*

When it comes to overcoming the consequences of global warming, millions of decisions will need to be made over the course of this century and beyond. Decision-makers will range from heads-of-state to heads-of-households. All of them will need accurate and understandable information in order to make the correct decisions. Attempting to achieve such a goal will require massive efforts. Much of the raw data will need to be gathered by experts using sophisticated equipment and techniques, such as the climate models run by supercomputers. The average person with minimal training will conduct other data collection efforts. But however it is collected, it must eventually find its way into the hands of everyone who needs it. And in order for such data to be useful, it will need to be transformed into information that the various types of decision-makers can understand and act upon – "actionable intelligence," to borrow a phrase.

The most basic function of a government is the protection of its citizens. As such, governments should ensure that every adult on the planet is armed with the information they need in a form they can understand to successfully adapt to climate impacts. People, wherever they are – even some place called The Big Ditch – have a right to know about the consequences of global warming that could threaten the well-being, the very lives of their loved ones. Governments should both protect and fulfill this right to know. If they fail to live up to this charge, their citizenry should avail themselves of the democratic dimension of freedom to encourage, and if necessary, demand that they do so – and replace them if they don't.

Right now even in a rich country like the United States, this goal is not being achieved. It can be if we dedicate the resources to do so. This is not the case for many poor countries, which lack the capacity to generate and transform the necessary information. The rich countries will need to assist the poor ones, whether through providing them the information or helping them to generate it.

A recent example demonstrates how providing poor farmers with the information and choices they need can reduce vulnerability and enhance resilience. In the rural, arid, and semi-arid lands of Kenya, home to approximately 10 million, over 60% of the population, lives below the poverty line.[749] Inhabitants already have enough problems without global warming making them worse. Fortunately, an adaptation pilot program begun in the fall of 2006, in an area called Sakai in the Makueni District located in eastern Kenya, is showing some promising results. It is giving poor farmers the "actionable intelligence" they need to make the changes necessary to adapt to the increased water scarcity and drought that global warming brings. This includes "downscaling climate forecasts [i.e., making the forecasts specific to their area] and presenting them to farmers in a format and language suited to their needs."[750] With a changing climate, this helps them know how to adjust their farming practices, making better informed decisions about when and what to plant. The project also includes information, training, and access to fast-maturing and drought-resistant crops, as well as how pest-management needs to adjust to the changing climate. Early results demonstrate that those participating in the program have higher yields than those who do not.[751]

While governments at all levels, from international to local, have a major responsibility to generate and to some extent transform the information into something that all decision-makers can successfully utilize, the Church and other non-profit and community-based organizations have a tremendous opportunity to help the poor understand and act upon such information. In this great cause of freedom, Christians and others of good will can provide information that will help "the least of these" make choices that will enhance their resilience and reduce their vulnerability to the impacts of global warming.

6. *AN INTEGRATED, COORDINATED RESPONSE*

As previously stated, adaptation is as much a process as it is a product. In Joseph's day the Pharaoh knew that the response needed to the approaching famine predicted by Joseph would require a tremendous amount of planning, organization, and cooperation, and the authority to get it done. That's why he put Joseph in charge.

I can't think of an institution of society that won't be involved in some way in climate adaptation, and every single person on the planet will have to adapt. While there won't be one overarching "International Department of Adaptation" that runs the whole show – and it would be a disaster if there were – we are going to need an intentional, integrated response between nations, across each society, and within institutions.

Adaptation is a crosscutting issue that will involve multinational organizations all the way down to the individual in the field or on the street. Without proper integration of the need for adaptation into the programs and procedures of society's institutions (applying a "climate lens," as one report put it[752]), and without proper coordination of adaptation activities, our efforts will be wasteful, sometimes counterproductive, and in many respects unsuccessful.

An example of successful integration and coordination comes from the project in the Sakai area of the Makueni District in eastern Kenya mentioned in the previous section. Here is a *partial* mapping of the various levels of integration for this one pilot project designed to enhance food security and water availability.

It was:

(1) initiated by the UN Environment Programme;

(2) assisted by an NGO called the International Institute for Sustainable Development, and by;

(3) African Centre for Technology Studies.

Funding came from:

(4) the Global Environment Facility, a multinational institution
supported by the United States and 177 other nations;

(5) Norway;

(6) the Netherlands; and

(7) Germany.

It involved the following agencies of the Kenyan government:

(8) the National Academy of Sciences and its Centre for Science
and Technology Innovation;

(9) the Ministry of Agriculture;

(10) the Agricultural Research Institute; and

(11) the community-based drought management program called the
Arid Lands Resource Management Project.

Implementation at the local level included:

(12) district-level government officials;

(13) community leaders;

(14) field staff of non-profits; and

(15) the local farmers themselves and their families.

As mentioned above, a crucial part of this project is the
downscaling of climate forecasting into forecasts for this specific area of
eastern Kenya, which in itself requires a massive amount of data collection
and analysis. Such forecasting relies on computer models that have been
developed since the 1950s and now involves scientists from around the
world, including the United States, the UK, Germany, France, Japan,
Australia, Canada, Russia, China, Korea, and Norway.[753]

But this particular project not only included providing local farmers
the information they needed in forms and languages they could understand.
It also involved introducing several types of drought-resistant crops,

implementing a number of water harvesting techniques, training on pest management and the proper storage of harvested crops, and help with starting small businesses through microcredit to diversify sources of income for local households.[754]

And guess what? It works!

7. COMMUNITY ENGAGEMENT

Let me sum up the discussion thus far. For the poor in developing countries to overcome the consequences of global warming, it will require

- the continued realization of the poverty-reducing and democracy-increasing dimensions of freedom;
- sufficient funding;
- good governance;
- the right policies;
- accurate and understandable information to decision-makers; and
- an intentional, integrated, coordinated response from international organizations and national governments down to those in poor villages and neighborhoods.

None of this will happen without

- sufficient moral and political will and a commitment to chart and stay the course.

Such summaries are helpful to remind us of the key points and provide scale and scope. By necessity they are generalized and abstract. Yet, by themselves they can mask something. As with all real change, there is an irreducible particularity and concreteness to adaptation that must not be forgotten for us to be successful.

Overcoming the consequences of global warming will require changes in the hearts and minds of particular individuals here in the United

States and all around the world. But such essential changes must also lead to specific changes in particular places, such as growing sisal in the Tanga region of Tanzania to enhance economic resilience, or capturing rainwater off roofs in Muthukandiya, Nandawaithe's village, or creating an early warning system for flooding that uses whistles in a Buenos Aires slum, The Big Ditch.

When efforts to change things get particular, as ultimately they must, it is going to require community engagement. This also is both process and product, but it is a process that involves particular people in particular places resulting in a product or outcome where they live.

Before you can have community engagement you actually need to have a community. People simply living close to one another do not equal a community. As important as the role of government is, it cannot mandate that people living in the same area create enough "social capital" to be able to work together for their common good. If the government is ready to play its part in overcoming the consequences of global warming, some semblance of community needs to exist where such efforts are to be attempted. Only people coming together and getting to know one another and discovering their common values and concerns will create the social capital necessary to make working together for the common good possible.

The city of Ilo in Peru, highlighted earlier, where there are no informal settlements in flood prone areas, provides an example. This was achieved because the government had civic associations and residents with a sense of community with whom they could partner.

But absent neighbors willing to work together for their common good, we can get the opposite of what we need. We can get maladaptation instead of enhanced resilience and reduced vulnerability.

Ad hoc, individualistic flood preparation offers some examples. It is very common in poor urban areas for there to be a household-by-household response to flooding as opposed to a coordinated, community-wide response. Far from creating community, this can lead in the opposite direction by increasing tensions and conflict. It also increases the overall

level of risk. For instance, in some poor neighborhoods in Buenos Aires each household uses debris to elevate their plot of land. With no coordination between neighbors, each plot ends up at a different height, with some becoming more flooded than their neighbors.[755]

Other examples involve the common practice of dumping waste in the streets, which ends up blocking what drainage channels might exist, thereby increasing flooding. A final example comes from Saint Louis, Senegal, where many households "construct flood defenses and embankments out of their household waste – thus, in effect, causing waste to be spread throughout a community when flooding occurs."[756] This has led to increased health problems, especially for the children. As one report summed up the situation, "Effective community-based pre-disaster measures to limit damage require levels of trust and cohesion – community social capital – that are often not present."[757]

So without some level of real community, more often than not the result will probably be the opposite of what we want.

There needs to be a semblance of community not only so that the government or an NGO can have one to partner with, but also so that the community itself can proactively engage the government or prominent businesses or NGOs – not only so that they can be engaged, but also so that they can do the engaging.

There are instances where the community and its allies will need to encourage the government to play its constructive role. In a good number of cases there are long-standing barriers that have impeded the effective delivery of government services to the poor. In many instances, especially in urban areas, the relationships between government officials and those living in poor neighborhoods, particularly slums, could use significant improvement. Sometimes there is outright hostility, with government officials seeing the residents as a problem and the residents in turn distrustful of the officials.[758]

Whether rural or urban, the poor can be underserved by their government if they are from minority or indigenous groups or different ethnic tribes. In addition, the government may not be protecting or recognizing the property and water rights of the poor. Lack of appropriate government personnel is also a serious issue. Finally, in many instances government officials simply do not know impacted neighborhoods and communities well enough to be completely effective. As one expert put it, reflecting upon post-disaster recovery efforts, governments have "thick fingers for such fine-scale work."[759]

An example is rain harvesting in Nandawathie's village in Sri Lanka. Initially the government had tried a rather top-down approach, imposed from the outside, which wasn't successful. A relief and development agency, Practical Action, tried again by engaging the community. Forty families agreed to participate. Those who did had twice as much water during the driest times compared to those who did not participate.

The facilitating and catalytic leadership provided by Practical Action in this particular case required care and patience. Such leadership will be needed repeatedly. As one report put it: "Facilitating and supporting truly constructive community engagement, especially with communities that may have little experience of joint decision making, can take skill and experience, as well as genuine commitment."[760]

Sounds like a job description for Christian servant leaders. This need for (1) real community, (2) having the community be engaged, and (3) the community doing the engaging, creates tremendous opportunities for Christians both here in the United States and for our Christian brothers and sisters in developing countries around the world.

As Christians we know two basic things when it comes to real community: (1) it is essential; and (2) it takes patient, persistent effort, genuine servant leadership. Whatever country we reside in, we are in the community-building business, which we realize to various degrees in our local churches, small groups, and ministries. We are also called to be community-building leaven where we live.

The need for community engagement for adaptation to be successful is a reminder to us as Christians of our calling to create real community. Our continuing efforts to build community within the Body of Christ and in the neighborhoods, villages, towns, and cities where we live, will provide part or all of the essential foundation for the community engagement necessary for successful climate adaptation.

Overcoming the consequences of global warming in this century will be close to impossible if we fail in this basic task of community engagement and if the LORD is not able to find others to make up for our failure in some way, shape, or form. It is there that the Risen LORD will be in the ditch with the least of these and watch as we pass by on the other side.

May it not be so! Woe to us if it is!

This is a reminder to us that as Christians, the most important ingredient in our faithfulness to overcoming both the causes and consequences of global warming is the strength of our relationship to the LORD, which in turn is nurtured in the local expression of the Body of Christ to which we belong. Is your church able to nurture you spiritually so you can play your part? Is it leading the way by its own activities? Are we supporting local churches of those in poor countries to do the same?

CONCLUSION

Floating gardens that provide food even when times are tough … Flood-resistant housing using readily available materials … Rainwater harvesting that doubles the amount of water available … Providing site-specific climate forecasts to poor farmers … Drought-resistant, year-round cash crops like sisal, whose by-products can be turned into an organic fertilizer that increases food crop yields up to 200%, and whose leftovers can be turned into biogas that both replaces firewood, thereby reducing indoor air pollution and deforestation, and powers sustainable economic progress … Successfully planning for floods and droughts in one of the poorest countries in the world, Mozambique, where they reduced flooding deaths by 89% and under-five mortality during severe drought to 41% below the national average … Benefits exceeding costs anywhere from 1 to 38 times for disaster risk reduction efforts … Slum dwellers in The Big Ditch getting their government to give them the flood warning information they need and creating a simple early warning system utilizing whistles to alert the community … The world well on our way to achieving the first Millennium Development Goal of reducing poverty by half by 2015.

These and many other examples help us see that in overcoming the consequences of global warming, challenges can become triumphs; adversities can be transformed into opportunities.

But I'll be honest with you. I'm more optimistic about the world overcoming the causes of global warming through mitigation than I am about the world helping the poor overcome the consequences through adaptation. In general, I think there will be many more people who get excited about creating a clean energy revolution that will protect us from climate impacts; far fewer, I fear, will be those who will care enough to see that the most vulnerable are helped to deal with the impacts that do happen. I have been studying this issue since 1992 and have been engaged full-time on it since 1997, and that has been my experience to date.

This is why I need you, the reader, to help me get the word out and help others understand what this is really about. It's about freedom and love: enhancing the freedom of the poor to create a better life for their families; and living out the love of Christ in our hearts.

Will we, in freedom, let Christ's love compel us to fight for their freedom? Will we pass by on the other side and let stand the injustice of their suffering from consequences they didn't cause? Or will our love for the LORD, who is with them in the ditch of global warming's impacts, compel us to stand with Him and them? Will we join our LORD in helping the poor overcome the consequences of global warming's tyranny?

We all have our particular roles to play. But as Christians we cannot simply focus on overcoming the causes.

What, then, is our role as Christians in the United States when it comes to adaptation of the poor in developing countries?

One thing we are definitely *not* to be are "adaptation heroes," riding in to save the day, spiritual superstar wannabes like Peter before the crucifixion, ready to go out in a blaze of glory with all eyes fixed on us and the sacrifices we are ready to make. That would be harmful to both them and us. As the saying goes, *it's not about us.*

We must remember that all human beings, regardless of economic status or station in life, have been made in the image of God and have been given the gifts of freedom and love just as we have. Whether they know it or not, they are also on their particular spiritual journey with the Risen LORD. Each one of them has particular gifts and abilities that can be actualized in addressing the challenge of climate adaptation.

Being ever mindful of Jesus' words to the disciples on the night he was betrayed will help us in framing how we should think about our role. After an argument broke out among them about who was the greatest, Jesus says

> The kings of the Gentiles lord it over them; and those who exercise authority over them call themselves Benefactors. But you are not to be like that. Instead, the greatest among you should be like the youngest, and the one who rules like the one who serves. For who is greater, the one who is at the table or the one who serves? Is it not the one who is at the table? But I am among you as one who serves.(Lk 22:25-27).

Even if we have the money or power to be a "benefactor," as followers of the Risen LORD, the One with the nail-scarred hands, we are to be the servants of others. It is especially important for us to keep this at the forefront of our minds when it comes to our role in helping people in poor countries overcome the consequences of global warming.

We are to be in the background playing a supportive role. It is not for us to do it for them, but rather to help create the conditions and opportunities they need to do it for themselves.

To help realize such outcomes, we must be advocates and supporters of **both** types of adaptation, and we must do so with **both** our government and the Christian organizations we support.

As citizens, we must be advocates on behalf of the poor for efforts that enhance the poverty-reducing and democracy-increasing dimensions of freedom. The Millennium Development Goals, to be achieved by 2015, offer an articulation of a good deal of what needs to happen in this regard – such as, the education of both boys *and* girls; reducing by two-thirds the under-five mortality rate and by three-quarters the maternal mortality rate; achieving universal access to drugs for HIV/AIDS; ensuring that 90% of the world's population has access to improved sources of drinking water; and making available new technologies such as cell phones and the Internet.[761] It is the enhancement of these poverty-reducing dimensions of freedom that will create what I called earlier a *resilience reserve* to help give them a fighting chance.

More broadly, we must support policies that help create the conditions for sustainable economic progress, the fostering of democracy and human rights, and the empowerment of local communities.

But as citizens we must also support funding and programs for targeted adaptation in keeping with our past generosity when compared to AIDS funding or responding to natural disasters.

Finally, as citizens we must not allow these two types of adaptation to be pitted against one another in terms of funding; we don't want a "robbing Peter to pay Paul" type of situation. Both must be fully funded because both are needed; to underfund one is to undermine the other.

As individual Christians, as Christian families, local churches, and Christian denominations, we must support Christian organizations working in poor countries to do four things: (1) share the Gospel, (2) build up local churches and communities, and (3) enhance the poverty-reducing dimensions of freedom. But we must also ensure that Christian organizations work with local families, churches, and communities to begin to (4) implement targeted adaptation projects (something that most have not started to do).

Adaptation is not only possible; it can be done in a way that improves the situation of the poor, helping them to create better lives for themselves and their families. The main thing needed? The will to make it happen. The love to not pass by on the other side.

Energized by our love for our LORD and with a servant's heart, let us use our freedom to fight for their freedom.

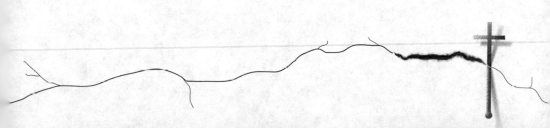

22

WALKING INTO THE FUTURE WITH THE RISEN LORD

IT'S OUR TIME

It is this decade, 2010-2020, a small sliver in time, which looms the largest in this great challenge to overcome global warming. What we do – or fail to do – will determine in large measure what global warming will do to the world in this century and beyond. Will our failure lead to a cascade of irreversible tipping points that result in a world unrecognizable to us? Or will we use our freedom to expand freedom around the world?

Will we use our freedom to follow the Risen LORD as He leads the way in overcoming global warming? Will we become Christian agents of transformation and help lead the clean energy revolution and ensure that the poor reap some of the benefits and opportunities? Will we fight to ensure that the poor have the resources they need to adapt and create a better life for their families? Will we not allow global warming to rob them of their dignity as fellow image-bearers with freedom and unique gifts, who, if given the chance, can make this world a better place? Have we allowed enough of the Risen LORD's love and spiritual presence to fill our hearts so that we will stand up and be counted and join Him in this great cause of freedom?

Or will we fail to follow and turn our faces away from the glory of the LORD's presence and sink into the shadow of our sinful finitude as our spiritual beauty fades and our freedom turns to dust.

It must not be so! In chapters 16-21 we have seen that it is indeed possible to overcome the causes and consequences of global warming. What we need is the will.

Brothers and sisters in the LORD, friends and colleagues, I hope that this book has helped you to see that *this is our time*.

Let me be frank. Many of us have been plodding along in chronological time on this great challenge, and have not awakened to the fact that we are now in *kairos* time when it comes to climate change. As used in the New Testament, the word *kairos* means a right or opportune moment usually associated with decisive action bringing about deliverance or salvation. If not acted upon, such moments can pass us by.

We are in the *kairos* climate moment because there is still time to overcome global warming. There is still time for us to be spared from many of its potential devastating consequences, for the poor to be delivered from even more destructive impacts, for less of God's other creatures to become extinct and be robbed of God's blessing of life.

If you are still operating in chronological time when it comes to overcoming global warming, **it's time to wake up**. Simply put: **our *kairos* moment on global warming has arrived, and it won't last forever**.

When Moses ended his final address to the people of Israel he said the following words that echo throughout the generations:

> *This day I call heaven and earth as witnesses against you that I have set before you life and death, blessings and curses. Now choose life, so that you and your children may live and that you may love the LORD your God, listen to his voice, and hold fast to him. For the LORD is your life, and he will give you many years in the land he swore to give to your fathers, Abraham, Isaac and Jacob* (Dt 30:19-20).

Global warming is in fact making the heavens (in this case the atmosphere) and the earth to be witnesses against us. They are pouring forth speech that unless we change our ways, curses and death will follow.

But if we love the LORD who is our life, listen to His voice, hold fast to Him, then curses can be transformed into blessings and death into life. Such is the way of the One who turns crucifixions into resurrections.

What is it that we must do? That each of us must do?

Choose life.

I'm going to put it to you straight out: will you follow the Risen LORD in this great cause of freedom, overcoming global warming? Will you make a solemn promise to the LORD that you will play your part and use the time, gifts, abilities, and resources He has given you?

My brother, my sister, now is the time. Today is the day to choose life. This moment is the moment to allow the Risen LORD also to be LORD of your response to this great challenge, to follow Him in freedom as He leads the way in glory.

BEING CHRIST'S AGENTS OF TRANSFORMATION

When we looked at how fully the causes of global warming are entwined in our lives and when we reviewed the magnitude of the impacts, especially on the poor, and when we grasp that this decade is our chance to turn the tide, it leads to the following conclusion: **all Christians have a contribution to make** in overcoming global warming.

Simply put: it is now part of our walk with the Risen LORD, as He is the One who is spiritually leading the way.

We are all called to be His agents of transformation in this great cause of freedom. We are all privileged with the opportunity to serve through our contribution to overcoming global warming.

Let us, therefore, explore together (1) seven things each of us must do; (2) what our gifts, talents, opportunities, and resources allow us to do; and (3) whether the LORD may be calling you to a particular climate focus.

But before doing so, let me highlight two important considerations. The first has to do with the need for leadership, which I am defining here as taking the initiative to practice continual creative engagement with what the LORD has blessed us with, and helping others do the same. In this way we are all called to be leaders in overcoming global warming.

And to demonstrate why our leadership is desperately needed, let me remind you of the following:

- ▶ Implementation will involve billions of people and millions of groups making decisions.
- ▶ The world needs to increase our carbon productivity tenfold in 40 years.
- ▶ Two-thirds of mitigation needs to occur in developing countries.
- ▶ Adaptation for the poor in developing countries is possible but in no way assured, either that the resources will be available or that such resources will end up helping the poor.

All of this cries out for creative Christian engagement and leadership. Christians should be helping to lead the charge, searching for creative ways to use our gifts, talents, opportunities, and resources to play our part in overcoming global warming and helping others to do the same. Simply having the right public policies in place and the right price signals to influence decisions in the marketplace are not going to be enough, given the scale and speed of the changes necessary.

Overcoming global warming is going to require individuals, primed to be change agents, who are actively searching out mitigation and adaptation opportunities. And we especially need people who are watching out for the best interests of the poor to ensure that their adaptation and mitigation opportunities are fully realized.

In helping to lead we must remember that as Christians we should have hearts of a servant, given that we are followers of the one who said, "But I am among you as one who serves" (Lk 22:27).

The second consideration is this: overcoming global warming will involve a grand and glorious mixture of both dramatic and mundane choices. There will be big showy choices such as passing laws and transforming a

company, and seemingly small, insignificant choices such as, "What type of light bulb do I buy?" or "Do I make the extra effort to recycle this aluminum can?"

In a profound and important sense, overcoming the causes of global warming is about individual choices, about each of us exercising our freedom to be a part of the Risen LORD's efforts in this cause. It doesn't matter whether we are the President of the United States, the CEO of General Electric or simply an average citizen and consumer.

But in an equally profound and important sense, overcoming the causes of global warming is about working together. It is about families working together, churches working together, communities working together, businesses working together, our country working together, and the nations of the world working together. Many of our most important choices as individuals involve choosing to work with others. So we must choose to play our part and work together.

We should ask ourselves, our family, friends, fellow churchgoers, work colleagues, online networks, and fellow citizens, "What's next? How can we do this in a climate-friendly way? How can we make this fun for everyone?" If we are to be Christian agents of transformation, helping to lead the way, it has to be about continual creative engagement in every area of our lives, including all the groups we belong to.

WHAT EACH OF US MUST DO

If we are to be faithful agents of transformation in overcoming global warming, there are seven things each of us must do: (1) keep spiritually healthy; (2) advocate for climate-friendly public policies; (3) reduce our own global warming pollution; (4) support the poor in their efforts to overcome global warming; (5) recruit and inspire others; (6) be a catalyst at our church, our workplace, and school; and (7) pray.

KEEP SPIRITUALLY HEALTHY

If there is one message I hope has come through in this book it is this: overcoming global warming is a marathon that will require spiritual stamina. It is part of our walk with the Risen LORD, who is the One who provides the spiritual sustenance for the journey. But we must choose to continue to follow Him as He leads the way in overcoming global warming. We must choose not to quit the race.

In the parable of the four soils (also known as the parable of the sower) found in Matthew, Mark, and Luke, Jesus compares the spiritual receptivity of those who hear the message about the Kingdom of God to four types of soil (Mk 4:1-20; Mt 13:1-23; Lk 8:4-15). Such an analogy can also be quite illuminating when applied to our subject.

When it comes to our receptivity to following the Risen LORD in overcoming global warming, what type of soil will we be? What type of harvest will be reaped?

The first type of soil is that along the path – in such soil the message never has a chance. If you've gotten this far you are not that type of soil.

For those who are rocky soil, they begin enthusiastically, but "they have no root" so "in the time of testing they fall away" (Lk 8:13).

For those whose soil has thorns, "as they go on their way they are choked by life's worries, riches, and pleasures, and they do not mature" (Lk 8:14).

But, as Luke puts it, "the seed on good soil stands for those with a noble and good heart, who hear the word, retain it, and by persevering produce a crop" (Lk 8:15). For those who do persevere, who do have "steadfast endurance" as another translation puts it,[762] Mark tells us that the message planted within such individuals can produce a crop that can be "thirty, sixty, or even a hundred times what was sown" (Mk 4:20).

So what type of soil will we be? Will we begin enthusiastically but quickly fade? Will we let life's worries and temptations keep us from full maturity? Or will we persevere with steadfast endurance to produce a bountiful harvest?

Thankfully, our spiritual situation is not fixed, not static. We can do things to become "good soil," and grow spiritually into a Christian that perseveres, one who has steadfast endurance, especially for marathons like overcoming global warming.

How? How can we help ensure that spiritually we become and remain good soil for producing these fruits?

There's no magic formula here. It's pretty straightforward, time-tested, standard stuff. Intentionally keep the LORD at the center of your life through constant prayer, daily study of the Bible (especially the Gospels), and regular communion with a local body of believers. Continually entreat the LORD to fill you with His empowering grace, to help you put Him at the center of your life, and to guide you with His Spirit in knowing what His Lordship is calling you to do in this great cause of freedom.

If we don't keep healthy and growing as Christians, we won't keep making and supporting the changes necessary. We must allow the LORD to enhance our spiritual freedom in the here and now by increasing our ability to do His will.

While there is no magic formula, that doesn't mean these proven methods for spiritual growth have to be boring. Be creative by adding your own special touches that enhance the experiences for you.

If you want to increase your spiritual stamina for this marathon, then do these practices creatively and listen for His voice so that you can stay centered in His Lordship. Remember: we are never alone in our overcoming global warming journey, because the Risen LORD is always with us.

ADVOCATE FOR CLIMATE-FRIENDLY PUBLIC POLICIES

For those of us in the United States, one of the great gifts of freedom we have been given is our citizenship in a representative democracy. We the people are ultimately sovereign. Just like any other gift we have been given, we have the opportunity and responsibility to exercise it in keeping with Christ's Lordship in our lives.

In this particular case, we must express our views to our elected officials about the need for strong public policies to overcome both the causes and consequences of global warming. (For those of you who may not have been involved in public policy engagement before, overcoming global warming provides you with opportunities to freely express your gift of citizenship.)

As was discussed in the mitigation and adaptation chapters, global warming will not be overcome unless we have the right public policies in place here in the United States and an international treaty that involves all countries.

The purpose of the climate mitigation policies is to set the "rules for the road" and harness private sector implementation. As I shared earlier, such policies are the "crack of the gun" to start the climate mitigation race, which will be run primarily in the private sector.

Such policies should have been adopted by now. Our elected officials have failed thus far to do so and time is running short. If those currently in office will not lead, then we must find others who will.

As for adaptation, in keeping with our past levels of generosity for major natural disasters and international health issues like AIDS, the United States should contribute 25% of what is needed to help the most vulnerable poor countries and communities adapt to the consequences of global warming. We should encourage our policymakers to support such funding levels, based on the reasons outlined in chapter 21.

Once a price has been put on carbon, adaptation funding commitments have been secured, and an international climate treaty ratified, these accomplishments do not signal the end of our public policy engagement. We must "stand watch" and ensure they are strengthened rather than weakened (1) in their implementation by administrative agencies, and (2) through subsequent actions by Congress.

As we do so, we must continue to be advocates for the poor in two respects. First, we must ensure that at a minimum they are not harmed by the implementation of such polices. Second, we must continue to search for

ways that mitigation and adaptation policies can provide opportunities for them to create better lives for themselves and their families.

A great way to know what to advocate for and when to do so is to join an organization/campaign you feel comfortable with that tracks such matters. This could be my organization, the Evangelical Environmental Network (EEN), or the National Council of Churches' ecojustice climate campaign, or the Catholic Coalition on Climate Change (which works closely with the US Conference of Catholic Bishops). A theme of this book has been that no one can overcome global warming all alone. We will need to be a part of various "teams," and one of these organizations/campaigns can be the team that helps you play your part in the area of climate advocacy.[163]

Our country has overcome great challenges before, such as during the civil rights movement, which resulted in important laws like the Civil Rights Act of 1964 and the Voting Rights Act of 1965. The environmental movement has secured important laws such as the Clean Air Act. Through it we've achieved tremendous success in reducing air pollution, with benefits exceeding costs by at least 12 times. Internationally, the Montreal Protocol has successfully addressed the problem of ozone depletion, and has cost 87% less than originally projected. We've done it before, and we can do it again.

We have the freedom to advocate creatively for the poor and to influence significantly the climate policies of perhaps the most important country in the world on this issue at this most crucial of times. This is an incredible spiritual opportunity! One for which we should be extremely grateful. It's our moment to make history.

REDUCE OUR OWN GLOBAL WARMING POLLUTION

Reducing our own pollution is essential for several reasons. First, it's an empowering way to train for our overcoming global warming marathon and have our values match our personal behavior in a tangible way. Second, it helps to invest us in the process, to strengthen our commitment by the very act of doing. It primes us for more, which is crucial. Third, it's a good witness to others. Finally, being part of a holistic approach can make a real difference. As shared previously, US households account for nearly 40% of CO_2 emissions. If a realistic number of households implemented a pollution-reduction program using available technology, we could reduce such emissions by 7% or more.[764]

Numerous books and Websites are devoted to helping us reduce our own global warming pollution.[765] The Energy Efficiency subsections on Home and Transportation previously covered provided numerous ways we, as individuals and families, can reduce our pollution, from purchasing energy efficient light bulbs and appliances to buying green electricity to purchasing a zero energy home near public transportation.

But what it means to reduce our own pollution will change over time (and thankfully so, as that means we're getting better). We can serve by staying abreast of the latest changes and how we and others can implement them.

In reducing our own pollution it will be important to pace things. Remember: it's a marathon, not a sprint. You don't want to burn out. Implement improvement projects on a realistic schedule. Purchase more efficient products when the ones you are replacing have run their lifecycle.

Reducing our own pollution is an empowering place to start and is an essential element of a holistic approach to overcoming global warming.

SUPPORT THE POOR IN THEIR EFFORTS TO OVERCOME GLOBAL WARMING

One of the truly great characters of the Bible is Barnabas, who we meet in the Book of Acts. His given name was actually Joseph, but the apostles nicknamed him Barnabas because it describes who he is, "Son of Encouragement" (4:36). Acts later describes him with high praise, saying, "He was a good man, full of the Holy Spirit and faith" (11:24).

Barnabas is one of the great role models in the New Testament for this reason: he doesn't have to be the star. Rather, he encourages others to

shine. This is especially the case with the Apostle Paul. Without Barnabas' facilitation and encouragement, there may not have been an Apostle Paul, as we know him today. When Saul tried to meet with the apostles in Jerusalem soon after his conversion they were all afraid of him, given his past persecution of the Church. "But Barnabas took him and brought him to the apostles. He told them how Saul on his journey had seen the Lord and that the Lord had spoken to him, and how in Damascus he had preached fearlessly in the name of Jesus" (Acts 9:27). In standing up for Paul, in facilitating his relationship with the other apostles, Barnabas put himself on the line.

But perhaps his greatest act of facilitative encouragement was to bring Saul from Tarsus to Antioch, which led to Saul becoming the Apostle Paul.

During Saul's first time in Jerusalem after his conversion, his preaching and debating were so successful that some plotted to kill him. Once the apostles heard of this they sent him to his hometown, Tarsus, in modern-day Turkey (Acts 9:28-30). Years pass, and in this period some of those who were scattered during the time of Stephen's martyrdom began converting Greeks in Antioch, Syria, the Roman empire's third largest city. "The Lord's hand was with them, and a great number of people believed and turned to the Lord" (Acts 11:21).

When the elders in Jerusalem heard about these conversions they sent Barnabas to Antioch to access the situation. "When he arrived and saw the evidence of the grace of God, he was glad and encouraged them all to remain true to the Lord with all their hearts" (11:23).

Barnabas saw that God was at work in the lives of the recently converted Gentile Christians in Antioch. He lived up to his nickname by encouraging them – which was no small thing given that the idea of Gentile converts to Christianity was still a new and radical concept. But he also saw that what this situation called for was the special gifts and talents of Saul of Tarsus. As Acts tells us, "Then Barnabas went to Tarsus to look for Saul, and when he found him, he brought him to Antioch" (11:25-26). Barnabas traveled over 200 miles round-trip to find the one man tailor-made to catalyze this new opening to those who had been excluded from hearing God's message in the synagogues and the Temple. The world would never be the same, because this leads to Saul becoming Paul, Missionary to the Gentiles.

When it comes to helping the poor overcome global warming, Barnabas can serve as a good role model for those of us in the developed countries. Like Barnabas, we should support, encourage, and facilitate their efforts and catalyze their gifts and abilities.

One of the ways we can do this is to once again become a part of a team effort. We can generously support Christian relief and development organizations like Food for the Hungry, MAP International, World Vision, and Plant With Purpose as they seek in a Barnabas-like fashion to help the poor enhance their resilience and reduce their vulnerability to climate impacts. They do this by facilitating sustainable economic progress and efforts to address water scarcity and sanitation, health concerns, refugees, and disaster risk reduction. They also do so with specific projects like more efficient cookstoves through *The Paradigm Project*, something individuals can support (www.theparadigmproject.org).

Be sure to let your team/organization know that you are supporting their relief and development efforts as part of helping the poor overcome global warming. Encourage them to help the poor implement *targeted* adaptation projects and plan for projected climate impacts in their area.

At times this work will require special expertise, knowledge, and resources that the poor simply don't have such as the Clean Development Mechanism reforestation project World Vision carried out in Ethiopia. Help will be needed to ensure that the poor have the "actionable intelligence" they need to be able to make informed decisions on when and how to adapt. But, again, this should always be done with the example of Barnabas as our guide. Rather than impede the free exercise of the creativity, gifts, and abilities the LORD has given to persons who happen to be poor, we must be encouragers and catalytic facilitators to let their lights shine.

When it comes to helping the poor overcome global warming it's ultimately about freedom and love: enhancing their freedom to create a better life for their families; and living out the love of Christ in our hearts.

RECRUIT AND INSPIRE OTHERS

There's an old saying, "May your tribe increase." In this instance what we want is to have our overcoming global warming team increase. We need to recruit others to this cause and inspire them to continue along the overcoming global warming path. Of course, such recruitment needs to be done in an inviting way.

The opposite of an inviting approach is one of self-righteousness. When it comes to new areas where we are changing our behavior, it can be very easy for any of us to slip into a self-righteous mode. But as we do make changes, we must find ways to make them a positive witness to others.

You can actively talk to others about overcoming global warming by looking for ways to naturally include it in conversations or activities. You can help others understand why it means so much to you. You can volunteer to help them implement projects that reduce their global warming pollution and explain why it makes a difference. You can let them know about opportunities to influence public policy. And if it feels right, you could even loan them your copy of this book.

One person you could make a special effort to recruit is your pastor, who you could encourage to join the Evangelical Climate Initiative (www.christiansandclimate.org), and other suitable efforts.

BE A CATALYST IN YOUR CHURCH, WORKPLACE, AND SCHOOL

Several areas ripe for continual creative engagement are where we go to church, where we work, and where we or our children go to school.

Imagining the possibilities is the best place to start. And remember: you don't necessarily need to be the one who does all of these things, but you can be the one who serves as a catalyst for them to happen.

For example, what if you helped your church to implement an energy efficiency program like the one Prestonwood Baptist did, which reduced their energy use by over 30%, saving approximately $725,000 a year? Another example is First Baptist Church of Springdale, Arizona, who in the first 14 months of their program saved $250,000 and reduced global warming pollution the equivalent of that produced by nearly 300 homes. They did so mainly by being aware of their energy use. As their project point person remarked, "The most significant things we did were to monitor our usage patterns and become aware of how to better use energy, and train both staff and members on how they could contribute to our energy savings effort."[766]

What if you also helped to ensure that your congregation understood how much global warming pollution was being reduced and why that was an important part of helping the poor?

What if you got the congregation to donate half of the savings to relief and development efforts that enhance the poor's resilience and reduce their vulnerability to climate impacts, and to invest the other half in future efficiency and pollution reduction improvements such as solar panels for your church's buildings? You would be a witness to the wider community of your overcoming global warming efforts.

You could help to organize a ride-sharing program that reduced by one-third the number of vehicles driven to church. By doing so congregants who lived close to one another would get to know each other better.

What if you helped to start an overcoming global warming small group to spearhead such efforts at your church, dedicated to (1) supporting one another, (2) educating the church about global warming and the positive Christian messages contained in this book, and (3) ensuring that pollution reduction projects are implemented?

If you work at a local grocery store similar to Vic's Market in Sacramento, what if you helped to convince the boss to upgrade the lighting and freezers, saving, say, $48,000 a year like Vic's did?

What if you worked in industry and convinced your company to implement a combined heat and power (CHP) system, doubling your energy efficiency? Or you convinced your plant to upgrade to more efficient boilers like the MinnTac plant of US Steel did, saving $790,000 a year, or to fix pipe leaks like at Dow Chemical's St. Charles Operations petrochemical plant in Hahnville, Louisiana, resulting in $1.9 million in yearly savings?

If you are the head of your own company, imagine if you decided to be like Lee Scott of Walmart and launch a major initiative with three overarching aspirational goals: (1) to be supplied by 100% renewable energy; (2) to create zero waste; and (3) to sell products that are produced sustainably.

If you are a student and want to educate your peers, you could be like Alec Loorz, who began to give presentations on global warming when he was 12 and has thus far educated over 10,000 students and has also founded his own organization, Kids vs. Global Warming?[767]

Or what if in all these areas of your life and more you came up with ideas that no one else could have and inspired others to work with you to make them a reality?

And what if in all these relationships you helped people understand that it's not really about lightbulbs and boilers, it's about creating a cleaner future for our kids and enhancing freedom for the poor, saving God's other creatures from extinction, loving God and our neighbors as ourselves, and becoming more spiritually beautiful?

PRAY

Prayer is one of the primal, irreducible elements of the Christian life. The Apostle Paul taught us to "pray without ceasing" (1 Thess 5:17, KJV), a seemingly, gloriously impossible command in our current sinful finitude – but one for which to aspire nonetheless. Scripture teaches us to pray for ourselves, pray for one another, pray for our Christian leaders, pray for our government officials, and perhaps most importantly, pray as Jesus taught, "Thy Kingdom come, Thy will be done on earth, as it is in heaven" (Mt 6:10, KJV).

Each of us should do all of these types of prayer when it comes to our current challenges, remembering that to pray "Thy Kingdom come, Thy will be done" today is to pray a prayer for God to overcome global warming. The impacts of climate change, such as those described in chapters 3-7, work against His will being done here on earth.

So when it comes to praying about overcoming global warming, here are some suggestions.

♦ Thank God for all of His many blessings, including the freedom and opportunities He has given us to help Him overcome global warming.

♦ Pray for the poor who have been and will be impacted by global warming. Pray that Christians and others help them overcome the consequences through targeted adaptation and by increasing the poverty-reducing and democracy-increasing dimensions of freedom. Pray that we find ways for the poor to benefit from climate-friendly technologies as well as policies that incentivize reductions in global warming pollution.

♦ Pray for ourselves that we are continuously filled with His love (which by its very nature must flow out of us to touch others). Pray that we remain centered in His lordship; that He continually gives us His empowering, forgiving grace and the guidance of His Spirit so that we can do His will in overcoming global warming. Pray for creativity in discovering new ways to overcome global warming and help with the activities and initiatives in which we are involved.

♦ Pray by name for those you know who are engaged in overcoming global warming. Pray for them the same things we prayed for ourselves (love; Lordship; grace; guidance). Pray also about specific activities and concerns that you are aware of (hint: this means you need to be aware of such things!).

- Pray for all leaders – Christian leaders and other religious leaders, government leaders, and business leaders. Pray the LORD gives them wisdom, courage, creativity, and whatever else they need to be the leaders He is calling them to be in overcoming global warming.
- Pray for our Christian leaders, collectively and by name, to help the Church rise to this challenge and support all the efforts to overcome global warming.
- Pray for our government officials that they adopt the policies necessary, both nationally and internationally, to create the rules and conditions that allow those in the private sector to run the climate mitigation race. Pray our government officials ensure that as a country we are contributing generously to help the poor adapt to climate impacts.
- Pray for our business leaders to encourage the creativity of their employees and make the investments necessary to spearhead our drive towards a tenfold increase in carbon productivity.
- Pray for those who are opposed to our efforts to overcome global warming, and that God would give them the blessings of love, Lordship, grace, and guidance.
- Finally, pray for God's will to be done in overcoming global warming. Pray we all understand His will and do it.

As we strive to achieve these tasks that all of us should do, let us remember the encouraging words of the Apostle Paul: "Stand firm. Let nothing move you. Always give yourselves to the work of the Lord, because you know that your labor in the Lord is not in vain" (1 Cor 15:58).

WHAT YOUR GIFTS, TALENTS, OPPORTUNITIES, AND RESOURCES ALLOW YOU TO DO

YOUR GIFTS, TALENTS, SKILLS, AND KNOWLEDGE

I can't think of a gift or talent that can't be used to help overcome global warming. Are you good with numbers? Are you good with your hands? Do you have the gift of hospitality? Are you good at teaching and communicating? Are you good with children? Are you a good project manager? Are you good at organizing people? Are you good at making money? Are you an artist or musician or writer? Are you good at figuring out what I forgot to put in here? You're all needed.

Overcoming global warming is going to require the talents, skills, and knowledge of engineers and scientists and welders and electricians and farmers and bankers and accountants and lawyers and nurses and doctors and homemakers and parents and students and, well, you name it.

We also need to break out of our "silos" and have intentional creative engagement across fields and disciplines that produces solutions never thought of before. For example, people in the field of information technology (IT) need to come together regularly with those from the field of clean energy technology to make the latter "smarter" and more efficient. And for this to occur we're going to need people who have skills at organizing people and projects, can manage new programs, and who can facilitate inter-disciplinary dialogue and catalyze collective innovation.[768]

A current example of how some cross-pollination is occurring is what is called the "smart grid" and its related enabling technologies. Electricity comes into our homes and businesses through something called the "grid," composed of power plants and transmission lines working as a system to bring us on-demand electricity. The twentieth century system that we live with now is a very centralized system, where the consumer is very passive with no real-time information on how much electricity is costing at any one moment.

Adding a "smart meter" and a "smart thermostat" would tell you how much your electricity is costing at any given moment. If you have "smart devices" (e.g., the washer and dryer, the dishwasher), you can receive information on when electricity costs more or less. They would turn themselves on or off accordingly and use less wattage during peak times. By having the option to program your integrated smart system to choose, say, the "eco" option, you can save money and overcome global warming simultaneously.[769]

However, we need to do even more than such things as a "smart grid." Whereas the smart grid is like two people – information technology and the electric system – coming together to form a marriage, where some integration occurs, what we need is the "offspring" of these types of marriages that combines the "DNA" of the parents to produce a completely integrated system. In other words, we need to creatively design things from their inception to deeply integrate the knowledge and potential of these fields. And the same needs to happen between many other areas of endeavor.

YOUR OPPORTUNITIES

What could you do to help overcome global warming with the leadership opportunities that will come your way? For example, if you become chair of your church's Worship Committee, you could talk to the committee about how the message of overcoming global warming might be introduced in worship. You could loan your pastor your copy of this book and talk to him or her about how to include this message in sermons.

If you are a scout leader you could help your scouts see the changes that global warming is bringing.

Or if at work you are put in charge of purchasing equipment, you could ensure that such equipment is as energy efficient as it can be.

Or if you are offered a new job you might be able to work it out so that a portion of the job is devoted in some way to overcoming global warming.

An example is my friend Larry Schweiger, a fellow Christian who was offered a job as the head of the country's leading conservation organization, the National Wildlife Federation (NWF). However, he took it on the condition that he could make global warming the organization's top priority. When he was being considered for the President/CEO position, Larry was perfectly happy as head of the Western Pennsylvania Conservancy. But the LORD had been working in his life to convince him that global warming was a challenge for which he needed to be a leader.

And so when Larry was being interviewed for the position, he told the search committee of his commitment to make global warming a top organizational priority due to the devastating impacts that global warming will have on wildlife. When describing the interview process, Larry said, "As I thought about climate change and I thought about [my daughter who was expecting her first child], it occurred to me that the question I never ever want to hear my grandchild ask is: 'Grandpa, where were you when they wrecked my planet?' I don't ever want to hear that. I told that search committee that."[770]

Thankfully the Board agreed with Larry's commitment to addressing climate change, and today NWF is one of the most important organizations of any kind engaged in overcoming global warming, helping to bring its membership of hunters, anglers, and conservationists to the cause.

YOUR RESOURCES

As I mentioned above, one of the ways we can help the poor in developing countries overcome global warming is by supporting the work of Christian relief and development organizations.

But some of us will have the financial resources to go further. Those of you who are seasoned investors could become what I will call "venture-tithers." After you have tithed 10%, consider investing an extra 2% in potentially risky but promising climate business solutions in developing countries (e.g., the Katani company described in chapter 20), and then reinvest any profits into the same or other ventures.[771]

I could see someone taking this a step further and creating a newsletter/blog that tells others about your venture-tithing investments. Or to go even one step further, someone could create and manage a venture tithing investment fund to provide investment opportunities for those of us who are willing to be venture-tithers.

CHRISTIAN CLIMATE ENTREPRENEURS AND FORWARD-LEANING TEAM-BUILDERS

I would like to lift up two types styles of leadership I think are especially needed to move us forward in overcoming global warming. Some of us are primarily one type or the other, but many of us have the capacity to practice one or the other style when they are called for. The first is a creative Christian climate entrepreneur, while the second is a forward-leaning team-builder. But regardless of what type we are or style we may be employing, we need both an "entrepreneurial spirit" and a "team spirit" as we work to overcome global warming.

CHRISTIAN CLIMATE ENTREPRENEURS

Creative Christian climate entrepreneurs with a servant-heart are those who take risks to overcome global warming. They put things together in new configurations; create new products and services and methods or approaches to doing things. They also take tried and true products and approaches and apply them in new ways. Christian climate entrepreneurs are ready to have their ideas stand or fall based on their value and effectiveness. They have the drive and tenacity to see things through even when on the edge of failure.

We need Christians who will be like those spotlighted in chapter 18, such as Bill Keith, a former roofer who invented a solar attic fan and took out a second mortgage to start his company, SunRise Solar.

Or Manoj Sinha, Gyanesh Pandey, and Chip Ransler, who founded Husk Power Systems, a private electric utility company that builds small power plants and local electric grid systems village by village, and utilizes rice husks as their power source.

Or Harish Hande, founder and CEO of Selco, an Indian company that currently supplies solar electricity to 105,000 households with plans to expand to another 200,000 over the next several years. Hande tells his employees that if a customer cannot afford their product, they have built the wrong product, that it should be paid for out of the new revenue or savings the customer is able to generate from it. Selco creates solar products that improve the lives of construction workers, rubber-tappers, flower-pluckers, and midwives.

We need Christians who will found companies or projects like Envirofit. Using a business approach they have created $25 cookstoves that the poor want to buy because it saves money and labor, reducing fuel use by up to 60% and cooking time by up to 50%. As an added bonus it cuts indoor air pollution, CO_2, and black carbon by up to 80%.

But we don't just need entrepreneurs in the marketplace. We also need "social entrepreneurs," like Alec Loorz who founded Kids vs. Global Warming, or the Rev. Sally Bingham, who founded Interfaith Power and Light, which has helped over 10,000 congregations purchase clean electricity and educate their members about global warming.[772]

FORWARD-LEANING TEAM-BUILDERS

A team-builder is someone who helps bind people together in a common cause and is able to call out the gifts of the members to achieve the group's goals. They help turn a group into a team. A team-builder is respected for his or her integrity and good intentions. The results of their leadership, groups with a team spirit, are part of the backbone of a healthy society, a healthy status quo.

But with overcoming global warming we need to create a new status quo, and do so again and again until the threat no longer exists. We need *forward-leaning* team-builders who not only keep the group/team together and working as a team – no small feat – but who help the team continue to see how they play an important role in overcoming global warming. This is what I mean by being forward-leaning – not just maintaining the health of the team, but leading the team to play its part in this great cause of freedom.

I'm sure that examples of such team-builders abound in your life. Who is it that keeps the groups to which you belong going? Who is it that helps create a team spirit at work and school? It may even be you!

Global warming will not be overcome unless groups working together as teams stay true to the cause. And for that we need forward-leaning team-builders.

Leadership of either style will bring challenges. And when they come, it will be helpful to remember these words of Paul: "For our light and momentary troubles are achieving for us an eternal glory that far outweighs them all" (2 Cor 4:16-17).

WHAT THE LORD MAY BE CALLING YOU TO

As I have been stressing, each of us has an important contribution to make; we all need to be "saints in the world" when it comes to overcoming global warming. Otherwise, we won't succeed.

But for some of you, overcoming global warming might become a full-time calling.

What would a "climate calling" entail? In the future there will be many types of full-time climate-related work. You may have gifts in the area of financial investing and may come to focus on clean energy, while keeping your eyes out for how the poor can benefit. Or you might be good at banking and pioneer new ways to extend microcredit to the poor so they can enhance their lives with efficient cookstoves or solar panels or biogas systems. Perhaps you become a carpenter or home builder who specializes in building zero energy homes. Or you might work for a Christian relief and development organization and help the poor implement targeted adaptation projects. You could start up a low-profit venture like The Paradigm Project or found a "venture-tithing" firm. You could be the Office Manager for an organization devoted to overcoming global warming.

In conclusion, the scale and speed of the changes necessary to overcome global warming provide us with an incredible challenge that can be turned into tremendous opportunities. As Christians we know this: "I can do all things through Christ who strengthens me" (Phil 4:13, NKJV).

OUR GREAT CLOUD OF WITNESSES

Because overcoming global warming is a marathon, there will be times when we grow weary. We might even want to give up when, in marathon-speak, we "hit the wall" – i.e., reach a point where we feel we can't go on. It is encouraging to remember that many have gone before us who have persevered in the faith, who have run their own races with the LORD, and we can envision them along the side of the road cheering us onward.

In faith Mary Magdalene first proclaimed the good news that Jesus was risen, even in the face of derision by the demoralized disciples (Mk 16:9-11, Lk 24:1-11).

In faith Peter persevered, even after denying Jesus three times, to boldly preach at Pentecost the good news about Christ's saving death and resurrection and to become the leader of the early Church.

In faith Thomas finally stopped his doubting and believed when he encountered the Risen LORD.

In faith the other disciples, even after failing to understand Jesus' teaching and ministry and abandoning him when it mattered most, persevered and allowed the Risen LORD to transform them into apostles.

In faith Barnabas, years after he first stood up for Saul, traveled hundreds of miles to bring Saul to Antioch and catalyze his calling.

In faith Paul, who once persecuted the early Church with murderous threats, persevered through tremendous suffering and opposition to become the Apostle to the Gentiles.

And there are many who have persevered closer to our own time and have displayed continued creative engagement of their faith with the great challenges and spiritual opportunities of their day.

In faith John Wesley, Anglican Priest and founder of Methodism, creatively wove together orthodox Christian theology, the experience of pietism, and the use of small groups for support and discipline, and as early as 1774 was a strong opponent of slavery.

In faith William Wilberforce, British politician and Christian, persevered for over 25 years to successfully abolish Great Britain's slave trade in 1807.

In faith Charles Finney became, during the first half of the nineteenth century, the father of modern revivalism, creatively perfecting many evangelism techniques. He was also a strong abolitionist and champion of the education of women and former slaves.

In faith a watchmaker named Corrie ten Boom and her family hid Jews in their home in Amsterdam during the Nazi occupation of Holland. They were eventually sent to concentration camps for doing so, where her sister (Betsie) and father died. Corrie persevered, and for over 30 years was a missionary and Christian speaker on the need for forgiveness and reconciliation.

In faith the Rev. Dr. Martin Luther King, Jr., a Baptist preacher, persevered as the premier civil rights leader of our time, with the movement's efforts leading to the 1964 Civil Rights Act and the 1965 Voting Rights Act.

These leaders weren't perfect, by any means. But that is part of the point. The Bible is filled with stories of sinners like us becoming instruments to do His will, flawed leaders who nevertheless were employed by the LORD to achieve great things. In their *kairos* moments, they answered the call, and then persevered in faith. They became "good soil" and produced bountiful harvests.

As the author of Hebrews entreats us, "since we are surrounded by such a great cloud of witnesses, let us throw off everything that hinders and the sin that so easily entangles, and let us run with perseverance the race marked out for us" (12:1).

Part of our race, the marathon marked out for us, is to play our part in this great moral challenge of our time. We can draw inspiration from this great cloud of witnesses.

But none of them had the opportunities you and I have to overcome global warming. We have the chance to be a part of an extraordinary worldwide effort to protect billions and enhance their lives simultaneously, to even lead the way.

So that we "will not grow weary and lose heart" (12:3), the book of Hebrews reminds us of our most important example of inspiration: "Let us fix our eyes on Jesus, the author and perfecter of our faith, who for the joy set before him endured the cross, scorning its shame, and sat down at the right hand of the throne of God" (12:2).

WALKING WITH THE RISEN LORD
IN THIS GREAT CAUSE OF FREEDOM

And so once again, let us fix our eyes upon Christ. As we do so, let us have eyes to see and remember who the Bible reveals Him to be.

Before He came to earth as a helpless baby refugee pursued by a murdering tyrant, before he pitched his tent amongst us, He was the pre-existent Son of God who dwelt in perfect love with the Father and the Holy Spirit. And as Their love expressed itself in creation, it was the Son through Whom all things were made. Christ is the Creator.

And what, then, does Christ the Creator see when He looks at what global warming is doing to His Earth? He sees it all.

But Christ is not just the Creator. Christ is also the Sustainer. He not only gives life to all of life, He sustains all of life. As John 1:4 says, "In Him was life." Hebrews 1:3 tells us He is "sustaining all things by His powerful word." And Colossians 1:17 says that in Christ "all things hold together." As the Sustainer, Christ has an intimate spiritual communion with all things. He keeps everything from slipping back into nothingness and from flying apart as He holds all things together.

And yet as Christ is sustaining all things, the life-denying effects of global warming are working against Him.

Christ is not just the Creator and Sustainer. As Jesus of Nazareth he lived his life as the true image of God, a true *imago dei*: free, beautiful, and glorious. Because of our sin, we are Venus de Milo, still beautiful, still made in the image of God, but spiritually warped, weathered and without arms. But Jesus freely and completely reflected his Father's will and thus was a true image or reflection of the Father's merciful and loving rule on Earth, the Kingdom incarnate.

And when global warming warps and distorts the beauty of the Father's creation, when it works against His merciful and loving rule, what then happens to our imaging the true Image and thus our relationship to him? We find that our actions push us away from him.

But Jesus Christ is not just the Creator, the Sustainer, and the true *imago dei* or Kingdom incarnate, he is also our Savior. It is precisely because we fail to be who God created us to be, true images or reflections of His will on Earth, that we need redemption. And it is precisely because Jesus of Nazareth is the true *Imago* that he can be our Savior. As Hebrews puts it, "For we do not have a high priest who is unable to sympathize with our weaknesses, but we have one who has been tempted in every way, just as we are—yet was without sin" (4:15). He was pursued even before his first breath by those who wanted to kill him, constantly criticized and opposed, misunderstood, and abandoned by his followers. And yet to save us he took the cup and was forsaken even by his Father as our sins poured into his heart and he became everything he was not. By his wounds we have been healed (1 Pet 2:24), healed to be transformed into true images of God as we reflect His glory, as we become spiritually beautiful in reflecting the will of the Father on Earth, as we fulfill the Lord's prayer, as we momentarily and fleetingly bring in His Kingdom when we do His will.

Through the transforming power of Christ's grace there is no problem we face, no problem the world faces, that cannot be overcome.

What, then, does global warming tell our Savior about us? That here is another spiritual problem for which we desperately need his empowering grace.

He is not just the Creator, the Sustainer, the true *Imago*, the empowering Savior, He is also our LORD. And yet as Peter and the other disciples help us understand, confessing him as Lord doesn't mean we fully grasp or completely fulfill his Lordship. We can even find ourselves in opposition to him, as when Jesus says to Peter not soon after Peter confessed him to be the messiah, "Get behind me, Satan!" (Mt 16:23; Mk 8:33). We must, therefore, continually strive through prayer and Bible study and communion with our fellow believers to have our hearts and minds transformed so that they become images or reflections of His heart and mind.

As Paul encouraged the Philippians, "Let the same mind be in you that was in Christ Jesus, who, though he was in the form of God, did not regard equality with God as something to be exploited" (2:5-6, NRSV). Jesus was the Son of God, the LORD, and yet here on Earth he did not try to exploit his Sonship or his Lordship. Rather, he "emptied himself, taking the form of a slave … he humbled himself, and became obedient to the point of death – even death on a cross" (2:7-8, NRSV). If his messiahship is one of service, how much more so should our discipleship be one of service?

And yet He is not just the Creator, the Sustainer, the true *Imago*, the Savior, the Lord, He is the *Risen* LORD.

It was the Risen LORD's *presence* that helped his frightened and doubting disciples see who He really was: "My LORD and my God." And from this confession they were given "the right to become children of God" (Jn 1:12) and "life in His name" (Jn 20:31). It was His presence that transformed wannabe-heroes-turned-deserters into bold apostles with servant-hearts willing to die proclaiming the gospel and building the Church. It was His presence that turned Saul the destroyer into Paul the Apostle.

And when it comes to overcoming global warming, it is His spiritual presence that will turn us from ignorance or denial or indifference or doubt or despair or fear into His agents of transformation with servant-hearts. We must remember Paul's words: "where the Spirit of the Lord is, there is freedom. And we, who with unveiled faces all reflect the Lord's glory, are being transformed into his likeness with ever-increasing glory, which comes from the Lord, who is the Spirit" (2 Cor 3.17-18). He frees us from what is holding us back from doing His will and being transformed into *imago Christi*, resulting in ever-increasing glory.

But if any of you harbor doubts, then it is time for you to **stop your doubting and believe** that the Risen LORD is LORD of overcoming global warming, that He is right beside you to empower and sustain you as you make your contribution to this great cause of freedom.

In following Him we will find ourselves … loving one another as our Savior loves us … loving our neighbors … loving our enemies by overcoming evil with good … loving the least of these … and loving God back, which is the purpose of our lives and the summation of these five great loves.

But does the Risen LORD see us following Him in overcoming global warming? He knows we have the potential, and he's beckoning us to join Him.

And where will we find the Risen LORD, that we might follow Him?

He is with those who find themselves in the ditch of global warming's impacts, urging us not to pass by on the other side.

He will be with poor children who formerly lived off food found at the local dump but during times of intensified drought will go hungry because no one will throw away scraps, as was the case in parts of drought-ravaged Guatemala in 2009-2010.[773]

He will be with those who have lost and will lose their land due to flooding, sea level rise, and hurricanes, such as Jyotsna Giri, who had lived on a small island in West Bengal, India. "I still remember that fateful day, when I lost everything …I suddenly noticed that my sheep were all drifting in the river … Slowly the entire island got submerged."

Jyotsna Giri, West Bengal, India[774]

He will be with those caught up in downward development spirals made worse by climate change. Alizeta Ouedraogo, a young woman from Burkina Faso, a poor country in Africa, describes what intensified droughts can lead to: "Every year, there is a food shortage. Children quit school because they cannot afford supplies and school fees. Girls sometimes prostitute themselves and may end up with an unwanted pregnancy or a sexually transmitted disease."

He will be with farmers like Pablo Huerta Mandez of Peru, who finds that "it barely rains now. Year by year it's less and less. I've farmed here for 10 years and there is more heat, which affects the plants and causes plagues … We're very worried about climate change."

He will be with those in the future who, because of global warming, will encounter situations like those we have met in earlier chapters, such as the father during the 2005 Niger famine, found with his family hundreds of miles from the nearest feeding station who said, "I'm wondering like a madman. I'm afraid we'll all starve."

Or the mother at a feeding station during the same famine who said, while watching her daughter die, "God did not make us all equal. I mean, look at us all here. None of us has enough food."

Or Bithi, the young girl from Bangladesh, whose family had to flee their home because of unusually heavy flooding and whose parents couldn't sleep because they were afraid that one of their children might fall into the floodwaters and drown.

Or 70-year-old Makhabasha Ntaote from Lesotho, South Africa, who, because a crazy climate has made it so difficult to grow crops, says, "I think God is angry at us, but I don't know why."

Or the young boys from Honduras, Olbin and Edgardo Casares, who, after their father was killed and their home destroyed by Hurricane Mitch and their mother's family didn't want them, became street urchins who had no idea where their mother was. "I don't know where she is anymore," said Olbin. "It's just me and my brother."

And of course He will be with those here in the United States who also encounter hardship and loss like the situation of Mike Johnson and his family during one of the recent Midwest floods who found themselves living in their Bronco SUV with Mike sleeping on the roof. Or that of Steve and Linda Hakanson of Northern California, whose house burned to the ground, "We basically all ran for our lives," said Steve. "I'm looking in my rearview mirror, and all I could see was red," recalled Linda.

Or Doug deSilvey of Mississippi, who lost his entire family to Hurricane Katrina. As his roof caved in and his lungs started filling with water, deSilvey remembers, "I just kept on asking Jesus not to let me go like this. I had to get my family out." But he didn't. "Losing a family – I don't think there's any words for it. Kinda makes you wonder what life is all about … I have nobody to plan for, or work on retirement for, or to save to buy a house for. It's just me."

But the Risen LORD will be with us also (Mt 28:20) as we strive to be His agents of transformation in overcoming global warming in this great cause of freedom.

Spiritually, whenever any of us make an effort to overcome global warming, the Risen LORD is there with us to encourage us forward. When we do things to keep ourselves spiritually fit to run this race and others, He is with us. When we pray for our leaders to play their part, He is with us. When we talk with our pastor or help to form a small group at church or do an energy efficiency program and donate the savings to helping the poor overcome global warming, He is with us.

The LORD is with us when we contact our members of Congress and urge them to support climate-friendly policies or when we decide to purchase clean electricity, LED lights, cold water detergent or a zero energy home close to public transportation. He is with us when we support Christian relief and development organizations that are helping those impacted.

Or if you decide to take a chance and found your own small company to make something like solar attic fans, like Bill Keith, the LORD will be with you.

Or if in playing your part to overcome global warming you decide to dedicate your skills in banking or investment or administration or project management or teaching or communication or hospitality, or if you provide leadership as a Christian climate entrepreneur or a forward-leaning team-builder, He will be with you.

And when we fail, He will pick us up, like He did Peter. And when we doubt, He will show us His nail-scarred hands, like He did for Thomas.

Truth is, like the disciples, we will be a motley crew of imperfect agents of transformation with a mishmash of motives. We are still just jars of clay with inestimable spiritual treasures inside from the LORD (2 Cor 4:7) that we clumsily put on at times like small children playing dress-up with their parent's clothes and jewelry.

But here's the thing. **We don't have to be perfect** to help the Risen LORD overcome global warming. Like the disciples, **we just have to be willing to follow**. In our climate *kairos* moments we have to choose life and be willing to stumble forward with the Risen LORD and allow Him to pick us up when we fall.

And it is precisely this marathon of overcoming global warming that provides us a key opportunity to become more spiritually mature, more fully who we were created to be. It is a great challenge like this that requires us to accept the spiritual food of Christ's grace as sustenance for the journey. The more we accept, the stronger and more spiritually fit and mature we become, the more we are able to do His will, which flows from His love. Both His grace and His will for us flow from His love. And it is our willingness to allow His love to transform us into who we were always meant to be that will make overcoming global warming be as natural as breathing.

So we see once again that it's ultimately about freedom and love. As we choose to love God back, as we exercise our spiritual freedom with steadfast endurance to do His will, to be His image, to become the righteousness of God, to embody His Kingdom as its citizens in these acts of loving others, we help free them from the tyranny of global warming's impacts. For the poor in developing countries, we help to create the conditions whereby the economic and political dimensions of freedom are enhanced rather than diminished so that they can create better lives for their families and loved ones. And in so doing, we become free, beautiful, and glorious as in freedom we reflect His beauty and glory.

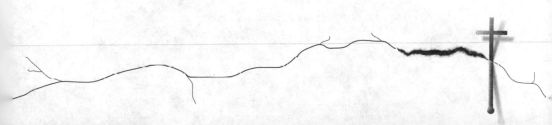

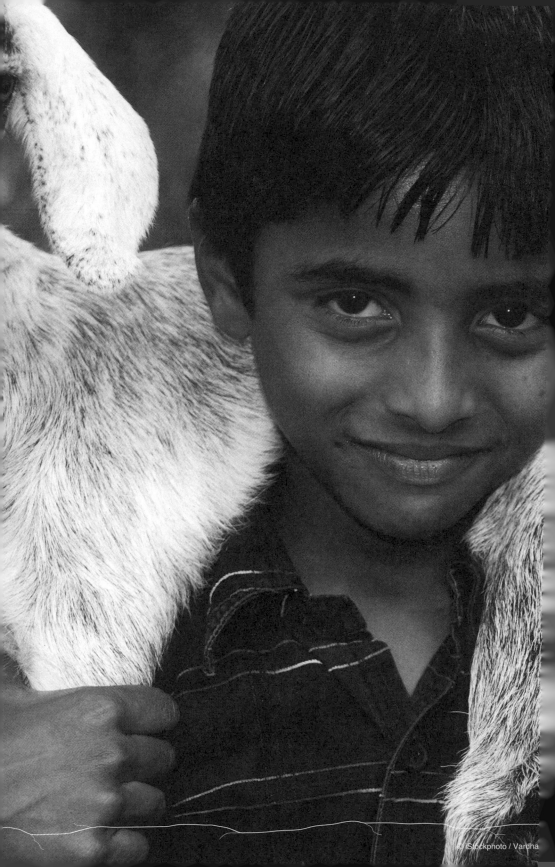

ACKNOWLEDGEMENTS

First and foremost I must thank my Creator and Lord, Jesus Christ, for such gifts and abilities that may be represented herein, and for His spiritual presence with me throughout the research and writing of *Global Warming and the Risen LORD*. Any glory is his; any defects are mine.

With a project like this there are too many people to thank (especially when considering how many shoulders one stands on). But let me start by recognizing the first person to encourage me to write a book on global warming, Scott Rodin. (For better or worse, here are the fruits of your encouragement!) Sir John Houghton and Dr. Katharine Hayhoe read relevant chapters to help keep me in line scientifically. The Rev. Mitch Hescox, my successor as president and CEO at EEN, has been one of the book's biggest champions, and it literally would not have been published without his support. Both Mitch and another EEN colleague, Alexei Laushkin, read the manuscript and offered helpful suggestions. Mark Russell, the book's publication manager, and his team at Russell Media, were absolutely terrific in helping to turn a manuscript into a book. I am also very grateful to the Energy Foundation (and especially Charlotte Pera) for their gracious support.

Finally, the person most responsible for this book becoming a reality is my wife, Kara. *Global Warming and the Risen LORD* has been a true Ball family project from start to finish. When I first began considering it, Kara was the one who insisted that I write the book even while recognizing the burdens this would place upon her. For nearly four years Kara supported me as I spent Saturdays, holidays, and vacations writing. She picked up the slack with household responsibilities, even while juggling her own full-time job at the National Wildlife Federation. Kara also patiently listened time and again when I read her something I just wrote, and put up with me when I became frustrated. She eventually read the entire manuscript, offering many suggestions that made the book much better. From A to Z, Kara has been my helpmate in ensuring that *Global Warming and the Risen LORD* became a reality. Words become empty ciphers when it comes to expressing love and gratitude to my Kara.

While numerous individuals read part or all of the manuscript, I take sole responsibility for the final product.

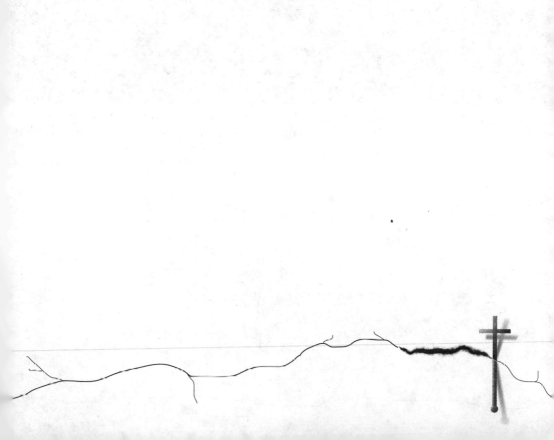

Jim Ball and His All Holiness the Ecumenical Patriarch Bartholomew
at the Patriarch's event on climate change for senior religious leaders

ABOUT THE AUTHOR

Nearly two decades ago when Jim Ball first began to study global warming he quickly realized what a tremendous impact the consequences would have on the world's poor – those who had done nothing to cause it. He knew then that Christ's Lordship was calling him to do something. Jim began to devote his life to overcoming global warming and rallying Christians to play their part. However, early on such efforts bore little fruit. No one showed up for his first scheduled presentation on global warming at a Christian conference focused on peace and justice issues. While finishing his Ph.D. in theological ethics in the mid-90s he wrote an initial primer for Christians on global warming. Few read it. Because there was no position in a Christian organization focused on climate change at that time, Jim took a job as the Climate Change Policy Coordinator with the Union of Concerned Scientists.

In late 1999 the Evangelical Environmental Network (EEN) recruited Jim and he made it clear then that his primary concern was global warming, while recognizing that it would take time before it could become a priority for the organization. Jim's first major opportunity to do so came in 2002 when he conceived of using the question, "What Would Jesus Drive?" in an educational campaign. This effort became an overnight media sensation, landing Jim on numerous national newscasts, including ABC's *Good Morning America*, where Diane Sawyer challenged him on how far Christ's Lordship could extend.

Jim's next big climate-related effort for EEN was to organize senior Christian leaders into the Evangelical Climate Initiative (ECI), which was launched in February 2006. It also garnered significant attention, including stories on all the major newscasts. The ECI is now highly regarded by policymakers and has established global warming as an important moral issue the evangelical community must address.

In recognition of his leadership on global warming, Jim has testified several times before Senate committees, most recently on why the U.S. should help the poor in developing countries overcome the consequences of climate change. Jim has been named by *Rolling Stone* magazine as one of their environmental "Warriors and Heroes," and *Time* magazine named him one of its five climate change "innovators" in its April 3rd, 2006 edition.

Jim currently serves as EEN's Executive Vice President for Policy and Climate Change. He is an ordained Baptist minister with a Ph.D. in theological ethics from Drew University, where his dissertation focused on the evangelical response to creation-care issues. He has a Master of Divinity from the Southern Baptist Theological Seminary and a BA from Baylor University. He was reared in Richardson, Texas, and currently lives in Virginia with his wife, Kara, her three cats, Emma, Midnite, and Spit (with whom Jim has an uneasy truce), and Iggy the iguana. Their beloved dog Mugsy went to be with the Lord several years ago, and Jim dreams that one day they will once again own a dog.

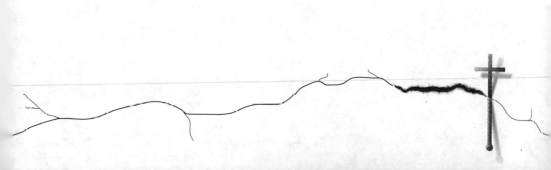

ABOUT THE PUBLISHER

The Evangelical Environmental Network (EEN) is a ministry that seeks to educate, inspire, and mobilize Christians in their effort to care for God's creation, to be faithful stewards of God's provision, and to advocate for actions and policies that honor God and protect the environment.

Founded in 1993, EEN's work is grounded in the Bible's teaching on the responsibility of God's people to "tend the garden" and in a desire to be faithful to Jesus Christ and to follow Him. EEN publishes materials to equip and inspire individuals, families, and churches; and seeks to educate and mobilize people to make a difference in their churches and communities, and to speak out on national and international policies that affect our ability to preach the Gospel, protect life, and care for God's Creation.

A PERSONAL NOTE from EEN's PRESIDENT & CEO

Many years ago, John Wesley described sin as an illness whose only healing is God's grace. In *Global Warming and the Risen LORD*, Jim Ball describes "the earth as having a fever." It's a fever caused by us, and like all sin, the only remedy is God's grace. *Global Warming and the Risen LORD* proclaims grace, transformation, and Christian discipleship. This book allows us to enter into a journey with the Risen Christ and reflect God's grace into our hurting, ill, and warming world.

Our world needs transforming grace. Like so many folks in the church, I grew-up with a Christianity that talked about "getting saved" and going to better place. It was easy to make that connection living in a western Pennsylvania village with un-reclaimed coal strip mines as my playground.

Why not grab the earth's resources, who needs them in heaven? However, as grace touched my heart and I took the time to read and understand Holy Scripture, it became very apparent that God's desire for us includes an abundant Christ-filled life now for all people.

Global Warming and the Risen LORD became my next conduit for God's grace. More than any other work, outside the Bible, Jim's discernment furthered my transformation in knowing that creation care, including climate change, is a matter of life for us today and certainly our children and grandchildren. I will be ever indebted to Jim for his passion, his teaching, his ministry, and especially the massive undertaking to co-create with God, *Global Warming and the Risen LORD*.

My experience led me to the decision to act as publisher. I wanted the world to read, ponder, and grow from each written word. *Global Warming and the Risen LORD* profoundly altered my life, and my prayer is that the same will happen to you. For *Global Warming and the Risen LORD* not only describes the fever we have given the earth, it proclaims the good news that is the sure and certain cure, God's grace.

In Christ,

The Rev. Mitchell C. Hescox
President / CEO
The Evangelical Environmental Network

NO TREES WERE HARMED
IN THE MAKING OF THIS BOOK

Truth be told, so a few
did need to make the ultimate sacrifice.

In order to steward God's good creation,
we are partnering with *Plant With Purpose*, to plant
a tree for every tree that paid the price for the printing of
this book.

PLANT WITH PURPOSE

Printed in the most climate-friendly manner available using 40% post-consumer recycled paper

FOR MORE INFORMATION

To find out more about the Evangelical Environmental Network (EEN) and how you can become involved, go to their main Website, **www.creationcare.org**.

For more on the Evangelical Climate Initiative (ECI), go to **www.christiansandclimate.org**.

To keep up with Jim Ball, visit his blog at **www.revjimball.com** or his Twitter account (**revjimball**).

For updates related to this book, including new stories and findings about overcoming global warming, the latest on impacts, or upcoming speaking engagements by Jim, visit the book's Facebook page at **www.facebook.com/RevJimBall**.

ENDNOTES

The book's endnotes are available in all e-book versions as well as at **www.creationcare.org** as a downloadable PDF document.